Veröffentlichungen aus der
Geomedizinischen Forschungsstelle
(Leiter: Professor Dr. med. Helmut J. Jusatz)
der Heidelberger Akademie der Wissenschaften

Supplement zu den Sitzungsberichten der
Mathematisch-naturwissenschaftlichen Klasse
Jahrgang 1985

Erhard Hinz

Human Helminthiases in the Philippines

The Epidemiological and Geomedical Situation

With 3 Coloured Map Plates and 83 Figures

Submitted to the Meeting of 30 June, 1984
by Richard Haas

Springer-Verlag
Berlin Heidelberg New York Tokyo

Prof. Dr. rer. nat. Erhard Hinz

Hygiene-Institut der Universität Heidelberg,
Abteilung Parasitologie, Im Neuenheimer Feld 324,
D-6900 Heidelberg

English Translation by

J. A. Hellen, M. A. (Oxon.), Dr. phil. (Bonn)
and Mrs. I. F. Hellen, B.A. (Newc.)
Newcastle upon Tyne

ISBN-13:978-3-642-70843-5 e-ISBN-13:978-3-642-70841-1
DOI: 10.1007/978-3-642-70841-1

Library of Congress Cataloging-in-Publication Data
Hinz, Erhard, 1931– Human helminthiases in the Philippines.
(Supplement zu den Sitzungsberichten der Mathematisch-Naturwissenschaftlichen Klasse ; Jahrg. 1985) (Veröffentlichungen aus der Geomedizinischen Forschungsstelle der Heidelberger Akademie der Wissenschaften)
"Submitted to the meeting of 30 June, 1984 by Richard Haas."
Bibliography: p. 1. Helminthiasis–Philippines. 2. Medical geography–Philippines.
3. Climatology, Medical–Philippines. I. Title. II. Series. III. Series: Veröffentlichungen aus der Geomedizinischen Forschungsstelle der Heidelberger Akademie der Wissenschaften.
RA644.H38H56 1985 614.5'52'009599 85-25003
ISBN-13:978-3-642-70843-5 (U.S.)

Softcover reprint of the hardcover 1st edition 1985

Printing of Maps: Justus Perthes, Darmstadt
2125/3140-543210

Foreword

For medical scientists, biologists and geographers interested in geomedical problems the helminthiases can be a fascinating object of research. Their distribution is due to the in part very complicated parasite life cycles which frequently depend on the presence of intermediate hosts. The search for the causes of the distribution of helminthiases requires to take into account not only such geofactors as affect the parasite developmental stages outside man but go beyond this and include the entire web of factors which contribute to the conditions for the distribution of their intermediate hosts. Last, but by no means least, it is, however, man who through his customs and habits, his settlements and dwellings, his population density and, above all, his interference in the environment, determines the distribution of helminthiases. The frequency, persistence and areal expansions are a consequence of the interplay and interaction of all the geofactors. The aim of every geomedical analysis must be to prove the causes of their distribution through a chain of causation which has no gaps. A classic example of such a chain had already been set out in the 1920s when Ernst Rodenwaldt investigated the occurrence of brugiasis in the Serajoe Delta on Java, and it is Rodenwaldt's analysis which has served as a model for this work.

The idea of producing the monograph presented here arose from the Geomedical Monograph Series edited by Helmut J. Jusatz. It was thanks to work on helminthological problems in South East Asia during the course of several research expeditions that the Philippine Archipelago – at the crossroads of South East Asia and the Pacific area and on the periphery of the palaeo-tropical faunal region – excited the author's interest above all. Several reasons led the author to attempt an analysis in the geomedical sense of the entire spectrum of human helminthiases in the Philippines: the occurrence of intestinal capillariasis as a human disease known for less than twenty years, the existence of a locally restricted *Schistosoma japonicum* strain having developed in co-evolution with its intermediate host as a result of geographical isolation, and the close connection of wuchereriasis with the cultivation of the fibrous banana, the source of Manila hemp.

Without the constant encouragement and furtherance provided by Professor Dr. med. H. J. Jusatz, professor emeritus of tropical hygiene and head of the Geomedical Research Unit of the Heidelberg Academy of Sciences, the realisation of this intention would have been impossible. To him I am indebted for the editing of and general concern for this monograph. It was through him that my interest in geomedicine as an interdisciplinary approach to research was aroused and, even more, it was he who critically watched over my efforts in this field for more than two decades.

Heidelberg, June 1984 Erhard Hinz

Table of Contents

List of Figures

1 Introduction

Extending over more than 1800 km from the Batan Islands to the islands of the Sibutu Group in the south, and a distance of almost 1100 km between the most easterly point of Mindanao to the Balabac Island situated at the southern tip of Palawan in the west, the Philippine Archipelago comprises a triangular area of around 1.3 million square kilometres of the Western Pacific. Broken up in part by Tertiary folds and partly by the post-glacial rise in sea-level, the surface area of the Philippines of almost 300,000 km^2 is divided up into more than 7,100 islands, only some of which are inhabited. Only 45 of them extend over an area of more than 100 km^2. Two-thirds of the total surface area alone are taken up by Luzon, the main island (105,708 km^2), together with Mindanao (95,687 km^2), and the main part of the remaining third by the Visayas lying between Luzon and Mindanao, and the islands of Mindoro and Palawan which, continuing on from Luzon, form a south-westerly axis. The population, which numbered 42 million in 1975 according to the 1980 Philippine Statistical Yearbook, and is likely to have passed the 50 million mark in the meantime, is equally unevenly distributed. Whilst Central Luzon reached population densities of almost 500/km^2 as early as 1975, other islands and provinces such as Palawan (20.1/km^2) and Kalinga-Apayao (23.1/km^2) were distinguished by a very sparse population.

In other respects the Philippines are also characterized by important regional differences. Besides mountain ranges, the highest peaks of which may come close to 3,000 m, hilly regions and low plains as well as isolated volcanoes towering above mark their heterogeneous morphology. Depending on the physiography, seasonally alternating air-masses of continental and maritime origin coming from the north-east or the south-west determine the amount and distribution of precipitation. On the other hand, altitude and precipitation determine the cultivation of plants, which in its turn exerts an influence upon settlement. These factors, whether taken singly or as the interplay of forces, are of decisive importance for the occurrence or absence of diseases, especially so in the case of endemic diseases such as the helminthiases, which are confined to certain areas.

The knowledge of the geographical distribution and the varying frequency of helminthiases on particular islands in the Philippines constitutes the pre-condition for effective control measures. An attempt is therefore made here to analyze the occurrence of the most important of the helminthiases in the geomedical sense and to describe the endemic areas on particular islands.

The climatic regions of the islands serve as an initial biogeographical basis, which is followed by a second survey in accordance with the administrative division into regions and provinces.

1.1 The Climatic Regions of the Philippines

The extent of the archipelago over almost seventeen degrees from north to south, and the largely parallel course of the mountain ranges "are the climatically most effective morphological factors", in which "it is not temperature that dominates the seasons which are important in terms of economic geography, but the distribution of precipitation" [439]. However, this annual precipitation regime is not only of significance from the viewpoint of economic geography, but also for geomedicine, for it is this which to a considerable degree decides directly and indirectly the distribution of helminthiases in the Philippines.

Three different air-streams – the south-west monsoon in summer, the north-east monsoon in winter, and the north-east trade wind – determine the distribution of precipitation over the course of the year, along with the depressions which arise in their marginal areas, which can when particularly strongly developed cross the Philippines (mostly in a west-north-westerly direction) as typhoons. This basic situation applies to the entire archipelago; it is, however, decisively modified by the mountain ranges which run counter to the direction of flow of the airstreams. In consequence four main types of precipitation distribution can be distinguished (cf Map Plates) in accordance with Coronas' classification which was also adopted by Kolb [439]: the monsoonal (west-side type), the permanently humid (east-side) type, and the two transition types.

The west-side type receives its main precipitation between May and October at the times of the south-west monsoon. As a rule no precipitation worth mentioning occurs there during the other half of the year as the then prevailing north-easterly airstreams have already shed their rain on the eastern side of the archipelago. This climatic zone is thus marked by alternating rainy and dry seasons; for a period of 4–6 months precipitation here tends to fall well below a monthly mean of 100 mm. This west-side type is in contrast to the permanently humid type as another extreme. In the east of the archipelago cyclonic rains fall even during the time of the south-west monsoon. The main quantities of precipitation in this area are, however, due to the airmasses approaching from a north-easterly direction; they reach their maximum of 400–500 mm and more a month in the period from November to Janurary. The east-side type is thus characterized by precipitation falling throughout the year, with maxima at around the turn of the year.

Situated between these two zones there is a transitional area of varying extent, where the prevailing conditions may be closer to those of the west-side type or the east-side type. In general, however, annual precipitation in this transitional area is markedly lower. In the case of the western transition type, it is due to the fact that it benefits little from the precipitation that falls during the time of the south-west monsoon. Against that this season with little precipitation (or a dry season) is reduced to from 1 to 3 months (February to April). The eastern transitional type, on the other hand, lacks the winter maxima which characterize the east-side type. Characteristic for this transition type are rains with slight monthly fluctuations falling throughout the year.

Within their zones, however, these precipitation types may locally be considerably modified, thanks mainly to the role of the luff and lee sides of the mountain ranges. In addition the precipitation total may fluctuate a great deal from year to year and deviate considerably from the long-term mean.

Besides the distribution of precipitation *temperature conditions,* too, are of significance for the occurrence of helminthiases. In this context it is less the minima than the maxima occurring in the course of a day, as well as during the year, which may have a restrictive effect on the occurrence of helminths as a result of their negative influence on those stages of their development which take place in the open. Due to latitude, but above all to air-streams that change with the seasons, as well as cloud cover and sunshine resulting from them, conditions vary according to the regions. In areas which belong to the west-side – or the western transitional type as regards precipitation distribution – the annual temperature cycle can be divided into three parts: the hot-humid one from June/July to October with medium temperatures; the cool one from November to February, and the hot and dry one from March to June, with absolute maxima of 40 °C and more being reached in places like northern Luzon. On the other hand, the east-side and eastern transitional type experience only two thermal seasons: a cooler one from November to March, and a hotter one from April to October, with absolute maxima possibly reaching 37 °C. During the hottest part of the year, this temperature difference between the two zones applies to the monthly mean maxima as well. These very high maxima tend to occur in low-lying plains, however; at higher altitudes in the mountain regions temperatures do not rise above 30 °C as a rule, a temperature which has no adverse effect on the development stages of helminths which take place in the open. On the other hand, temperature minima, falling below 8–10 °C, may have an adverse effect on the development here.

Temperatures also play a part in the speed with which helminth larvae develop inside the transmitting insect. In the case of *Wuchereria* larvae there is a clear positive correlation between temperature and speed of development. But if temperatures exceed the optimal value of c. 27 °C deformation sets in and increases, leading to irreversible damage at temperatures of c. 33–34 °C.

1.2 The Administrative Units, Regions and Provinces of the Philippines

With statistical data in particular being based on administrative units, it is necessary to know their particular locations if existing material and the author's own data are to be worked up. The section which follows is therefore intended to present a brief survey of these circumstances.

Since 1975 the 75 provinces and sub-provinces of the Philippines have been combined in 12 regions of varying extent, with the National Capital Region of Manila forming an additional thirteenth (Map Plate 1). This division presents the basis for the following survey of the different parts of the Phillipine state. The population data referred to in the text are based on those of the 1980 Statistical Yearbook for the year 1975.

1.2.1 The Ilocos Region

Situated in the north-west of Luzon, the Ilocos Region, which is divided into the provinces of Ilocos Norte, Ilocos Sur, Abra, Mountain, La Union, Benguet and Pangasinan, extends over an area of 21,568 km² with a population of 3.27 million people.

Related to the total area, the population density of 151.6/km² conceals the real problematic nature of population pressure in this region, since only 42% of the country is described as being "alienable and disposable"; Wernstedt and Spencer [831] had already given the physiological density of the provinces of Ilocos Norte, Ilocos Sur, Abra and La Union as being around 750 persons per km² of land under cultivation. Large numbers of the indigenous Ilocanos of this region were thus forced to leave their home country and to migrate to the western highland or to move south along the coast or to other parts of the land.

In terms of the landscape areas the Ilocos Region presents a number of different components. Following the division suggested by Wernstedt and Spencer [831] the Ilocos Provinces, together with Abra and La Union belong to the Ilocos coast region, Benguet and the Mountain Province to the North Luzon highlands and Pangasinan in part to the Central Plain region and in part to the Zambales-Bataan region. The entire region is subject to the south-west monsoon, and consequently receives 90% of its precipitation in the period from May to October (sometimes attaining extreme values thanks to typhoons crossing over North Luzon in particular).

The main crop is rice, although maize, vegetables, fruit, coconut palms and tobacco are also cultivated, the latter as a cash crop. The agriculturally usable area being small, and the population density, however, being high (save for that part of Pangasinan belonging to the Central Plain), the Ilocos Region is regarded as an area with a deficit in regard to food supplies. Supplements to the diet in the form of animal foods such as fish, snails etc., which the Ilocanos tend to consume raw, are the reason for the frequent occurrence of food-borne helminthiasis in this region (cf above all 4.1., 4.5 and 4.6).

1.2.2 Cagayan Valley Region

The north-eastern provinces of Luzon, including the Batan and Babuyan Islands, are combined in the Cagayan Valley Region (Map Plate 1). Their total area is 36,403 km². The actual valley of the Cagayan River forms the core of the region (Isabela and Cagayan Provinces); it is one of the large low-lying plains of the Philippines. Open to the Babuyan Channel in the north, it is screened from the Pacific by the mountains of the Sierra Madre in the east; the western edge is formed by the Cordillera Central with the provinces of the Central Highlands, some of which – Kalinga-Apayao and Ifugao – belong to the Cagayan Valley Region, whilst the barrier to the south is formed by the Caraballo Mountains in the Nueva Vizcaya and Quirino Provinces region. In regard to a population of 1.93 million, the region shows a population density (53.1/km²) far below the average for the country, chiefly as a result of the negligible amount of settlement in the Sierra Madre, but also on account of the small number of people living in the Cordillera Central. Even in the 'sixties two-thirds of the population of the Cagayan Valley proper were made up of immigrants from the Ilocos Region and the smaller part of the indigenous Ibanags, Gaddangs, Itavis, Yogods and Isnegs. Other ethnic groups such as the Kalinga, Ifugao, Kankanai, Inibaloi etc. from the highlands of the Cordillera Central, as well as the inhabitants of the Batan and Babuyan Islands complete the heterogeneous ethnic mix of the Cagayan Valley Region [cf 831].

The Cagayan Valley Region, too, is one of the areas of the Philippines that are more severely affected by typhoons, which in their turn influence the frequency and distribution of precipitation. The western part tends to receive its precipitation from the south-west monsoon in close relationship to its season. In the Cagayan Valley itself the season with less precipitation is less marked and of shorter duration, whereas the Sierra Madre and the Batan and Babuyan Islands belongs to the zone of precipitation that falls throughout the year (cf Map Plates).

The Cagayan Valley is an area of surplus production in so far as agriculture is concerned. Rice growing is most important, the area under cultivation being around 360,000 ha in the early seventies, although it is followed closely by maize with c. 261,000 ha. As in the Ilocos Region tobacco presents the most important cash crop.

1.2.3 The Central Luzon Region

Central Luzon is characterized by a lowland plain that extends between the Lingayen Gulf and Manila Bay, and is flanked by mountain ranges on both sides. It includes the provinces of Bataan, Bulacan, Nueva Ecija, Pampanga, Tarlac and Zambales (Map Plate 1) with a total area of 18,231 km^2 and an overall population of 4.2 millions, giving it a position well above the average where population density (231/km^2) is concerned. Situated entirely in the lowland plain, the Pampanga Province (478/km^2) is the most densely populated one, whereas the hills of Zambales Province present the most sparsely populated area of this region (112/km^2). In the north the part of the plain belonging to Central Luzon is formed by the watershed between the river Agno (Pangasinan), which flows into the Lingayen Gulf, and the River Chico Pampanga. The latter is a tributary of the Pampanga River, which drains into Manila Bay after dividing up into numerous arms. The inhabitants of Central Luzon belong to a variety of ethno-linguistic groups. In the Central Plain Pampangan predominates, whereas Tagalogs are in the majority in the Bataan Peninsula. Originally inhabited by tribal groups known as Zambals, the province of Zambales experienced an influx of Ilocanos as early as the 19th century, and these came to be the dominant group in the population from early in the 20th century.

The climate of Central Luzon is determined by the south-west monsoon, which brings 80–90% of the precipitation in the period from May to October. On the leeward side of the mountains the total amount of precipitation may be less than 1800 mm in the course of the year, but it can be much higher on the windward one. The distribution of precipitation throughout the region nonetheless follows the seasonal change of the dry and the rainy seasons.

Central Luzon is the main rice-producer of the Philippines. In the early 'seventies rice was grown on more than 600,000 ha – that is about 20% of the total rice-growing area of the Philippines [676]. Maize, taking up 77,000 ha, was much less important. The cultivation of sugar cane (an area of 70,000 ha) represented, on the other hand, an economically important factor, although it tends to be produced on large haciendas – unlike rice, which is grown on small farm holdings. North of Manila Bay, where there are swamps of brackish water, an additional area of more than 30,000 ha was utilised for fish ponds.

1.2.4 National Capital Region (Metro Manila)

Located on the eastern side of Manila Bay, the National Capital Region occupies an area of 636 km². It includes Manila itself at the centre and sixteen other towns grouped around it in a semi-circle. The total population being 4.97 million in 1975, the average density of population was 7,814 per square kilometre. In this conurbation the city of Manila represents the most densely populated area (38,619/km²), with extreme values far in excess of the average in the over-populated Tondo District. As the most important centre of the Philippines for administration, education, commerce and industry and equipped with a port which handles the greater part of imports and a considerable percentage of the exports, Metro Manila presents all the contrasts that are characteristic of metropolitan cities in tropical developing countries. Here it is chiefly the varying living conditions and life-styles, ranging from the upper-class residential quarter to those in the so-called squatter areas, which are reflected in the difference in the pattern of disease frequency in general and of the helminthiases in particular.

1.2.5 Southern Tagalog Region

The Southern Tagalog Region includes eleven provinces and sub-provinces, which must be associated with different landscape regions. The part on Luzon includes the south-west of the island between Manila Bay and Tayabas Bay, with the Rizal, Cavite, Laguna and Batangas provinces, as well as that section of the Eastern Cordilleras which extends southwards from the border of the Cagayan Valley Region to the Bondoc Peninsula, and is largely identical with Quezon Province (including Aurora). In addition the islands of the Sunda Shelf (Mindoro and Palawan) and those of the Sibuyan Sea (Marinduque, Romblon, Tablas, Sibuyan) are included in the Southern Tagalog Region (Map Plate 1) as well.

In 1975 the total population of the region amounted to 5.214 million inhabitants, living in an area of 46,924 km². Admittedly more than half the people (2.878 million) lived in the provinces of south-west Luzon, where the population density was 383/km². All the remaining parts of the region are much more sparsely populated: Marinduque: 170/km²; Romblon: 134; Quezon: 93; Mindoro: 56; Palawan: 20.

The climate of south-west Luzon, Occidental Mindoro and Palawan belongs almost exclusively to the area affected by the south-west monsoon, which tends to produce precipitation in the period from May to October; a seasonal change of rainy and dry seasons is therefore a characteristic of these areas as it is for Central Luzon. Whilst the islands of the Romblon Province still experience a brief dry season, Oriental Mindoro, Marinduque and the larger parts of Quezon Province are predominantly affected by rains falling throughout the year, which reach their maxima in the east coast region of Quezon at around the turn of the year.

Depending upon the orography and the amount and distribution of precipitation, land use varies greatly within the region. With 406,805 ha of the total area under cultivation, rice comes first [676], followed by maize (183,938 ha), coconuts (104,344 ha) and sugar cane (45,142 ha).

1.2.6 The Bicol Region

The provinces of Camarines Norte, Camarines Sur, Albay and Sorsogon, which are situated in the extreme south-east of Luzon, together with the island provinces of Catanduanes and Masbate, form the Bicol Region, which thus extends over a total area of 17,633 km^2. Numbering 3,194 million, the population shows a population density well above the national average as well with 181 persons to the square kilometre being attained. Apart from the fringe areas the region enjoys a very uniform structure in the ethno-linguistic sense; according to Wernstedt and Spencer [831] Bikol was the mother-tongue for 93 per cent of the inhabitants.

Much of the Bicol Region is mountain and hill country, above which rises a string of volcanoes. Between Mount Mayon (2,421 m) and San Miguel Bay there is the extensive lowland plain of the Bicol River, in the main given over to rice cultivation. Since humid air-masses from the north-east as well as from the south-west have access to the entire region, there is no season without precipitation; the Bicol region is thus characterized by rains falling throughout the year, reaching their maxima in the east between October and January, but chiefly between June and October when exposed to the west. Besides permitting the cultivation of rice (315,302 ha), maize (117,866 ha) and coconuts (226,018 ha), climatic conditions also guarantee the growing of abaca (*Musa textilis*) which supplies the so-called manila hemp. With more than 65,710 ha planted with abaca, the Bicol region in the early 1970s supplied more than 42% of the total cultivation area of the Philippines. However, abaca cultivation is not only an important economic factor but is also closely allied to a certain worm infection, wuchereriasis (cf 2.2.3).

1.2.7 Western Visayas Region

The four provinces of Panay (Aklan, Antique, Capiz, Iloilo) together with Guimaras and Negros Occidental form the Western Visayas Region (Map Plate 1), an ethno-linguistic unit with Hiligaynon as a common language. The total area is 20,223 km^2, the population numbers 4.146 million people, and is thus, with a population density of 205/km^2, distinctly above the national average.

On Panay the mountain range of the Western Highlands running in a north-south direction, and the chains of hills radiating from it to the north-east, divide the island into three lowland areas; the Iloilo Basin, the flats on the north coast and those on the west coast. The island of Negros, too, is traversed by a volcanic range of mountains that follows a north-south direction and merges with the Tablas Plateau in the south west. The north and west includes a plain that rises gradually from the coast to the mountains.

The Western Visayas Region receives its main precipitation from the south-west monsoon in the months from June to October, but the north-east of the region in particular is also hit by the north-east monsoon, so that this region has got only a very short period with little precipitation, by contrast with the south-west, where it may continue much longer, i.e. from December to May (cf Map Plates).

Although one of the most important rice producers of the country, the Western Visayas Region derives its chief importance from the cultivation and refining of sugar-cane. More than half the total area under cultivation in the Philippines –

446,908 ha in the early 1970s [676] – is, on average, found in the province of Negros Occidental alone. The dominance of sugar-cane cultivation is, on the other hand, also the reason for Negros being one of the deficit areas in respect of food crop production which must rely on imports from other parts of the country.

1.2.8 Central Visayas Region

Comprising the provinces of Cebu, Negros Oriental, Bohol and Siquijor, the Central Visayas represent the region with the smallest surface area (14,952 km²); the inhabitants are the ethno-linguistically uniform group of the Cebuanos. The large number of inhabitants (1975: 3.387 millions), however, makes the Central Visayas one of the most densely settled areas of the Philippines (227/km²). More than half the population (1.818 million) is concentrated on Cebu, where the density attains the very high value of 357/km². This has long been the reason for Cebu remaining at the top of the list for internal migration, which is chiefly directed at Mindanao.

Protected from all sides from rain-bearing air-masses, the location of the Central Visayas is the reason for it being one of the regions with the smallest amount of precipitation in the Philippines. On average 67% of the precipitation falls in the period from June to October, but the western parts are characterized by a more distinct separation of dry and rainy seasons than the east (northern Cebu and Bohol), where precipitation falls more evenly throughout the year.

Partly as a result of the small amount of precipitation, maize is the main agricultural product of the Central Visayas. In the early 1970s an area of 240,000 ha was planted with maize; that is about 20% of its total in the Philippines. Rice cultivation is only of importance on Bohol and in the south-west of Negros Oriental. Due to the high population density the Central Visayas are one of the areas with a deficit in basic food production, so that staple foods have to be imported from other parts of the country. On the other hand, the deficit is also brought about by the fact that a not insignificant part of the agriculturally utilisable area is reserved for cash crops, such as sugar-cane, coconuts, bananas and mangoes, as well as to a lesser extent abaca.

1.2.9 Eastern Visayas Region

Administered as six provinces or sub-provinces, the islands of Samar, Leyte and Biliran form the Eastern Visayas Region (Map Plate 1). Their combined surface area is 21,432 km². With a population of 2.6 million (1975), the general population density of 121/km² is below that of the national average. The population is very unevenly distributed, however: whilst Eastern Samar is very sparsely populated (66/km²), 46 per cent of the population of the entire region is concentrated on North Leyte (192/km²), which is one of the provinces with significant internal migration. There is no ethnolinguistic unity in the Eastern Visayas in so far as Waray-Waray is spoken in Samar and North-east Leyte, while Cebuano, the language of the Central Visayas, predominates in the west and south of Leyte.

In the Eastern Visayas there is no season with little or no precipitation at all. Even April, the month with the least precipitation, contributes 162 mm to the mean annual total of 3,086 mm. Apart from some more sheltered areas on the west coast of Samar and the western third of Leyte, the Eastern Visayas belong to that climatic zone which is characterized by rains falling throughout the year and additional maxima at the turn of the year. Agriculture is therefore not tied to seasons. Rice and maize, both occupying c. 200,000 ha, are almost equal in importance among the food crops. Coconuts are the most important cash crop. In addition, climatic conditions permit the cultivation of abaca, around 30% of the total crop of which was produced on Leyte and Samar in the 1970s. However, the distribution of precipitation and abaca cultivation are also the decisive precondition for the persistence of schistosomiasis and wuchereriasis.

1.2.10 Western Mindanao Region

Western Mindanao divides into two sub-regions: sub-region 9 A with the provinces of Basilan, Sulu and Tawi-Tawi, and sub-region 9 B with the two provinces of Zamboanga, which are part of Mindanao (Map Plate 1). Basilan and the Sulu Archipelago together occupy 21.5% of the total surface area of 18,685 km², whereas the Zamboanga Provinces take up 78.5%. A somewhat different relation (27.1% and 72.9%) occurs in the distribution of the population, so that a somewhat higher population density results in sub-region 9 A (138/km² in contrast with 102/km²), the total population figure being 2.048 million. In the Sulu Archipelago, on Basilan, and the south-west orientated section of the Zamboanga Peninsula the muslim population dominates, with the Tausug Moro as one of the important ethnic groups, whereas in the eastern part of Zamboanga immigrants from the Central Visayas and their descendants form the majority of the population.

The greater part of the region receives precipitation throughout the year, but the south-west section of the Zamboanga Peninsula is characterized by a season with little precipitation, lasting from January to April. With an annual mean of 1125 mm Zamboanga City even ranks as one of those localities with the least precipitation in the Philippines. As a rule, however, precipitation is sufficient for the cultivation of the most important food crops. On the whole rice plays the dominant role as a food crop, followed by maize; in large parts of the Sulu Archipelago, however, cassava (manioc) is the more important staple food. Coconuts and abaca are the most important cash crops.

1.2.11 Northern Mindanao Region

The coastal provinces of Misamis Occidental, Misamis Oriental, Agusan del Norte, Surigao del Norte, the island province of Camiguin and the adjoining landward provinces of Bukidnon and Agusan del Sur together form the Northern Mindanao Region (Map Plate 1), a heterogeneous area which is part of such varied regions as the Pacific Cordillera, the Davao-Agusan Depression, the Central Highland and the Bukidnon-Lanao Highland. The Northern Mindanao Region extends over 28,328 km², but considering that the total population numbers 2.314 million, the majority of which consists of immigrant Cebuanos and their descendants, it is relatively sparsely populated, the average being 82 persons/km². There are considerable differences within the region in regard to population density, with very low values occurring in the mountains (e.g. Bukidnon Province: 64.3/km²) and especially in the area of the Agusan Depression (Agusan del Sur: 23.8/km²) and high figures in the coastal lowlands (Misamis Occidental: 230.6/km²), so that the latter are among the emigration centres.

Northern Mindanao belongs to three different climatic zones: the eastern part (Surigao and Agusin Provinces) is subject to precipitation falling throughout the year with extreme maxima at the turn of the year in some parts. In the sheltered situation of Misamis Oriental, on the other hand, the influence of the south-west monsoon serves to concentrate the precipitation predominantly in the summer and autumn months, whilst the period from February to April may well have little rain or be quite dry. The remaining area (Misamis Occidental and large parts of Bukidnon) are characterized by a largely uniform distribution of precipitation falling throughout the year. Depending on the amount and distribution of precipitation, on the altitude and soil conditions, maize and rice predominate in the region as the main agricultural product. The most important cash crops – again in different key areas – are coconut and abaca. Moreover the cultivation of certain crops intended for export, such as pineapples, has increased in economic importance.

1.2.12 Southern Mindanao Region

Extending over 31,693 km², the Southern Mindanao Region occupies about one third of the total area of Mindanao. Surigao del Sur, Davao Oriental, Davao del Norte, Davao del Sur and South Cotabato, the five provinces of the region together have a population of 2.715 million and an average population density of c. 86/km². The Pacific Cordillera, the Davao-Agusan Depression, the Central Highland, the south-western Coastal Range and foothills of the Cotabato Basin divide the region into sections with varying possibilities for settlement, which consequently results in an uneven population distribution. The majority of the present population are descendants of immigrants from the Western and Central Visayas.

The climate of Southern Mindanao is characterized by rains falling throughout the year, which reach their maxima in the region of the East Coast around the turn of the year. In sheltered locations (e.g. in low-lying areas protected by mountains) agriculture may well experience a shortage of precipitation, which can only be made good by additional irrigation. In most parts of the region maize dominates over rice, with different kinds of root crops forming an important supplement to foodstuff production. Among the cash crops coconuts rank highest, with abaca in the next place. Admittedly in part abaca cultivation has shifted from its original centre around Davao City to other areas in the region. In the early 1970s Southern Mindanao with an area of 22,000 ha under cultivation with abaca had only a 14% share in the total area of cultivation. Other cash crops, such as coffee and pineapple, on the other hand, increase in importance.

1.2.13 Central Mindanao Region

At present the provinces of Lanao de Norte, Lanao del Sur, Maguindanao, North Cotabato and Sultan Kudarat are included in the region of Central Mindanao (Map Plate 1); with the Maguindanao and Maranao Moro they have a significant proportion of Moslems among the population. As with other parts of Mindanao, so, too, the Central Region had been subject to substantial immigration from the Visayas, so that Cebuanos and their descendants predominate on the coasts of Lanao del Norte and in the west of North Cotabato. In 1975 the total population was about 2.07 million, the surface area 23,293 km^2, putting the population density (89/km^2) clearly below the national average as in other regions of Mindanao as well.

Cotabato Basin with the Tiruray Highlands and the Central Cordillera as the western and eastern fringes, parts of the Bukidnon-Lanao Plateau, the Lake Lanao area and the lowlands in the region of Panguil Bay are the most important topographic components of the region. Its climate is that of the zone characterized by precipitation falling throughout the year. Central Mindanao is one of the most important maize-producing areas of the Philippines, though rice is also cultivated on a fairly large scale except at higher altitudes. Abaca and coconuts as cash crops play a lesser role than in other regions of Mindanao, whereas the cultivation of coffee, ramie, pineapple and other fruit is increasing in importance.

2 Insect-Borne Helminthiases: Filariases

In the Philippines two species of filariae pathogenic to humans are to be found: *Wuchereria bancrofti* and *Brugia malayi*. Belonging to the nematodes, both species are parasites of the lymphatic system and can cause blockage of vessels and a stagnation of lympth, thereby leading to the clinical picture of elephantiasis; this chiefly presents itself in the swelling and thickening of the lower extremities, sometimes extending to the scrotum and less often to the arms, vulva or mamma and considerably impairing the ability to work. The filaria-♀♀ are ovoviviparous; the sheathed microfilariae set down by them pass into the blood vessel system, from whence – having spent some time in the capillaries of the lungs – they tend to be carried into the peripheral bloodstream; this process follows a diurnally fixed rhythm and is known as periodicity. In the Philippines the two species of filariae are distinguishable from one another by the periodical occurrence of their microfilariae in the peripheral blood of the final host (cf 2.2.1.2 and 2.3.3). Whenever mosquitoes, which are suited to becoming vectors, take up these microfilariae as part of their blood meal, further development takes place inside them; first of all, having lost their embryonic sheath, they enter their first larval stage (L_1), which is followed by a first ecdysis and a second larval stage (L_2). Finally, after another moulting the development inside the mosquito leads to the infective larva (L_3). Such infective larvae enter the proboscis of the vector, leave the labium of the insect at the next sting and penetrate the final host by way of the channel made by the sting. Having settled in the lymphatic vessels they remain there, grow to maturity, and mate. For *W. bancrofti* man is the only final host, whereas *B. malayi* finds a reservoir in a number of mammal species; this results in differences in the epidemiology of both species.

Rozeboom and Cabrera [668] attribute the existence of both the species in the Philippines to introductions from Melanesia during the low-water periods of the Ice Age, whereas Pichon [627] maintains that the Austronesian region itself (of which the Philippines forms a part) constitutes the area of origin. Pichon considers the sub-periodic form of *B. malayi,* which is limited to Austronesia, to be the forerunner of the periodic variant; even *W. bancrofti* is thought to stem from a sub-periodic mammalian filaria of this region, which is closely related to *B. malayi.*

2.1 Short Historical Review

Filarial infection of man in the Philippines was first proved shortly after the transfer of power in that island country to the United States of America when Strong [742; cf 129] diagnosed microfilariae in the blood sample of a European who had settled at

Iloilo. A little later Calvert [187, 188] reported further cases of filariasis among Filipino prisoners of war, two of whom came from Albay, a third from northern Luzon. In this the two microfilariae carriers from Albay were probably subject to nocturnal periodicity.

In the following years filariasis research received significant impulses through the setting-up in 1906 of the Board for the Study of Tropical Diseases in the Philippine Islands [64][1]. The first result was the description of *Filaria philippinensis* as a new aperiodic species of filaria in the Philippines, and their development in the vector by Ashburn and Craig [65–72]. But this new species was not to stay: Low [468], Fülleborn [307] and Walker [821] refuted the justification of its separation from *W. bancrofti.* (This may also apply to the *Microfilaria philippinensis* described at a much later date by E. Y. Garcia [321, 322] in 1964 and again in 1966.)

Until then filariasis in man was regarded as insignificant in the Philippines. This view only changed when Phalen and Nichols [617–620] carried out extensive surveys. Their first investigation included a number of cases of elephantiasis in the Sorsogon Province, Bicol Peninsula, which continues to be one of the most important areas of endemic wuchereriasis. Already at that time elephantiasis, locally referred to as "titibac", was considered by the native population to be predominantly connected – directly or indirectly – with the cultivation and processing of abaca (manila hemp). However, the proof that there is indeed an important relationship between the growing of abaca (*Musa textilis*) and the distribution of wuchereriasis was only achieved at a much later date (see below).

Though in the main based upon surveys of Philippine Scouts, Filipinos employed at military bases and prisoners [cf 120, 622–625 a], and therefore not considered representative, the results published by Phalen and Nichols [618–620] nevertheless permitted conclusions to be drawn as to the chief endemic areas. These were above all the Bicol Peninsula centred on the Kinali Valley (Albay Province), followed by Leyte, Bohol, Samar and various provinces in Mindanao, which continue to be regarded as endemic for *Wuchereria bancrofti.* Moreover Phalen and Nichols succeeded in proving that it was a filaria of exclusively nocturnal periodicity, a statement largely confirmed by other authors subsequently [22, 72 a, 73, 166, 467, 790]. Investigations carried out in Taytay showed that isolated autochthonous *Wuchereria* infections occurred on Luzon, even to the north of the Bicol Peninsula [561 a].

This state of knowledge about the geographical distribution of wuchereriasis has only been insignificantly extended by occasional surveys over the decades that followed [73, 353, 467, 790]. But the opinion did evidently gain ground that filariasis was an important health problem for the Philippines, which required greater attention. To begin with, the Division of Malaria examined blood smears from 19,345 persons in 18 provinces over the period 1955–1957, establishing a prevalence of 8.8 per cent [129]. This result led to the development of a five year programme, in the course of which microfilaraemia rates of 0 to 10.18% were found among the blood smears from 182,712 persons in 49 provinces [290; cf Table 1[2]]. This established the extent of the problem for all parts of the country except Panay and Negros Occidental, so that measures for the control of filariasis could be undertaken (cf 2.4).

---

[1] According to Simmons [723] this institution, which was originally known as the "U.S. Army Board for the Study of Tropical Diseases as They Exist in the Philippine Islands", had been established in 1900 already, but had not been working from 1902 to 1906

[2] Table 1 and further tables are to be found at the end of the book on page 234

On the evidence of the investigations carried out so far, *Wuchereria bancrofti* could be taken as being the sole species of causal agent. It was only in 1964 that *Brugia malayi*, with its nocturnal-subperiodic variant, was also found to occur endemically in the Philippines, though never on its own but rather together with *W. bancrofti* (cf 2.3).

The question of which species of insects functioned as vectors of *W. bancrofti* remained open for a long time. In the light of the laboratory experiments carried out by Ashburn and Craig [71], as well as by Africa [7], *Culex quinquefasciatus* was at first incriminated until B. D. Cabrera and Tubangui [172] showed in field investigations on the Bicol Peninsula that *Aedes poicilius* was the only possible natural vector. This mosquito species had an infection rate of 35%, with a considerable proportion of infective *Wuchereria* larvae. *Culex quinquefasciatus* on the other hand, although also infected at a rate of 12.7%, never harboured infective larvae.

Aedes poicilius is one of the axil-breeding mosquitoes. The leaf-axils of a variety of plants may serve as breeding places, but the *Musa* species play an especially important role, particularly the fibrous banana (*Musa textilis*), referred to as abaca or manila hemp; on the Bicol Peninsula, the Visayas and Mindanao its area of cultivation is likely to be about 250,000 ha (cf 1980 Philippine Statistical Yearbook, p. 239). This results in a close connexion between abaca cultivation and the distribution of *Wuchereria*, which will be treated in greater detail at a later stage.

Outside the areas of abaca cultivation, that is on Palawan, the Sulu Archipelago and in Kalinga-Apayao Province (northern Luzon), *Anopheles flavirostris* takes over the function of transmitting filariasis [127, 663, 664, 667], being at the same time the most important vector of malaria in the Philippines.

The mosquitoes transmitting *B. malayi*, the second species, were also first described by Rozeboom and Cabrera [666] in 1965, and later on confirmed as species of *Mansonia*, albeit requiring quite a different ecosystem as the basis for their distribution (cf 2.3.2).

2.2 Wuchereriasis

2.2.1 The Significance of Abiotic and Biotic Factors for the Ontogenesis of *Wuchereria bancrofti*

In the ontogenesis of *Wuchereria bancrofti* two factors stand out as of geomedical and epidemiological significance. In the one case it is the geofactors which affect the development of the filaria larvae in the transmitting insect, with temperature being of pre-eminent importance. The second factor concerns the periodicity, that is the rhythmic occurrence of the microfilariae in the peripheral blood.

2.2.1.1 Temperature, Atmospheric Humidity and Development of Larvae

What applies to other helminths also holds true for *Wuchereria bancrofti:* the duration of the larval development in the vector depends upon the temperature (cf Fig. 1). At a temperature of 16 °C only a few larvae attain the second stage (L_2), and development to the infective larva (L_3) does not occur. At 19 °C the development to L_2 is

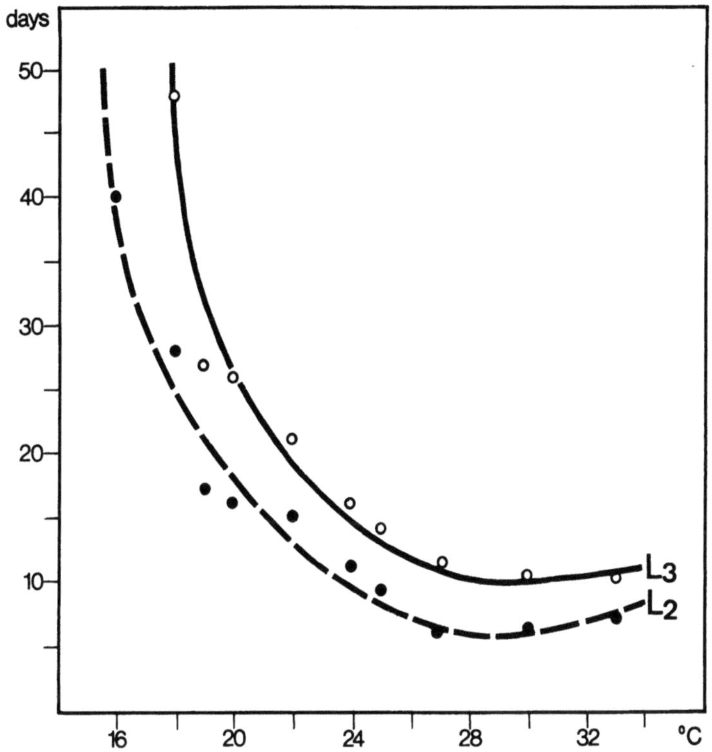

Fig. 1. Development of *Wuchereria bancrofti* larvae in *Culex pipiens pallens* as a function of temperature. L_2 second-stage larvae, L_3 third-stage larvae [Source: 578]

completed after 17 days, that to L_3 after 26 days. A rise in temperature by 3 °C to 22 °C leads to a further shortening of the development period: after 15 days there are L_2 and after 21 days infective (L_3) larvae. Conditions are at their optimum at about 27 °C, a temperature which enables the full development to be completed within 11 days. Temperatures of 30 °C and 33 °C, though shortening the development time by another day may allow the larvae to come to grief: in some larvae a temperature of 30 °C leads to deformative swellings of the caudal body section; at 33 °C most larvae appear to be deformed [cf 578].

Thus a constant temperature of about 27 °C is of special significance; without having a damaging effect upon the larvae it guarantees the shortest development term in the vector; what is more, it presents the optimum for the cultivation of abaca (*Musa textilis*), the leaf-axils of which the chief vector of *Wuchereria bancrofti* in the Philippines – *Aedes poicilius* – prefers as breeding places. A very short development duration is not of so much significance in respect of a rapid succession of parasite generations – time does not matter much where the average prepatence is one year – as regards the limited life-span of the transmitting mosquitoes. The shorter the development period of the larvae, the greater the probability that one individual vector, having taken on microfilariae in the course of biting an infected person, can, in turn, transmit L_3 at its subsequent blood-meal.

The relative atmospheric humidity is also of significance for the development of the *Wuchereria* larvae in the vector. Values in excess of 90 per cent are considered optimal [cf 556]. High atmospheric humidity also plays a role in the process of penetration of the final host by the infective larvae (L_3), in so far as damp skin protects them from drying-out[1]. Thus temperatures of about 27 °C and high atmospheric humidity present optimal conditions for the development of larvae in the vector and for the percutaneous penetration of L_3 in man.

2.2.1.2 Microfilarial Periodicity of Wuchereria bancrofti

The rhythmic occurrence of microfilariae in the peripheral blood at the very same time of day represents an adaptation of the parasite to the activity-phases of the transmitter, thereby optimizing the maintenance of the cycle and in turn serving the survival of the species. The nocturnal periodicity of *Wuchereria bancrofti* in the Philippines is an adaptation to the predominantly nocturnal activity of *Aedes poicilius* and *Anopheles flavirostris.*

Initially, when research into filariasis in the Philippines was beginning, it had been thought to be an aperiodic form of *Wuchereria bancrofti* [71], but later investigations by Phalen and Nichols [617, 618], J. W. Smith [728], Lopez-Rizal and Padua [467], Africa [7] and R. V. Guzman [353] showed that it was a nocturnally periodic form. These observations were later confirmed by Africa et al. [22] on the basis of repeated counts of microfilariae in a patient (cf Fig. 2). Their counts showed that blood samples taken at 10 a.m., 12 noon and 2. p.m. never produced microfilariae in the peripheral blood, whereas the mean maxima of the number of microfilariae occurred in the time between 10 p.m. and 2 a.m.

The final confirmation that *Wuchereria* bancrofti in the Philippines was to be classified as a typically nocturnal periodic form was ultimately obtained by investigations carried out in the Calaccad Valley (Kalinga-Apayao), on Palawan and with prisoners in Muntinglupa (Rizal) [166, 664], the results of which it was later found applied to other endemic areas as well, such as Bongao, Tawi-Tawi [158], and Agusan del Norte [752]. In the case of all these investigations the mean maxima of microfilaraemia adjusted to the average percentage of peak count occurs at night between 10 p.m. and 3 a.m. (Fig. 2).

Consequently the variability of the maximum density of microfilariae in the peripheral blood is distinctly less well marked in persons infected with *Wuchereria bancrofti* than in patients infected with that form of *Brugia malayi* which is to be classified as sub-periodically nocturnal in the Philippines (cf Fig. 3 and 2.3.3). Such differences between the two species of filariae can only be confirmed by calculating the so-called periodicity index. First introduced by Sasa and Tanaka [699, 700] for characterizing the rhythmic occurrence of microfilariae in the peripheral blood, taken over later by Aikat and Das [37] on an alternative basis of calculation, this index has been accepted as a basis for co-ordinating variations in the periodicity of *Wuchereria bancrofti* and *Brugia malayi* as they occur in different endemic areas across the globe. In this regard *W. bancrofti,* attaining indices of 85.79 to 112.36 (Sasa and Tanaka) or 107.81 to 141.78 (Aikat and Das) in the Philippines, turned out to be markedly periodic (Table 2); so far as calculating the time at which the peak values occur, it proved to be nocturnal. For Palawan the so-called "peak hour" is given as being between 23.7 and 0.1 hours on the 24-hour clock (Aikat and Das) or as between

[1] Transmission of an animal filaria (*Brugia pahangi*) to experimental animals by *Aedes aegypti* at a relative humidity of 90–100% showed, for example, a three to seven-times higher rate of penetration for L_3 than at a relative humidity of 40–50% [462]

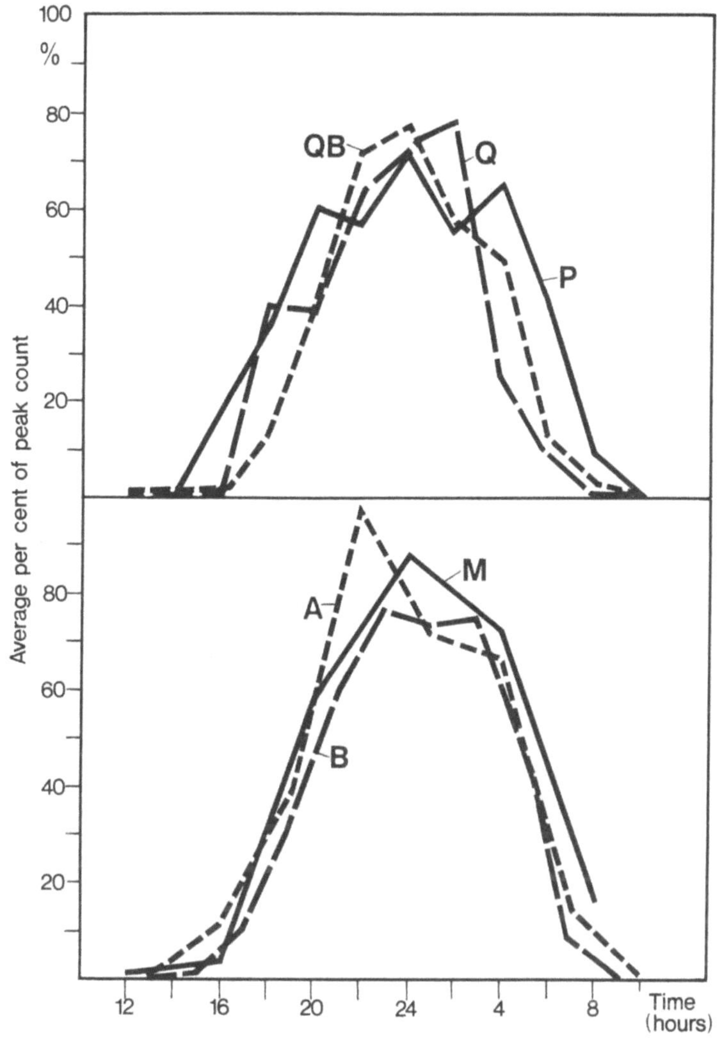

Fig. 2. Microfilarial periodicity of *Wuchereria bancrofti* in different areas of the Philippines. *A* Agusan del Norte; *B* Bongao, Tawi-Tawi; *M* Muntinglupa Prison, Rizal; *Q* Quezon, Palawan; *QB* Quezon, Palawan (patients with *Brugia malayi* double infections); *P* counts of a single patient on 8 different days [Sources: 22, 158, 166, 664, 752]

22.9 and 23.7 hours (Sasa and Tanaka), for Agusan del Norte as 24.0 hours and for Muntinglupa as 0.8 hours [Aikat and Das: cf 750]. Accordingly one may expect the mean maximum of the microfilarial density at about midnight (fluctuating by ±1 hour). To what extent this constitutes evident adaptations to the activity phases of the carriers will be discussed in the following section (2.2.2).

2.2.2 The Vectors of *Wuchereria bancrofti* and Their Bionomics

On the evidence of experimental findings [7, 66, 67, 69, 72, 172, 660, 661] *Culex quinquefasciatus*, probably the most important transmitting species throughout the world,

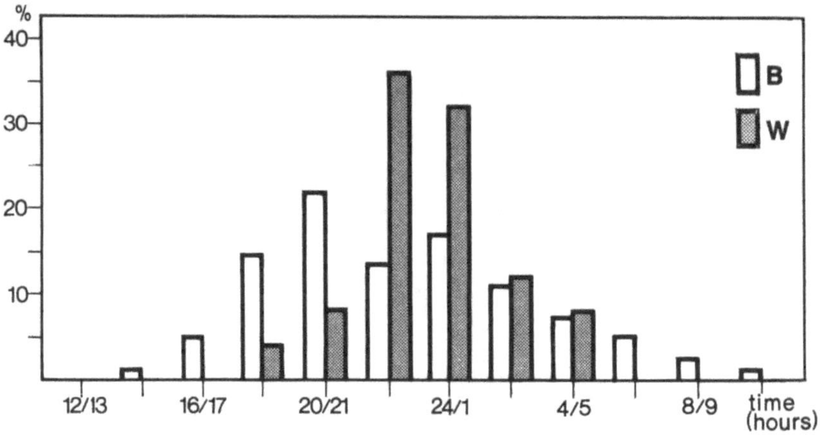

Fig. 3. Percentage distribution of maximum microfilarial densities in wuchereriasis (*W*) and brugiasis (*B*) cases during the course of the day [Sources: 158, 166, 752, 830]

was initially also incriminated as the chief vector of *Wuchereria bancrofti* in the Philippines. But already in 1951 B. D. Cabrera and Tubangui had reported on their field investigations in the Bicol Peninsula where infective larvae of *W. bancrofti* were never found in *C. quinquefasciatus*. A high percentage of another species of mosquito, namely *Aedes poicilius*, proved, however, to be infected with larvae of this sort. Further investigations in the Sorsogon Province, on Catanduanes, Samar and Jolo (cf Table 4) led to a far-reaching confirmation of these findings. However, in other endemic areas like the Calaccad Valley, the only *Wuchereria* focus in northern Luzon, as well as Bongao and Palawan, it was possible to establish *Anopheles flavirostris* as being the vector of *W. bancrofti*. Although *Armigeres baisasi* and various *Aedes, Culex* and *Mansonia* species among all the other Culicidae that have so far been examined proved to be infected as well (cf Table 4), they are likely to be of either no or only rather minor significance as vectors of *Wuchereria*, since they were never (or in only very few cases) host to infective larvae and often achieved only a limited population density. On the evidence of the findings so far presented, it can consequently not be denied that *Culex quinquefasciatus* for instance and *Aedes niveus* – possibly even *Armigeres baisasi* – could under certain conditions and to a limited extent perform the function of additional vectors. Whether this applies to some *Mansonia* species as well must remain doubtful since the findings presented by Estrada and Basio [290] refer to the Quezon and Brooke's Point municipalities in southern Palawan, which later turned out to be endemic for *Brugia*. Without being confirmed for all endemic areas, the general opinion now accepts *Aedes poicilius* and *Anopheles flavirostris* as playing the dominant role in the transmission of *Wuchereria bancrofti* [cf for example 151].

2.2.2.1 Aedes poicilius

Once it had become clear that *Aedes poicilius* functions as the chief vector of *W. bancrofti* in the Philippines [75], research into the bionomics of this mosquito species became of especial significance. It was known that it belongs to the axil-breeding species, i.e. those species whose breeding-places are the water that collects in the axillary joints of plant leaves. It is in these leaf axils that the eggs are deposited and here that the larvae emerge and develop by way of the pupae into the imago.

In principle the leaf axils of a wide variety of plant species can act as breeding-places as long as a sufficient amount of water remains there throughout the period required for the development of *Aedes poicilius.* But in order to establish wuchereriasis, the density of vectors has to pass a certain threshold value; again, this is only possible if there is a sufficiently large number of breeding-places. Either there is a considerable quantity of plants available in the vegetation where water can collect and be preserved in the axils, or cultivated plants meet these requirements.

Baisas et al. [75] have presented a thorough study of this problem in an endemic area of the Sorsogon Province. The following plants were initially considered by them:

Anahaw (*Livistonia rotundifolia)*
Banana "saba" (*Musa sapientium compressa*)
Abaca (*Musa textilis*)
Biga (*Alocasia machrorrhiza*)
Gabi (*Colocasia esculenta*)
Galiang (*Cyrtosperma merkusii)*
Nipa (*Nypa fruticans*)
Pandan "bankoang" (*Pandanus simplex*) and
"various small gabi-like ornamentals and wild plants".

As a result of preliminary investigations, which proved *L. rotundifolia, C. merkusii, N. fruticans* and *Colocasia*-like plants to be lacking in any significance, the study proper concentrated upon the two *Musa* species besides *A. machrorrhiza* and *C. esculenta.* The leaf axils of the plants served as breeding-places for 18 different mosquitoes, among which *Aedes poicilius* and *A. ananae* clearly surpassed the other species in respect of frequency (cf Table 3). *A. poicilius* quite evidently prefers the axils of the "saba" banana as its breeding-place, for the entire collection from 165 of these plants amounted to 2849 larvae and pupae, compared to the 951 found on 150 abaca plants. The same preference has been observed by Kurihara et al. [448], by B. D. Cabrera and Valeza [177, 178] and by Valeza and Grove [803]. It must be attributed to the fact that the axillary joints of the "saba" banana are able to contain more water (an average of 36.23 ml (20–80)) than those of the abaca plants (average 23.95 ml, varying from 15 to 33 ml), and that shrubs of the edible banana have more leaves and thus more axils than abaca plants. However, this evidently only applies to the saba banana, for in Cebu, where the kardaba banana predominates, abaca is the preferred breeding-place. There 37.5 per cent of the leaf axils of the abaca contained larvae of *A. poicilius,* compared to a mere 12.8 per cent of the axils of the bananas [708].

Marked differences have also been observed in the distribution of larvae and pupae over particular axils of a plant, though the investigations of Baisas et al. [75] do not specify whether this is also due to differences in the quantity of water, something confirmed later on by Kurihara et al. [448]. Axils of the abaca at a height of 1–2 m above the ground contain an average of only 8.0 ml, increasing to 25.9 ml at a height of 2–3.5 m, and even to 32.4 ml of water at a height of 3.6–4.5 m. The density of larvae rises accordingly from 0.94 to 7.89 and, finally to 9.39, with *A. ananae* being the only one in evidence.

In contrast to the two *Musa* species, other plants which serve as breeding-places for *A. poicilius* are distinctly less prominent in their importance. This applies even to *Alocasia machrorrhiza*, which favours sites on damp sections of river banks next to running waters. Although Baisas et al. [75] on occasion found larvae in their leaf axils it plays a generally subordinate role as a breeding-place for this species of mosquito because it is periodically cut off from the population and the water contained in its leaf axils frequently turns "soapy". This quality of the water appears to favour the development of *Armigeres baisasi*, which then proceeds to crowd-out all other species. The similarly small importance of *Colocasia esculenta* finds its explanation in its low number of small leaf axils (containing an average of 15 ml of water) and in its being harvested within three months of planting. *A. poicilius* larvae form only some 3 per cent of the total mosquito fauna developing in leaf axils of *C. esculenta* [75], or are totally absent [803]. The same is true of breeding-places in the leaf axils of *Pandanus*. On Catanduanes the axils of this plant never contained the larvae of *A. poicilius* during the period of low precipitation, and at times of ample precipitation Valeza and Grove [803] only found 25 larvae of this species when examining 155 *Pandanus* plants.

Whether the species of plants mentioned above are really able to ensure a vector density sufficient for the maintenance of the *Wuchereria* cycle does not, however, depend upon the degree of their suitability as breeding-places but rather upon their numbers. Suitability as a breeding-place for *A. poicilius* and mass cultivation (1978: 243,800 ha) together therefore make abaca the most significant factor in the epidemiology of filariasis on the Philippines. This fibrous banana (*Musa textilis*) is cultivated in plantations as well as smallholdings. The "saba" banana, however, must also be taken into account as an important breeding-place; it cannot be estimated, however, just what proportion of the total area cultivated with bananas (1978: 284,400 ha) is given over to this variety, nor whether single bushes standing in the immediate vicinity of houses have been included in the official statistics. But thanks to the short flying-distance, such free-standing single bushes may be the very ones responsible for frequent contact between vector and man. In line with the present state of knowledge, however, it is the cultivation of abaca, not of the "saba" banana, which is coupled with the distribution of wuchereriasis.

Nor is it the number of plants alone which cause a high density of vectors, for it also depends to a significant degree upon the frequency and distribution of precipitation in the course of the year, as high precipitation falling throughout the year is the only guarantee for the constant presence of sufficient quantities of water in the leaf axils for development of the mosquitoes. Biotic and abiotic factors thus determine the density of vectors and with it the occurrence and degree of endemicity of wuchereriasis. (In addition the cultivation of abaca itself depends upon environmental factors, which result from the locational and climatic requirements of the plant, so that a more detailed study of this seems to be called for: cf 2.2.3.)

Even in those areas of abaca cultivation which concentrate upon the two climatic zones without a dry season but with ample precipitation (cf Map Plate 1), a distinct seasonal dependence may be observed in respect of the density of larvae in the leaf axils, as well as in the population density of the imagos. This can be proved unambiguously by reference to investigations in the Sorsogon Province [75, 85, 177, 178, 290]. As a rule the density of larvae in the leaf axils of abaca attains its maximum values during the particularly wet months of November and December, its minima in the rather drier months from May to July. If, for example, November and December yielded about six larvae each per abaca bush, their numbers fell to 0.2 to 0.5 larvae per plant between May and July [75: cf Table 5]. A particularly marked correlation between the density of larvae in the leaf axils of abaca and the annual course of precipitation was noted by Basio [85] on the basis of investigations carried out in Barrio Bentuco, Gubat, over the period from June, 1960 to May, 1961 (Fig. 4). Although the quotient calculated by him does not permit conclusions concerning the absolute density of larvae, the seasonal dependence it expresses may nonetheless reflect the actual conditions, since the proportion of leaf axils used by *A. poicilius* for breeding fluctuates by less than a factor of two, i.e. between 36.4 and 60.9 per cent in the course of a year [75]. This small degree of fluctuation can therefore exert only an insignificant influence upon the great difference between the density of larvae – which Basio [85] calculated – of 3.1 and 109.1 (factor 35). It must therefore be assumed that the quantity of water in the axillary leaf-joints, which depends in turn upon precipitation, determines the density of larvae.

The "saba" banana as a breeding place is, on the other hand, probably subject to other conditions (cf Table 5). Apart from the findings in the case of abaca, the maximum for the number of observed larvae on the plant was 17.24 in September, the minimum of 3.05 per bush being in February. Here the increase in precipitation that commences in June appears to be the cause of the high density of larvae, which maintain their above-average values until January and drop markedly only then.

The seasonal fluctuations in the population density of the adult *A. poicilius* on the other hand follow quite a different rhythm, for by comparison with earlier months

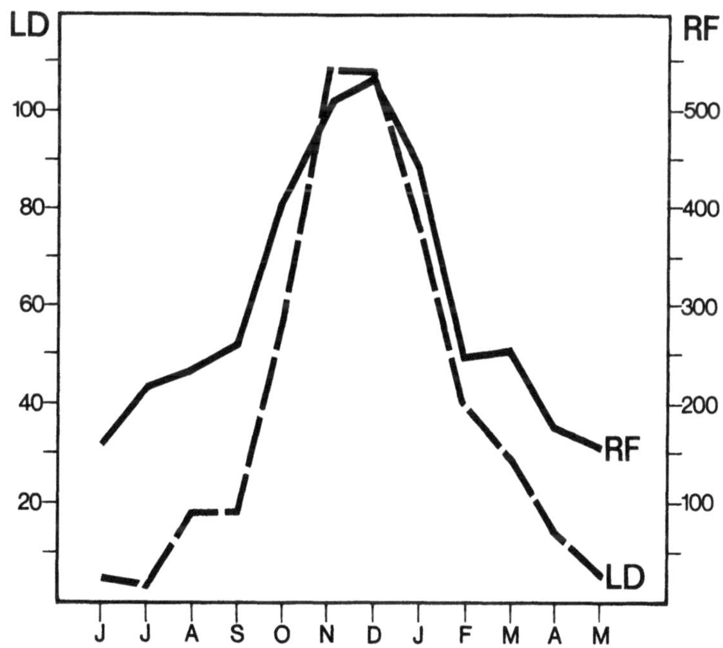

Fig. 4. Monthly densities of *Aedes poicilius* larvae and rainfall in Sorsogon Province
LD larval density in Gubat = $\frac{\text{no. of larvae collected}}{\text{no. of axils with water}} \times 100$
RF rainfall (mm) in Juban (20 years' average) [Sources: 85, 827]

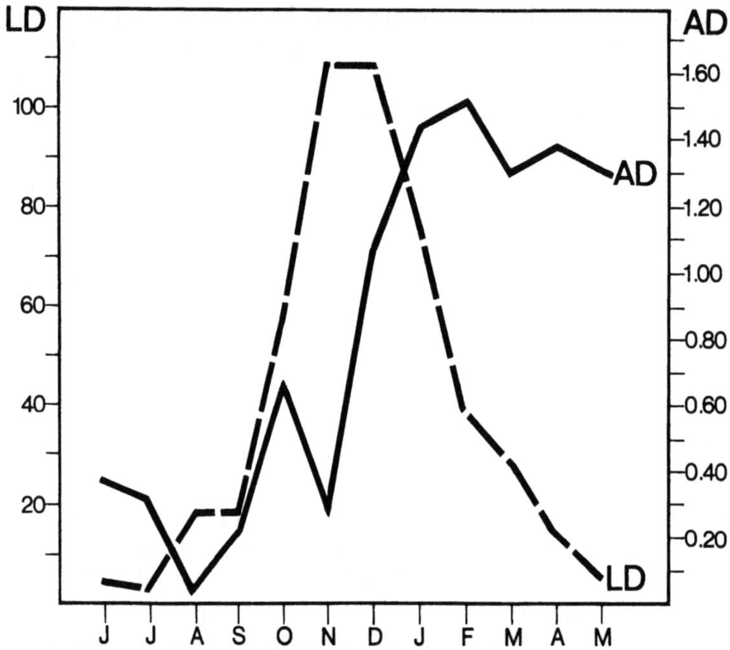

Fig. 5. Monthly densities of *Aedes poicilius* larvae and adults in Gubat, Sorsogon

LD larval density $= \frac{\text{no. of larvae collected}}{\text{no. of axils with water}} \times 100$

AD adult density per 10 minutes [Source: 85]

the number of imagos increases noticeably only in December or January; the density remains at a high level until and including April or May, before suddenly dropping in May or June [85, 177, 178]. This seasonal rhythm is, however, coupled throughout with the density of larvae: a high population density of adult mosquitoes presupposes an earlier high density of larvae; due to the development period from egg to imago a shift of phase of about one month must be reckoned with (cf Fig. 5). A drop in larval density, however, has no immediate effect upon the population density of the imagos since their life-span exceeds the term required for the development of the imago. The life-span of the imagos is finally the reason for their persistence of a high-level of their numbers until well into May. It is no contradiction that the density of larvae decreases continually during this time, in spite of the presence of numerous adult ♀♀, but this can be explained by the decreasing quantity of precipitation. It is not the diminished availability of breeding-places that matters, but the smaller quantity of water in each leaf axil. The intra-specific competition this causes – together with the competitive pressure exerted by other axil-breeding species – must necessarily lead to a decrease in the density of *A. poicilius* larvae.

In this context the question remains to be discussed, too, of whether the seasonal fluctuations in the density of adult *A. poicilius* is in any way accompanied by differences in the risk of infection for man. Basio [85], who had obtained 1,902 *A. poicilius*-♀♀ by "human bait trapping" in Gubat (Sorsogon) and examined them for *Wuchereria* larvae during the period July, 1961 to June, 1962, found an average infection rate of 4.47 per cent, with monthly fluctuations of 1.71 to 17.04 per cent. However, these percentages are of little value by themselves as no allowance is made for the seasonally-related population density in this. Basio's data [85] on the monthly density of the imagos and the corresponding infection rates, however, permit observations as to fluctuations, possibly even regular fluctuations, in the number of infected mosquitoes over the course of the year. This is evidently the case (cf Table 6), since the population density of infected mosquitoes is clearly below the monthly average (= 100%) during the months from June to November, and is followed by an almost explosive four- to five-fold increase. These conditions thus largely agree with those of the overall population (cf Fig. 5). During the months December to May, people in the endemic areas are quite clearly more frequently exposed to infectious bites than during the other half of the year. (For the actual number of infectious bites in the course of the year cf 2.2.5).

Besides the indispensable condition which *A. poicilius* fulfills that the main vector must be an anthropophilic species, a distinct concordance between the periodicity of microfilariae and the activity of vectors in the course of the day is required. Since *A. poicilius* is predominantly active at night [cf for example 85, 290, 803, and Fig. 6], though it will also bite during the day in abaca plantations, if it is rainy and the sky covered by cloud [151], such an accord also exists with the nocturnal periodicity of *W. bancrofti* in the Philippines (cf 2.2.1.2). As a rule the mosquitoes begin to enter the houses after sunset, reaching their maximum density towards midnight, and leave the houses again before sunrise, having rested before and after their blood-meal. During the day they live in the open in cool places, which offer protection against wind and

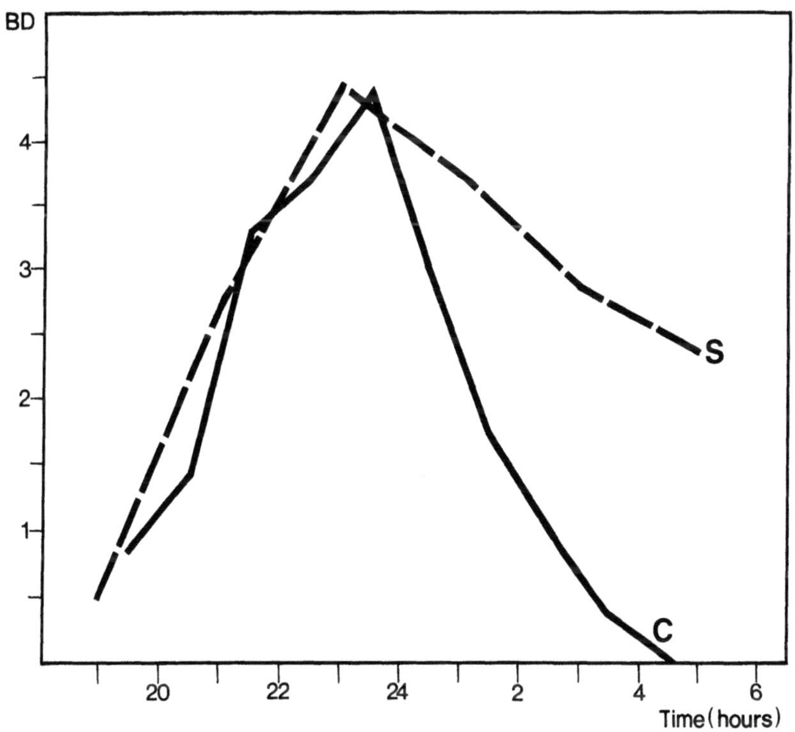

Fig. 6. Hourly and bi-hourly biting densities (*BD*) of *Aedes poicilius* in Sorsogon (*S*) and Catanduanes (*C*) [Sources: 85, 803]

direct solar radiation; when at their greatest density they keep a distance of about 10 m from the houses [cf 290].

The activity phases of infected mosquitoes do not differ seriously from those of the overall population; for man the risk of infection also concentrates upon the time between 10 p.m. and 2 a.m. When calculating the mean transmission index from the data given by Estrada and Basio [290] for the areas of Sorsogon Province (Areas I, II, III), the percentages for the different hours of the night are as follows:

6– 8 p.m.: 2.3%
8–10 p.m.: 14.2%
10–12 p.m.: 27.8%
0– 2 a.m.: 24.0%
2– 4 a.m.: 18.8%
4– 6 a.m.: 13.0%

Thus not only its breeding habits but also the distribution of its biting frequency makes *Aedes poicilius* an optimal vector, especially in those areas of the Philippines where abaca is cultivated.

2.2.2.2 Anopheles flavirostris

Although there is little evidence available (cf Table 4) *Anopheles flavirostris* is considered the main vector wherever *Aedes poicilius*, the axil breeder, is absent or reaches only a low density. This applies above all to those endemic areas where abaca is not cultivated, and ought therefore to be in the main limited to the Calaccad Valley and Palawan. The bionomics of *A. flavirostris* is very well documented, as this mosquito species also represents the most important vector of malaria in the Philippines [cf for instance 152]. This fact also results in a problem which has hitherto not yet been entirely solved, however: namely, why is *A. flavirostris* not thought to function together with *Aedes poicilius* as vector in areas where malaria is endemic and abaca is cultivated, or, in other words, in areas where abaca cultivation takes place and malaria occurs endemically. This contradiction can unquestionably not be resolved by appropriate investigations in the south of the Bicol Peninsula, an area considered largely free from malaria [cf the map in 152], although it has attracted a concentration of investigative work on filariasis in the Philippines. If there are no different strains of *A. flavirostris* on the various islands of the Philippines, which differ in their vectorial capacity, it does not preclude both these mosquito species from functioning as vectors of filariasis, at least not in the case of Mindanao. Appropriate entomological investigations are not so far available, however. If the population density of *A. flavirostris* in the abaca-growing areas of Mindanao were similar or as low as had been established for the Sorsogon Province, Catanduanes and Samar – to name but a few examples [175, 177, 178, 803, 830] where this species was frequently completely absent from the mosquitoes collected – *A. poicilius* will have to be regarded as the sole vector in Mindanao as well. Moreover, the correlation established there between the cultivation of abaca and the prevalence of Wuchereria (cf 2.2.3.4) points to the dominance of *Aedes poicilius* as vector.

An essential difference between the two species, however, is that *A. flavirostris* presents a predominantly zoophilic species [151, 152], though Bruce-Chwatt et al. [110] in their list of host preferences of this *Anopheles*, give a figure of 47.1 per cent – out of 140 positive tests – containing primate blood (Bovinae: 48.6%; Suidae: 4.3%).

On the other hand this result is evidently to be attributed to the fact that the examples under investigation had been indoor catches, for only 7 per cent of the 1,433 *A. flavirostris* caught outdoors contained human blood, compared to 57 per cent (out of 122) which had been caught at night with a human-baited trap [231, 232].

So too in their breeding habits the two species differ: unlike *Aedes poicilius*, *A. flavirostris* does not breed in leaf axils but as a stream-breeder prefers clear, cool and slow moving waters, on the edge of which larvae and pupae develop in the shade of bamboo, grasses and other plants. "Slow moving waters such as streams, seepages and rivers with grassy edges" [231, 232] are regarded as typical breeding-places of *A. flavirostris;* they are chiefly found in the "foothills and mountain slopes" [152].

This species of mosquito occurs in all the provinces of the Philippines [671] and occasionally even up to altitudes of 1500 m above sea-level [152]. As to the flow-velocity of the waters, the speed of 1.5–16.5 cm/sec seems to be optimal for achieving the greatest larval density; the critical point appears to be at 30 cm/sec [410; for further results on the physico-chemical qualities of breeding-places preferred by *Anopheles* see 411–414].

The seasonal fluctuations of the biting and population density of *A. flavirostris* evidently depend upon local conditions and the methods of trapping them. The greatest density both of larvae and imagos was observed mainly during the months December to February [cf for example 155, 231, 232, 799], but investigations in the context of transmission by *W. bancrofti* are not available.

Belonging to the nocturnally active mosquitoes, *A. flavirostris* shows a distinct similarity to the periodicity of the microfilariae of *W. bancrofti* in the Philippines. Seen from this point, *A. flavirostris* can be considered a suitable vector. Its pronounced exophilia and zoophilia present qualifications. At nightfall it leaves its resting places, which tend to be in the vegetation on the banks of waterways, and looks mainly for carabaos (water buffalo) as hosts. In this regard the greatest density of bites is observed between 10 p.m. and midnight. Those entering houses and biting man attain their maximum density a little later between 11 p.m. and 1 a.m., a period that coincides with the greatest density of microfilariae in the peripheral blood of man (cf 2.2.1.2).

In conclusion, it can be established that knowledge of the bionomics of *A. flavirostris* is based almost exclusively on the investigation of malaria epidemiology in the Philippines and has been placed on record in numerous publications (cf the bibliographies of Russell [670] and D. W. Walker [820]). Thanks to its distribution throughout the country, this species of mosquito may be regarded as a potential vector for all the areas in the Philippines with endemic wuchereriasis, but due to marked zoophilia and exophilia their significance as actual vectors pales considerably when compared with that of the anthropophilic *Aedes poicilius.*

2.2.3 Abaca Cultivation and Its Importance in the Distribution of Wuchereriasis

2.2.3.1 Musa textilis and Its Locational Conditions

Very similar to edible bananas, *Musa textilis*[1] is a member of the Musaceae, a bush of about 3–6 m height with an inedible fruit, 5–10 cm long, rich in tannin, and leaves which attain lengths of 90–240 cm and a width of 30 cm.

[1] The information of *Musa textilis* and its cultivation in the Philippines presented in the following sections was taken from the publications by Copeland [237], Kolb [439], Huke [395, 396], G. Franke [303], Berger [92] Elsdon & Pirson [289], Purseglove [633] and W. Franke [304].

The cultivation of abaca requires the following conditions:

- The presence of deep, loose soil such as occurs in volcanic loams and alluvial soils.
- The soil must manifest good water storage, without however permitting the water to become impounded, i.e. the soil structure must not be too fine because inadequate drainage could then lead to the rotting of the roots. Too coarse a soil, on the other hand, drains too fast and does not allow the plants to develop a firm anchorage.
- The ground-water level must not exceed 90 cm (root rotting!).
- The climatic requirements are a tropical lowland climate with average temperatures of 27 °C, a relatively high atmospheric humidity of 78–86%, and a high level of precipitation, distributed evenly throughout the year and reaching a total of at least 2500 mm.
- The cultivation areas must be situated outside the main typhoon zone, since *Musa textilis*, being a shallow-rooting plant, is susceptible to wind-breakage. (On the Bicol Peninsula in particular attempts are made to meet the danger of wind-breakage by planting shelter-trees. Leguminosae plants are preferred for this, as they add the necessary nitrogen to the soil at the same time.)

These conditions are met in the more southerly parts of the Philippines, especially in the river plains and coastal areas at altitudes of up to 500 m. Abaca will flourish there, and it is precisely under these conditions that *Aedes poicilius* can develop in great numbers and transmit *Wuchereria*.

2.2.3.2 The Increase in Abaca Cultivation and Its Present Importance

The Philippines may be regarded as the country of origin of *Musa textilis;* it has supposedly been grown there for its fibres from ancient times. The manufacture of fabrics from abaca was already practiced when Magellan "discovered" the Philippines in the 16th century. Until the beginning of the 19th century, however, cultivation took place exclusively for the home market. A systematic expansion of the area under cultivation only began after a sample of fibres had reached the U.S.A. in the year 1820, and where their great strength and resistance to seawater was recognised, along with a low specific gravity as another important advantage. In the period that followed abaca fibres were preferred for the manufacture of ships' ropes and fishing nets.

Abaca was first exported in 1818; the evidence is of a total of 41 tonnes. A gradual increase in production was followed by an expansion of abaca processing (cf Table 9 and Fig. 7), eventually leading at the turn of the century to abaca constituting almost 68 per cent of the Philippines' total exports. The peak of the development was reached in 1929, when a cultivation area of 500,000 ha yielded more than 200,000 t of fibres. Fluctuations in the price on the world market, but above all World War II, led to a drastic reduction in cultivation. It only gradually recovered later – all the more since competition brought about by the development of man-made fibres necessitated a shift in the use of abaca to other areas. It now includes the production of paper (for instance for cigarettes and paper money) as well as the manufacture of handbags and other similar products. The data for 1978 show a cultivated area of 243,800 ha, producing 129,800 tonnes.

Nonetheless significant regional changes in regard to abaca production have taken place in the Philippines in the course of time, and undoubtedly had their effect upon

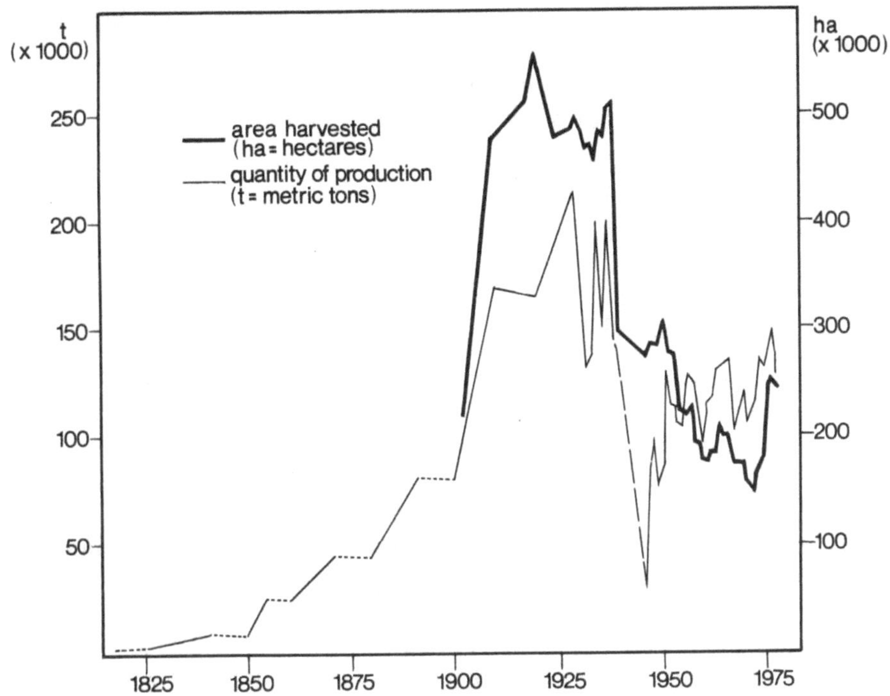

Fig. 7. Abaca in the Philippines: area harvested, 1903–1978, and quantity of production, 1818–1978 [Sources: 395 and 1980 Philippine Statistical Yearbook]

the endemicity of wuchereriasis. At the beginning of the 20th century Camarines Sur, Albay and Sorsogon, three provinces of the Bicol Peninsula, predominated. In 1903 they contributed about 63 per cent of the area under cultivation (= 137,738 ha). Against that other also important areas of cultivation like Leyte (22,038 ha), Samar (12,368 ha) and northern Mindanao remained quite unimportant. At that time Davao with 2,499 ha under cultivation supplied less than 1 per cent of the overall Philippine production. There was a distinct shift in the years to come. By 1918 the cultivated area in the Davao region had increased tenfold by comparison with 1903; by 1937 it had reached a total area of 108,820 ha, contributing 31 per cent of the country's overall production. At that time the share of the Albay and Sorsogon provinces (including Catanduanes) had decreased from 46 to 30 per cent of the total area, whereas the cultivated area on Leyte and Samar had roughly trebled (forming about 20 per cent of the total area). Over and above this there were further important abaca plantations at Cavite (> 20,000 ha) and Laguna, on Romblon and Bohol after World War I, which had to be almost totally abandoned later on when the plants became diseased. Since then the situation has largely stabilised at this level. In the early 'sixties the areas under cultivation were divided among the regions roughly as follows:

Davao	40%
Albay, Camarines Sur, Sorsogon, Catanduanes	32%
Southern Samar and Northern Leyte	9%
Cotabato, Surigao, Bukidnon, Camarines Norte	12%
Other provinces (30)	7%

Davao also maintained its position at the top with regard to its share in the production (1960: 45%).

A distinctly different distribution of cultivated areas is, however, furnished by Salita and Rosell [676] for the period 1970–74:

Southern Tagalog	0.1%
Bicol Region	40.8%
E. Visayas	31.1%
W. Visayas	2.4%
N. & E. Mindanao	11.9%
S. & W. Mindanao	13.7%

This change from the situation in 1960 does not reveal whether this is a passing or permanent one. For example, the possibility cannot be excluded that abaca cultivation on Mindanao has only temporarily declined as a result of disease such as Baisas et al. [75] observed in the Cotabato provinces, where mosaic disease had brought about a calamity that destroyed numerous abaca plantations.

Though the cultivation of abaca has markedly declined in significance by comparison with the 1920s and 1930s – as can be demonstrated with a number of parameters (cf Figs. 8 and 9) – there are still an estimated 2.5 million persons in the Philippines engaged in the cultivation, harvesting and manufacture of this fibrous banana, both as peasants or as plantation workers. These people are thus to a considerable degree exposed to the risk of *Wuchereria* infection[1]. Thanks to the agricultural-industrial development which began in 1904, this applies above all to the Davao area. The example of this particular area is to demonstrate the circumstances which set the development going there, thereby causing the continuous increase in the risk of infection.

2.2.3.3 The Davao Case

The flourishing of abaca cultivation in Davao began as a result of Japanese development work, with the construction of a road that was to connect Manila with Baguio – which is situated in the mountains to the north. When attempting to carry out this project, which the then American colonial administration had initiated, the indigenous Philippine labour proved inefficient. Eight hundred Japanese workers from Okinawa were therefore hired for the continuation of the road works. Having completed the road ahead of time, about 500 of these workers lacked the necessary means for the return journey to their homes, and remained in the Philippines. Under the leadership of K. S. Ohta 150 [439] or, according to other sources, 350 of these former road labourers [395, 396] left for Davao, where there was a shortage of labour in the abaca plantations. Kolb [439] depicts the further development as follows:

> All of them found employment in the vicinity of Davao, then still a small capital, some on the Juan-Awad Plantation in Lapanday, others on the Sanchez Plantation in Barong, or on the Libby Plantation in Libby. Ohta himself did not work as an agricultural labourer but opened a grocery store at Talomo. On coming to Davao none of these Japanese had intended to stay for good. But as a number recognised the great agricultural possibilities, the fertile volcanic soils and the guaranteed

[1] For instance, taking into account the usual intervals between plants of 3.70 m x 3.70 m to 2.80 m x 2.80 m, a cultivated area of 243,800 ha would provide a total of about 250 million bushes of fibrous bananas as breeding-places for the vector *Aedes poicilius*

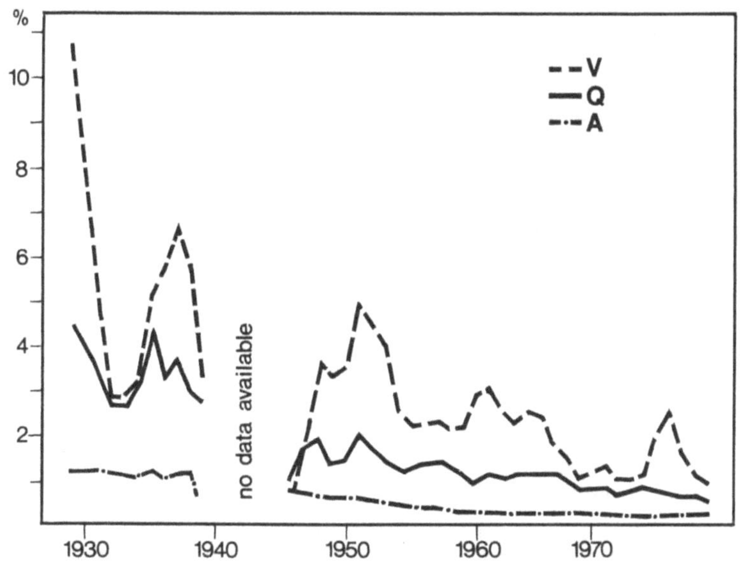

Fig. 8. Abaca in the Philippines 1929–1978: area harvested (*A*), quantity of production (*Q*) and value of production (*V*) as percentage of all crops [Source: 1980 Philippine Statistical Yearbook]

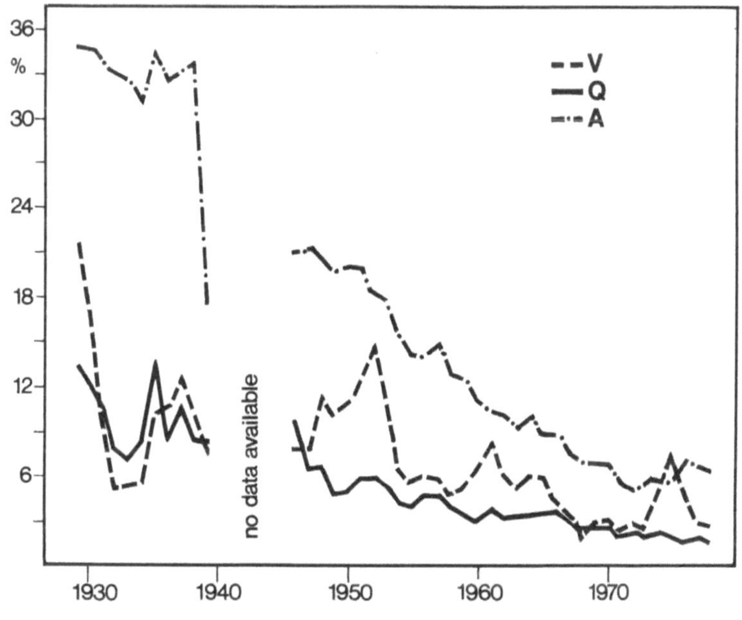

Fig. 9. Abaca in the Philippines 1929–1978: area harvested (*A*), quantity of production (*Q*) and value of production (*V*) as percentage of commercial crops [Source: 1980 Philippine Statistical Yearbook]

> sales of manila hemp offered, they themselves became producers and as such became involved in the economic development of Davao. On May 3rd, 1907, Ohta, together with a number of friends, founded the Ohta Development Company at Talomo, which was to develop into one of the largest Japanese companies in the years which followed . . .
> Clearance and planting was mostly carried out in corporate co-operation by the partners who had founded the company. The Ohta Plantation became a model for quite a number of Japanese company foundations which followed . . .

By 1918, when the area under cultivation with abaca amounted to 24,432 ha, the number of Japanese companies had risen to 60 and the number of Japanese living at Davao to about 10,000. Though at the time Davao had not yet caught up with any of the traditional abaca-growing provinces in respect of the extent of the cultivated area, these others were then already outclassed in terms of quality and productivity. After World War I falling prices for raw materials led to stagnation and decline in abaca production and a recovery in the market only came about in 1927; the Japanese companies in particular took advantage of it. By 1939 the Japanese population had increased to 17,888 in the town of Davao alone. Davao had become a Japanese abaca colony, holding 40 per cent of the entire Philippines area under abaca cultivation.

Even in the early 1960s the Davao area still occupied the first place among the abaca growers. Ten years later, however, a distinct change had taken place: due to plant diseases and attacks by pests there was a severe recession; Davao lost its position as the greatest abaca producer. Burley [123] comments on this:

"However, in recent years failure to control pests and diseases not only has forced farmers northward from the original centre of production around Davao but also has reduced the overall importance of the region as a whole."

On the evidence of our knowledge of the close connection between abaca growing and the prevalence of wuchereriasis it may be assumed that the Davao development described above was followed by changes in the occurrence of wuchereriasis. For this the following causal chain can be established:

- construction of the Manila-Baguio road
- inefficiency of the indigenous labour
- recruitment of Japanese labour from Okinawa
- lack of funds for their return on completion of the road
- shortage of labour at Davao
- transfer of the Japanese to Davao
- foundation of Japanese abaca companies
- great demand for manila hemp before and during World War I
- continuously increasing extension of cultivated areas
- implicit creation of breeding-places for *Aedes poicilius*
- high population density of the vector
- increase in wuchereriasis
- diseases and pest attacks affecting the abaca bushes
- reduction in cultivated areas
- reduction in breeding-places for *A. poicilius*
- decreasing population density of vectors
- fewer cases of wuchereriasis
- shifting of the overall problem to areas further north

2.2.3.4 Wuchereriasis Prevalence and the Area Under Abaca Cultivation

In conclusion it remains to be discussed whether there is quantitative proof of the significance which is ascribed to abaca growing for the epidemiology of wuchereriasis. However, a more precise analysis cannot rest on the knowledge of prevalence rates and of the area under abaca cultivation alone; one would rather more require data on the absolute number of infected persons in each endemic area and their microfilarial density too, since they are the only ones to characterize the actual adverse impact upon a population group. Moreover, it ought to be known in every single case whether cultivation takes place on a smallholding or on a large plantation, as the two differ in their labour requirements and thus in the number of persons exposed to risk. Another important factor to be considered would be the distance of domestic dwellings from the abaca fields, and the number of people living in the houses. Besides this, it is important whether saba bananas are grown in the vicinity of the homes or not, as they may also figure as breeding-places for *Aedes poicilius.*

Such detailed investigations, which include all the factors listed above, have so far not been carried out in the Philippines. Only B. D. Cabrera and Tamondong [169, 171] had, in the case of Mindanao and Jolo, sought to establish whether, on the basis of prevalence rates and the area under abaca cultivation, a demonstrable relationship exists between the two variables. Based upon their calculations, statistically significant co-efficients of correlation were finally arrived at for Agusan, Bukidnon and Davao provinces, as well as for the whole of Mindanao. By ordering the prevalence rates given by B. D. Cabrera and Tamondong [171] for 128 municipalities on Mindanao in groups according to the extent of their particular area under abaca cultivation, the dependence of wuchereriasis on abaca cultivation can be made graphically clear as well (Fig. 10); an increase of the area under abaca cultivation from 0 to 1600 ha is coupled with a rise in the average prevalence from 0.73 to 2.60 per cent. But where the abaca cultivation area is regarded as a percentage of the total available area for agricultural use, a relationship is at first scarcely recognisable; an increase in the proportion from 0 to 3.9 per cent results in a slight rise from 0.73 to 0.97 per cent; only a proportion up to 7.9 and then again to 15.9 per cent causes an increase in prevalence by about 1 per cent. in each case. Once the proportion of land under abaca cultivation rises to 16 per cent and more, a decrease in prevalence by 1.2 per cent sets in. The then lower rate of microfilaraemia (1.73%) in municipalities where abaca plays a particularly large role could be attributed to the fact that this work is done mainly on large plantations. But only four out of 128 municipalities belong to that category presenting a proportion of 16 per cent or more under abaca cultivation, so that the small number alone distorts the true picture.

To summarize, it must be stressed that the case of Mindanao at least permits the establishment of a quantitative and statistically secure correlation between abaca cultivation and wuchereriasis prevalence. This does not seem to hold true for the situation on Jolo [169]. There, as in other endemic areas, further investigations are required with the parameters mentioned above being borne in mind if one is to arrive at a final judgement.

2.2.4 Frequency and Transmission Index

At present a total of about 18 million people are exposed to the risk of infection *by Wuchereria,* all of whom live in the rural areas of the Philippines' 42 endemic prov-

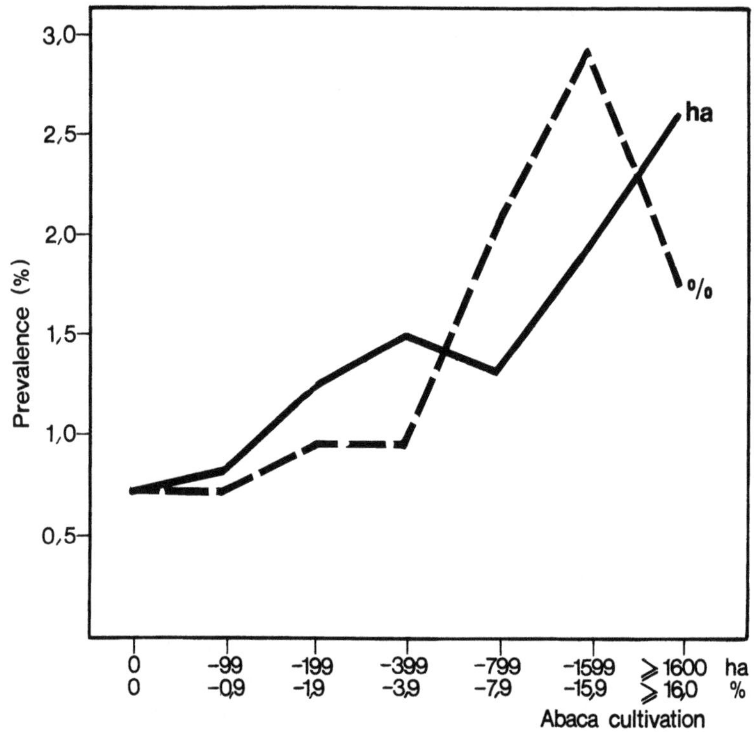

Fig. 10. Wuchereriasis and abaca cultivation in Mindanao. *ha* abaca-cultivated area in hectares, % abaca-cultivated area in percent of available farm land [Source: 171]

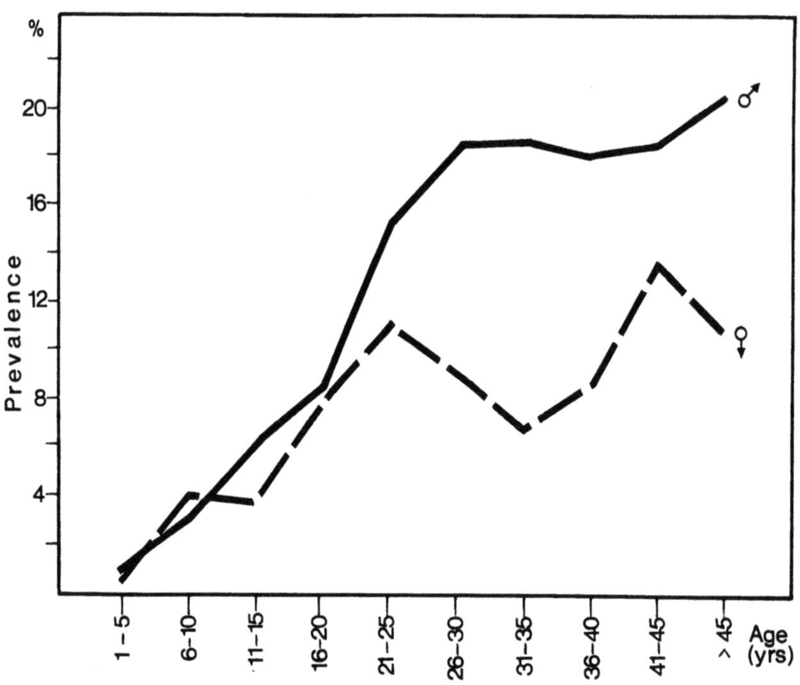

Fig. 11. *Wuchereria bancrofti* microfilaremia rates in the Philippines by age and sex (average of data) [Sources: 74, 168–171, 175, 665, 667, 830]

inces. Assuming an average rate of prevalence of 3.7 per cent means that more than 660,000 Filipinos are actually infected. This average prevalence naturally remains distinctly below the values in the endemic areas where – as on the Bicol Peninsula, on Mindoro, Samar, Mindanao and the Sulu Archipelago – rates of 20–30 per cent and more have been noted in some places (cf Table I[1]). Almost all the mass examinations carried out so far are based on the evidence of microfilariae in the blood smears. As has been shown in the case of the Philippines as well [720, 721] this method does not detect patients with a low density of microfilariae in the peripheral blood. Thus the rate of microfilaraemia in the Barrio Salvacion (Sorsogon) amounted to 11.5 per cent where blood smears only were evaluated, but to 16.1 per cent where additional millipore and nuclepore techniques were applied. If the overall situation in the Philippines were based on this result, the average prevalence would have to be around 5.2 per cent and the number of all wuchereriasis cases about 920,000. Whether this is actually so must remain an open question until a larger number of comparative investigations becomes available.

For the exact assessment of the situation in the endemic areas themselves, however, average prevalence is insufficient as a sole measure, since the structural selection of examinees tends to differ more or less markedly from that of the actual population. It is therefore absolutely essential – especially for the working of the control programmes – to analyse in greater detail the conditions of various age groups of both sexes (cf Fig. 11 and Table 7).

Assuming the equal disposition and exposure of all population groups to the bites of the transmitting insects, the prevalence of wuchereriasis (together with the numbers of infections) would be rising with increasing age so that the highest rates (linked with the greatest parasite density) occurred amongst the oldest people. Such age-dependencies are not, however, solely a consequence of disposition and exposure. On the contrary, they are not insignificantly modified by the age of the parasite as well as by the progressive condition of the host's immunity. Admittedly too little is known about predisposing factors favouring or impeding an infection with *Wuchereria*, i.e. about varying susceptibility of sections of the population and about the immunological process in an existing wuchereriasis and its effect upon successive superinfections, to be able to include these factors in the analysis. In general, differences in prevalence are chiefly attributed to differences in exposure.

Thanks to a fair number of mass examinations (cf Tables 11 and 12), from which mean values have been calculated (Fig. 11, Table 7) this appears to apply up to the age group 26–30 years (♂♂) and 21–25 years (♀♀). In both sexes an almost continuous increase takes place until then, though distinct differences in prevalence can already be observed in the 21–25 year age group (♂♂ 15.1%, ♀♀ 10.9%). Whereas the increase continues still further among the men, probably due to exposure, prevalence among women gradually decreases to 6.8% among the 31–35 year-olds. As the prevalence rates given above were based exclusively on the evidence of microfilariae in the blood, they do not of course permit the conclusion to be drawn that the decrease in the rate of microfilaraemia implies a decrease in the rate of infection by adult worms, for it is known – evidently under the influence of immunological processes – that adult worms continue to thrive, though possibly limiting or ceasing with the production of microfilariae.

[1] Tables I–IV are on the back of the Map Plates

The drop in the rate of microfilaraemia in later life may therefore well be attributed to various causes. A mere reduction in exposure may be responsible, though it would have to date back a long way considering the life-span of adult worms of 10–15 years [cf 280] – in the period of puberty and the early marrying-age in the case in question. This in turn touches upon another factor: the hormonal change of the female organism, which takes place at that time. It is known from experimental investigations with animals that it can influence the susceptibility to helminth infections. But it could also be a matter of the interaction of reduced exposure, relative immunity and hormonal change that explains the drop in the rate of microfilaraemia. Likewise only assumed but unproven is the later rise in the number of infections to 13.5 per cent among women of the 41–45 year age group as being caused by a reversal of these factors (increased exposure, reduction of relative immunity, menopause).

Conditions are quite different among the male sex. After a continuous increase the rates of infection with microfilaraemia reach a maximum (18.4%) among 26–30 year-old men, and remain more or less at that level. There is no doubt that here, too, exposure plays an important role in the course of the graph. Besides the relative immunity, which prevents a further rise in prevalence, it must be borne in mind that among 26 to 30 year-old men the reservoir of exposure may already be almost entirely exhausted. In an area of abaca cultivation, for instance, the section of the population actively engaged in cultivation – and thus in the front-line so far as bites by *Aedes poicilius* and the risk of infection are concerned – only constitutes part of the overall population, whilst other groups are less infected, or not at all. Other additional causes may, however, also be responsible for the marked difference among adult men and women, as experimental investigations of animals have shown the distinctly higher susceptibility of males for some helminth infections (sex hormones play an important role in this).

As far as man is concerned – and he as the sole final host is the only one of epidemiological significance besides the vector – it is not only the different rates of microfilaraemia in the various age groups of the two sexes which determine the valency as transmission reservoir; for a more exact evaluation the age structure of the population and the density of microfilariae in the blood of the different sections of the population need to be taken into account as well, for the high prevalence in a small section of the overall population may well be of much less epidemiological relevance than a low prevalence in a relatively large section of the population. In other words, it revolves around the question of the distribution of the microfilaria carriers, which are actually present, among the different age groups of the two sexes – i.e. prevalence rates or, as the case may be, microfilaraemia, are to be weighted according to the population structure.

This approach shows that carriers of microfilariae are clearly over-represented in the 21–30 year-age group of men, as 16.8 per cent of the microfilariae-positive cases are found in this population group, which constitutes a share of a mere 7.9 per cent of the population (Fig. 12). Thus the very same young men are affected who have to undertake the hardest work in the fields, including above all the stripping of abaca, which requires the greatest application of energy. At the same time they constituted the most important potential for the infection of vectors, and thus for the maintenance of the cycle. In all, 50.4 per cent of all microfilaria carriers are adult men (> 20 years) who, on the other hand, constitute only 22.4 per cent of the total population.

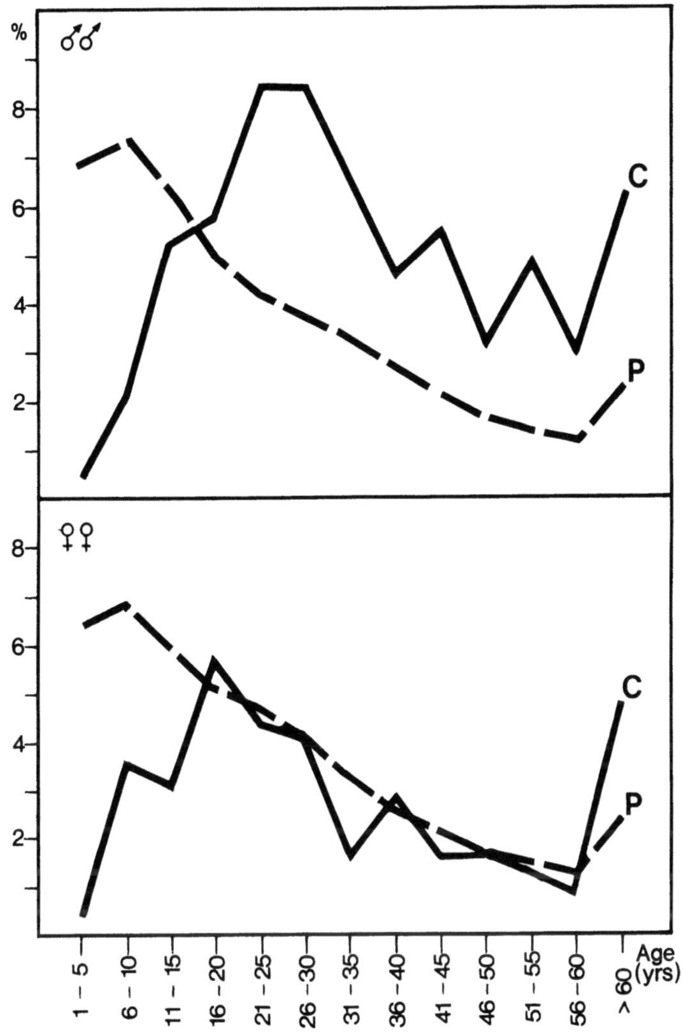

Fig. 12. Age and sex distribution of cases positive for microfilariae of *Wuchereria bancrofti* (*C*) and population structure (*P*) in the Philippines (average of data) [Sources: 158, 168–171, 175 and 1980 Philippine Statistical Yearbook]

Among the adult women of the same age, however, conditions appear to be perfectly balanced (23.4% of the microfilaria carriers in 23.8% of the population).

Estimating the epidemiological importance of sections of the population on the basis of microfilaraemia rates and population structure of course only applies when the density of microfilariae, which indicates the level of infection, too, can be assumed to be equally high – i.e. when quite independently of the donor the vectors take up an equal number of microfilariae when feeding on blood. This, however, is not the case, as a number of investigations have shown (cf Fig. 13, Tables 7 and 13). Assuming the quantity of blood taken in by the transmitting mosquitoes in one bite to be on average 2 µl, 2 to 4 microfilariae on average (with a range of < 1 – 19) will be contained in it. As a rule this average figure should neither impede the insect's ability to fly nor the period of its survival, but it must be taken into account that up to twelve times more microfilariae can be absorbed by the vector than are contained in the corresponding quantity of blood of an infected person [cf for example 523]. This is a concentration effect, which is attributed amongst other things to "bait" contained in the insect's saliva. Though observed among various species of mosquito, this high concentration capacity does not seem to apply to *Aedes poicilius* and *Culex quinquefasciatus* in the Philippines [cf 405]. In how far a high number of microfilariae or an accumulation of microfilariae have a damaging effect upon the vector due to the concentration effect, or as a result of repeatedly feeding on blood, has so far remained largely unknown for open air conditions, but it may be assumed that where too small a number of microfilariae have been taken up, the result will be a failure to form infective larvae, whereas too large a number will lead to the death of the vector [cf 302][1]. The epidemiological significance of the density of microfilariae in the different age groups thus eludes exact analysis – though certain estimations can be made. A definitive statement is, however, possible in respect of the distribution of microfilariae in different age groups: higher rates of microfilaraemia and higher densities of microfilariae indicate that about 70 per cent of the microfilariae of *Wuchereria* are to be found among the male population.

The transmission index proper, which allows conclusions as to the infection potential in the various endemic areas as well as among the different age groups of both sexes, is composed of three factors quoted above: the rate of microfilaraemia, the population structure and the density of microfilariae in the peripheral blood. It makes it possible to balance the significance of one area against that of another, as well as to recognize differences in the significance of sections of the population within one endemic area. Such absolute total indices for the various endemic areas may be found in Table 8.

A true comparison of these values is never admissable unless the investigations give a clear indication of the size of the basic entity from which the sample has been drawn. But information of this sort tends not be contained in the publications. This applies to other geomedically relevant factors as well, such as population density, the extent of the area concerned, as well as the distribution in the area under investigation – all of them factors which are to be taken into account – in the planning and carrying out of control programmes.

Thanks to the investigations so far carried out, the comparison of indices of population sub-groups within a given area is a reliable instrument for picking out those groups which constitute the main potential for transmission. Taken as a relative transmission index, it also permits the comparing of conditions in a number of endemic areas (cf Table 13). In respect of the overall situation (mean relative index) it shows that the largest transmission potential of 14.14 per cent (population share: 4.14%) is in the group of 21–25 year-old men (Fig. 14 and Table 7). The indices for adult men (21–40 year age range) add up to a sum of 40.12 per cent (population share: 13.77%). It also means that a systematic treatment of a bare 14 per cent of the infected population would reduce the risk of transmission by about 40 per cent. As the part of the population in question works the hardest, wuchereriasis presents an important economic problem as well (cf para 2.4).

---

[1] Having been experimentally infected, 94 per cent of *A. poicilius* survived for a period of two weeks after taking the blood of donors with a low density of microfilariae. If it was blood from patients with a high density of microfilariae, this period of survival was attained by only 67 per cent. Giving corresponding values of 90 per cent or 83 per cent respectively, *C. quinquefasciatus* appears to react less sensitively [405]

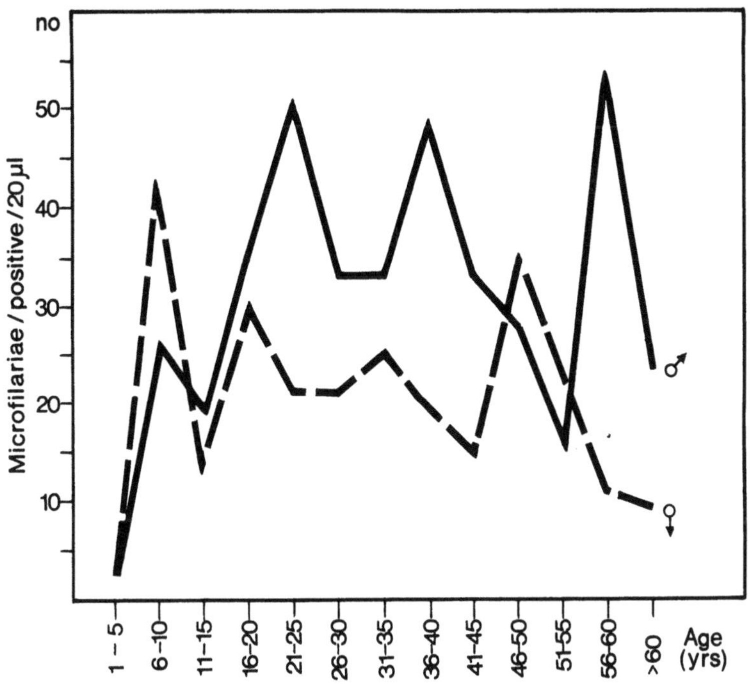

Fig. 13. Density of microfilariae of *Wuchereria bancrofti* in the Philippines by age and sex (average of data) [Sources: 158, 168–171, 175]

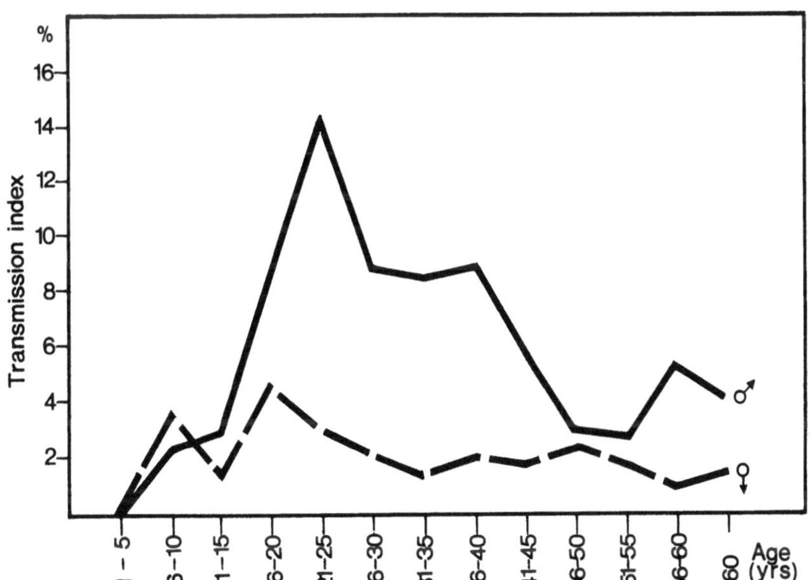

Fig. 14. Relative transmission indices of *Wuchereria bancrofti* in the Philippines by age and sex [Sources: 158, 168–171, 175]

2.2.5 The Endemic Areas

The data on the occurrence of *Wuchereria bancrofti* in the Philippines which has been made available so far has been brought together in Table I. The quantitative interpretations for the various endemic areas are based upon this table, though figures showing up inexplicable discrepancies in the original publications concerning the prevalence calculated from the numbers of infected and examined persons were not included. Neither does the list of endemic areas which follows include parts of the country or provinces where only isolated cases have been noted, without being able to ascertain that they were autochthonous infections. The endemic areas are discussed in a north to south sequence (cf Map Plate 2).

2.2.5.1 The Calaccad Valley

Closer investigations concerning microfilaria-positive blood samples containing the marker "Bontoc" in the description of their place of origin, and at first entered under this allocation as well [661] led to the discovery of the only *Wuchereria* focus in northern Luzon. It is essentially limited to the Calaccad Valley. The Barrio Calaccad and the Parasiles area proved to be endemic, with prevalences of 10.7 and 3.6 per cent [662, 664, 665].

Calaccad in the Kalinga-Apayao Province is a village situated on both sides of the Malig river, 20–25 km south of Tabuk; its inhabitants are chiefly members of the Kalinga tribe. Parasiles lies about 8–11 km south of Calaccad. During the dry season, which extends from November until well into June, the Malig river is relatively shallow and carries clear water like its numerous tributaries. Thanks to this low depth of water and the prevailing high temperatures, especially towards the end of the dry season, all the running waters in the Calaccad area are exceptionally well suited as breeding places for *Anopheles flavirostris;* 1.3 per cent of these were host to infective *W. bancrofti* larvae. On the other hand, *Aedes poicilius* is only seldom found there, even during the rainy season, because the bananas grown there store very little water in their leaf axils; thanks to its low population density – Rozeboom and Cabrera [664] counted only one *A. poicilius* for every 137 *A. flavirostris* in their adult catches – this species cannot play a role as a vector of *Wuchereria* in the Calaccad Valley.

The various sitios forming the Barrio Calaccad show distinct differences in their prevalence rates. Admittedly this is in part due to some sitios having had only a very few inhabitants examined (although also because two houses with accordingly few inhabitants may well be described as a sitio if they are fairly close to each other). All *Wuchereria* cases live in houses in the immediate vicinity of the Malig river, its tributaries or streams flowing into it, and thus of breeding-places of *A. flavirostris* (cf Fig. 15). On the other hand the population of sitios like Tanduyong, Dalanacan, Angalipan and Ban-ing, which are further away from the barrio centre, are not affected.

As a result of the same exposure-risk there are practically no differences in the prevalence rates between the sexes in Calaccad (♂♂: 11.4%; ♀♀: 9.7%). But there is a marked dependence upon age, with the younger age groups (up to 20 years) yielding only two 12–15 year old boys as carriers of microfilariae (=2.2%), indicating a very limited exposure to infectious *Anopheles* bites. Then the prevalence rises, reaching 10.8% among the 21–25 year olds and a maximum of 24.1% in the 41–50 year age group, dropping again after that [cf 664]. The density of microfilariae in the peripheral blood is also described as very low, with 20 out of 30 persons examined having only

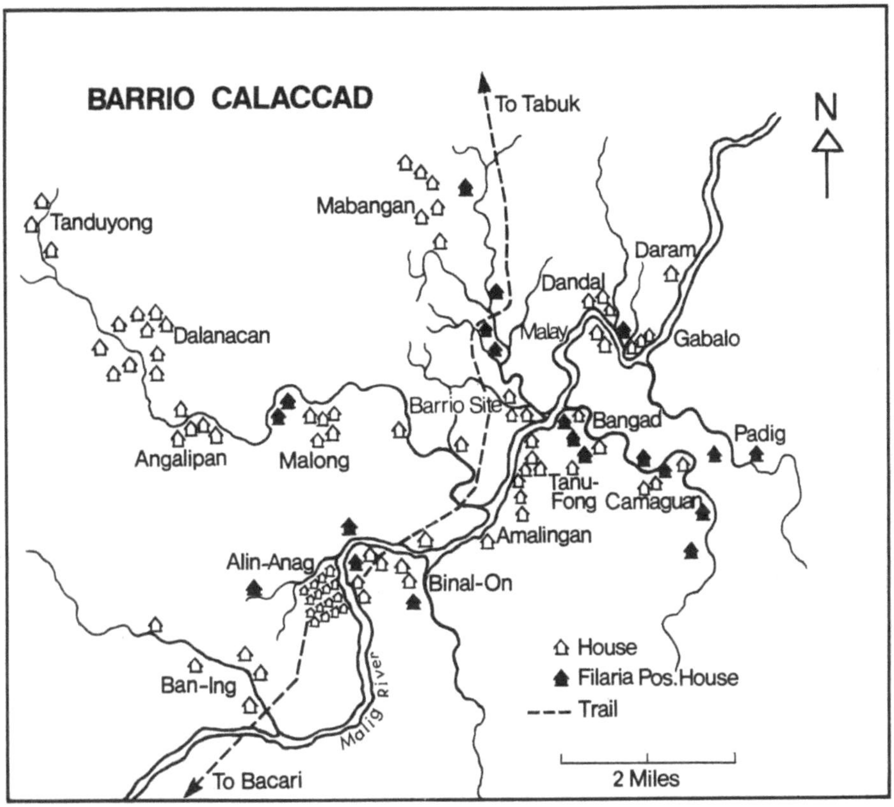

Fig. 15. Location of *Wuchereria bancrofti* positive and negative houses in Barrio Calaccad, Kalinga-Apayao [Source: 664]

1–5 microfilariae/20 μl blood. Nonetheless, as the blood samples were taken between 20.05 and 23.45 at night, and *W. bancrofti* shows a distinctly marked nocturnal microfilarial periodicity, a higher average density of microfilariae than that recorded must be assumed. Admittedly even so low a microfilaraemia is sufficient for transmission, and thus for maintaining wuchereriasis as an endemic disease, since a concentration factor – until now not exactly defined – results in mosquitoes taking in 8 to 12 times more microfilariae than can be shown to exist in the corresponding quantity of the donor's blood [cf 523]. The question, however, of why wuchereriasis has remained confined to this narrow endemic area in northern Luzon, must remain open.

2.2.5.2 The Bicol Region

The Bicol Region, a string of peninsulas and islands, forms the outermost south-east of Luzon and includes the provinces of Camarines Norte, Camarines Sur, Albay, Sorsogon and Catanduanes, as well as a small part of Quezon Province in the north-east. Lacking a dry season altogether, it is characterized by high precipitation, which exceeds 2000 mm a year everywhere [831]. In conjunction with evenly high temperatures and high atmospheric humidity, this creates favourable conditions for the

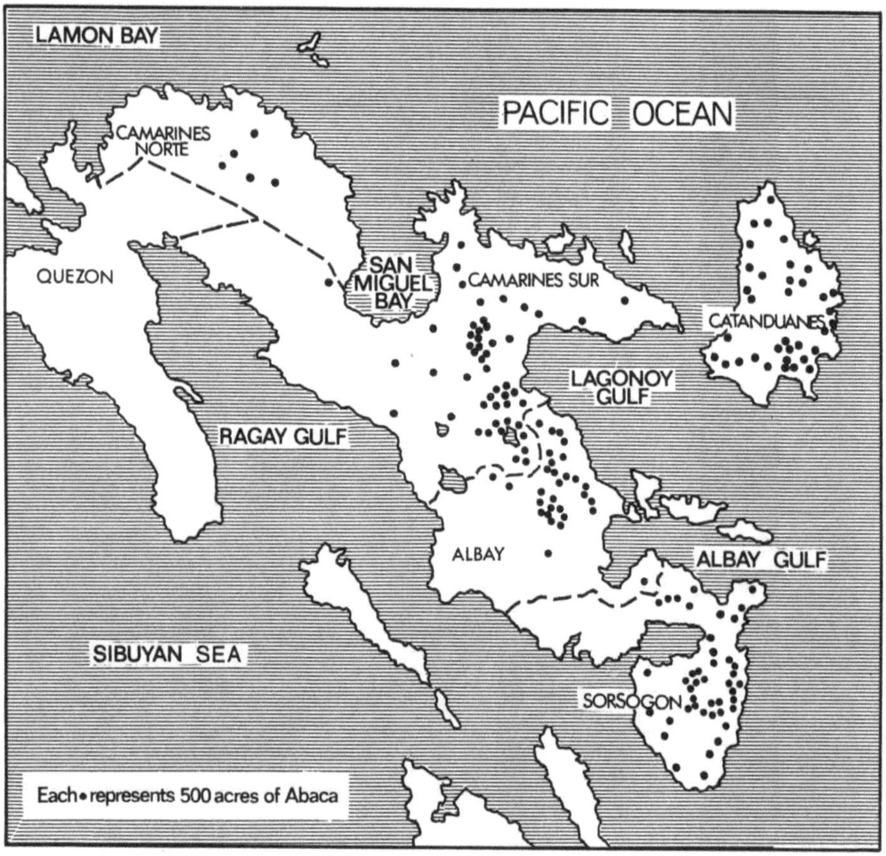

Fig. 16. Abaca cultivation in the Bicol Region [Source: 831]

cultivation of abaca, and thus for *Aedes poicilius,* the main vector of *W. bancrofti,* which belongs to the axil-breeding mosquito species. On average about 32 to 40 per cent of the total area of abaca cultivation in the Philippines fall to the share of the Bicol Region, especially to the provinces of Albay, Sorsogon, Camarines Sur and Catanduanes (about 60,000 ha with about 60 million abaca bushes). In particular, the relatively fertile volcanic loams at the foot of mountain ranges and volcanoes, as well as the lower slopes, are used accordingly. Areas with near-monoculture of abaca are found between Mount Isarog and Mount Mayon, i.e. on the eastern side, following the Bicol river and its south-east to north-west-sloping plain, on the slopes of Mount Bulusan (Sorsogon) and Mount Labo (Camarines Norte), as well as the Central Uplands of Catanduanes (cf Fig. 16). These areas in particular provide great quantities of breeding places for *A. poicilius,* so that its high population densities guarantee the continued persistence of *Wuchereria* infections.

Though long recognised as endemic, as a result of the explorations by Phalen and Nichols [617], systematic analysis of the extent of the wuchereriasis problem in the Bicol Region only took place after World War II. Sorsogon Province with its Filariasis Unit at Gubat has, of course, become the most important area for the inquiry into wuchereriasis and the development of measures for its control (cf 2.4).

The decision to conduct the research carried out there as a pilot scheme has undoubtedly been influenced by the high prevalence rates which Baisas [74] illustrated by extensive data (cf Table I). He found an average prevalence of 9.8 per cent for the total of 110 towns and barrios in nine muncipalities altogether. In this Sta. Magdalena, a muncipality on the south-east coast of Sorsogon, showed the highest rates of microfilaraemia with 19.7 per cent., Barcelona, a municipality further north, the lowest with 7.0 per cent. (cf Fig. 17). Still more marked, however, were the differences between the various barrios within the municipalities, with the administrative district itself – the so-called poblacion of the muncipality, tending to register distinctly below average values. No positive cases at all were diagnosed in three barrios, whereas 21 had prevalence rates of 20 per cent and more.

In this way the population of the abaca-growing areas of the Bicol Region was exposed to considerable stress upon their health merely as a result of wuchereriasis. The mass examinations carried out from 1959 to 1963 again showed Sorsogon as having the highest percentage prevalence rates (8.40%). The rates for Albay (6.40%), Camarines Sur (7.66%) and Camarines Norte (6.09%) are slightly lower, whilst those for Catanduanes (3.43%), and especially for Quezon (0.85%) are significantly less [290; cf Table 1, and Fig. 17]. In any case Quezon presents an exception in so far as only the far south-east, which is affected by wuchereriasis, belongs to the geographical area of the Bicol Region.

The relative frequencies quoted above do not, of course, permit the true distribution of cases to be recognised; this is only achieved by the reduction of prevalences to the population concerned (cf Table 15). It then transpires that around 1960 about 140,000 persons in the Bicol Region must have been infected with *Wuchereria bancrofti*. In this the overwhelming majority of 124,000 occurred in Camarines Sur, Albay and Sorsogon, the chief abaca-producers among the Bicol provinces. They also showed a very high density of wuchereriasis cases per spatial unit, i.e. 11.9 to 13.7 per km^2. This implied the concentration of a large number of infected people in the very area of Camarines Sur, Albay and Sorsogon, i.e. a relatively small area, which constitutes the important epidemiological and geomedical prerequisites for persistence of high endemicity. Due to the population growth, the resulting undiminished prevalence rates ought by now to have resulted in a noticeable increase in wuchereriasis cases (i.e. about 173,000 by 1970, about 186,000 by 1975, and approximately 200,000 in 1980). As the results of large-scale examinations from the recent past are not available, the present extent of the wuchereriasis problem in the Bicol Region cannot be estimated more precisely. Entomological studies carried out in the 1970s [175, 803] revealed the infection rates to be still very high among *Aedes poicilius;* with an average of three infective larvae per mosquito, they amounted to 2.6 and 4.0 per cent; these values indicate a great number of infectious bites to which the population of that region is constantly exposed.

For the Sablayan Island lying in Sorsogon Bay B. D. Cabrera and Valeza [175] established an average prevalence of 11.4 per cent (the variability between the diverse sitios being 8.3 to 16.3%). Though predominantly engaged in fishing, even the limited cultivation of abaca (1 ha), bananas and taro appears to be sufficient for an adequate supply of *Aedes poicilius* and the endemic persistence of *W. bancrofti* among the roughly 900 inhabitants. Indeed, *A. poicilius* was the predominant species: the mosquitoes caught in carabao- and human-baited traps were made up of *A. poicilius*

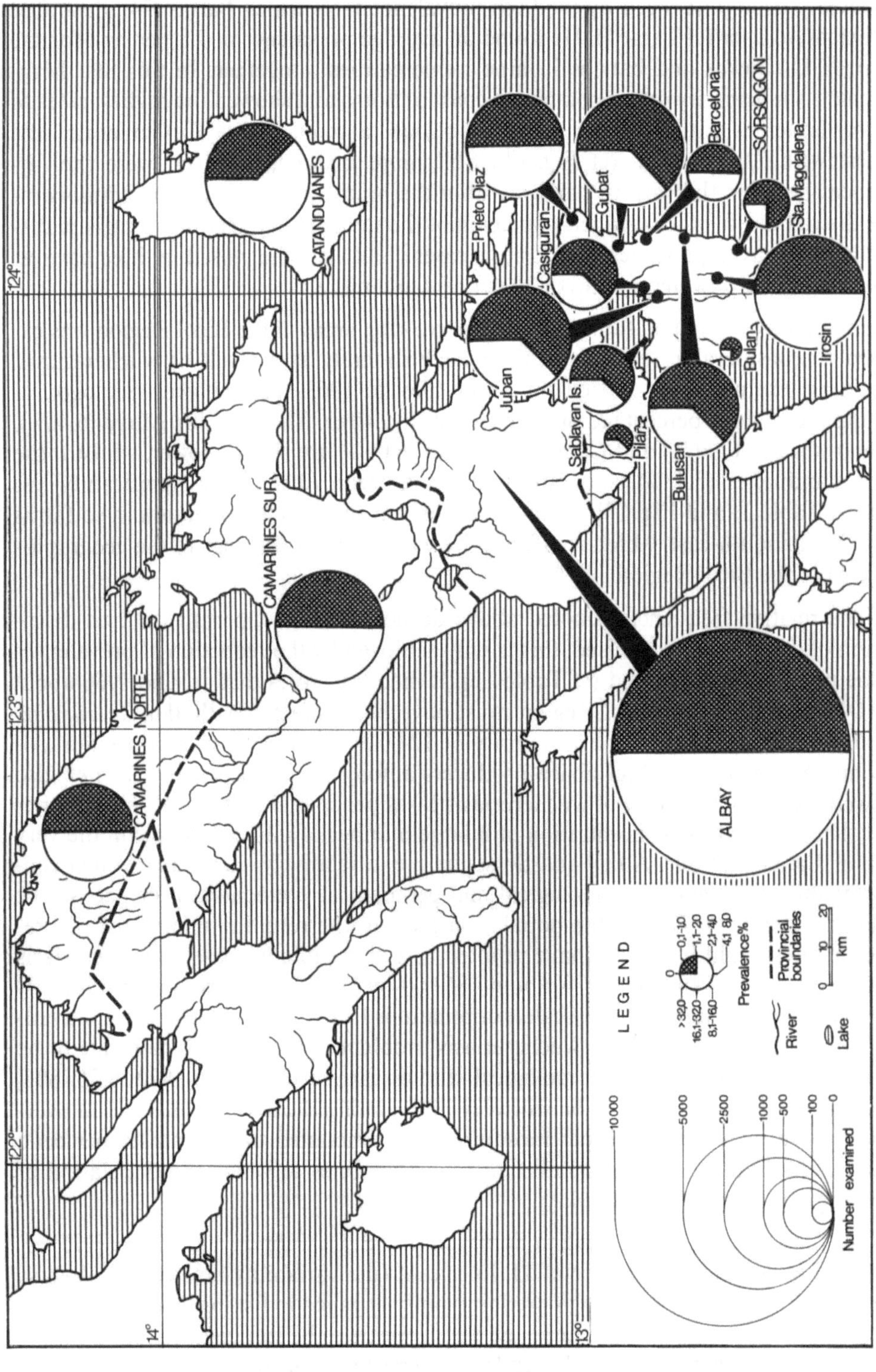

Fig. 17. Wuchereriasis in the Bicol Region [Sources: 74, 290, 661]

(39.1%), followed by quite a large margin with *Anopheles maculatus* (18.7%), *Armigeres baisasi* (9.8%) and various other species.

That a very high prevalence is indeed the characteristic of a population predominantly engaged in cultivating abaca was also confirmed by investigations on Catanduanes [349, 803]. At Barrio Paraiso (about 30 km north of Virac), 328 inhabitants (= 86%) were examined for microfilariae; 37 per cent of the male and 17 per cent of the female population proved to be infected, with the maximum rate of microfilaraemia of 67 per cent occurring among the 51–60 year-old men. The chief occupation there was growing abaca; as many fields are situated at a considerable distance from their homes, the men frequently spend the night in the fields. The intensive exposure to infectious mosquito bites linked with this accounts for the high prevalence rates among the men. It probably explains the differences in prevalence between men and women which have been observed in almost all the endemic areas. The proportion of *A. poicilius* in the entire mosquito fauna was noticeably still higher than on Sablayan Island. Using human bait trapping, 60.6 per cent were found to belong to this species, and in indoor collections even 80.2 per cent. Taking into account the rates of infectiousness, and the density of bites which had been observed, Valeza and Grove [803] calculated that every inhabitant of this endemic area is exposed to an average of 80.3 infectious bites per year.

2.2.5.3 Mindoro

As a consequence of the central mountain range, which crosses this large island of the Sunda Shelf in a NNW–SSE direction, Mindoro has clearly differentiated climatic conditions: the areas in the east belong to that type of precipitation which is characterized by rainfalls of 2000 mm or more throughout the year, whereas the west – although also affected by very high precipitation – receives this in a concentrated period during the months of June to September, and has a 4 to 5 month dry season. Correlated with this are the rates of prevalence, which added up to an average of 1.5 per cent in Oriental Mindoro but to only 0.2 per cent in Occidental Mindoro. Although possible from the aspect of precipitation distribution, abaca is not grown in Oriental Mindoro. It is not known whether bananas therefore take over the role of a breeding place for *A. poicilius*, nor whether other mosquito species function as vectors. With malaria being widely spread on Mindoro it is of course feasible that *Anopheles flavirostris* transmits *W. bancrofti* as well.

Due to the low population density (1960: 30.6/km^2, 1975: 56.1/km^2) and the paucity of prevalence on Mindoro, the density of *Wuchereria* cases is very small:

1960:	Or. Mindoro	0.79/km^2
	Occ. Mindoro	0.03/km^2
1975:	Or. Mindoro	1.34/km^2
	Occ. Mindoro	0.06/km^2

On the other hand, Enarson and Enarson [285] showed only very recently that locally very substantially diverging conditions may occur. These authors established age-adjusted microfilaria carrier rates of a maximum 28.7 per cent among the Mangyans, the originally indigenous population in the area of San Teodoro in the north of Oriental Mindoro. A second important focus with an average rate of 15.8 per cent was discovered by them in the south-eastern part of the Bongabon-Mansalay-Bulalacao district. With 17.6 per cent the area of Mamburao (Occidental Mindoro) was considerably affected as well. Enarson and Enarson [285] estimated the total figure of cases in the population investigated by them to be 2,780. They are of the opinion that wuchereriasis on Mindoro is in the main restricted to the Mangyans; the large number of inhabitants from other Philippine islands is said to be scarcely affected at all.

In accordance with observations made in the majority of the other areas of endemic *Wuchereria* the male Mangyan population proved to be more often affected (27%) than the female one (17%). In respect of the microfilarial density the differences were slight (♂♂ 20.5, ♀♀ 17.5 per 20 μl blood) so that the relative transmission indices (♂♂ 65%, ♀♀ 35%) diverge less strongly than in Mindanao for example (♂♂ 79%, ♀♀ 21%) or Sorsogon (♂♂ 76%, ♀♀ 24%). It is striking how the indices of 35–44 year old men (28.8%) and women (16.5%) markedly exceed those of the other age groups. This means that 37 per cent of the entire quantum of microfilariae are found in 9.4 per cent of the population.

The dissimilarity between these conditions and the situation in the abaca-growing areas indicates a different transmission in the *Wuchereria* foci on Mindoro. Nevertheless, the basic fact remains that prevalence rates and infection levels in men exceed those among women. Whether differences in exposure are again the sole explanation for this cannot be gathered from Enarson and Enarson's [285] data.

2.2.5.4 Palawan

Palawan, too, is one of the islands of the Sunda Shelf. The climatic conditions are characterized by alternating wet and dry seasons, the dry season being less marked in a broad coastal strip between the Green Island Bay and San Antonio Bay on the east coast than in the remaining parts of the island. Since abaca is not grown, *Anopheles flavirostris* is regarded as the vector of *W. bancrofti* in the light of evidence supplied by Rozeboom and Cabrera [667].

Wuchereriasis occurs widely in Palawan; its prevalence rates vary, so that the declaration of a mean value of 2.4 per cent [290] does not permit the true distribution pattern to be discerned. The investigations by B. D. Cabrera and Tamondong [168] in all parts of Palawan Province, together with the findings of B. D. Cabrera and Jueco [163] allow, on the other hand, a more detailed description. They indicate an increase in prevalence in a north to south direction (cf Table 14). North Palawan (including the Calamian and Cuyo Islands, as well as Linapacan and Dumaran) can be considered *Wuchereria*-free. The endemic areas begin roughly at the northern latitude of 10°15′N. in the southern part of the Roxas Municipality and on the west coast in the north of the Puerto Princesa Municipality. The average prevalences for the two municipalities are about 2–3 per cent. The adjacent southern district of Aborlan follows with an average rate of 7.6 per cent, and the Brooke's Point Municipality in the extreme south presents the maximum with 12.9 per cent. The south-west coast (Quezon Muncipality) in turn has a smaller frequency of *W. bancrofti* infection (7.2%), but there is an additional *Brugia malayi* focus with an average prevalence of 24.3 per cent (cf 2.3.1).

Although the wuchereriasis frequency fluctuates considerably within the municipalities mentioned above (cf Table I) there does indeed seem to be a south to north gradient, which is also brought out by the fact that the microfilarial density in the peripheral blood of the infected person, as well as the relative frequency of barrios with positive diagnoses, decrease from south to north, parallel to the prevalence (cf Table 14).

This gradient can have different causes. As far as the vector is concerned, the hilly-mountainous character of Palawan is likely to offer suitable breeding places for *A. flavirostris* throughout the island. A high density of the mosquito species must be assumed for large areas as a correlation of the evidence on malaria endemicity [cf 152]. It is true that too little is known about it to permit appropriate, regionally differentiated conclusions to be drawn. The more densely-wooded central and southern areas of Palawan possibly provide a better-suited breeding habitat for the transmitting *Anopheles* than the northern part. It ought, however, to be considered that the north-south gradient described above might be a diffusion process, which is not only caused by the vector-related differences, but – emanating from a centre of endemicity in the south – by a northward-bound expansion.

In Palawan the infection frequency is also higher among the male population. In respect of microfilarial density there is a noticeable difference between the sexes in Brooke's Point, although not in Quezon, Puerto Princesa and Roxas. The question whether the presence of such differences is to be attributed solely to exposure must remain open for the wuchereriasis foci in Palawan as well. Since *A. flavirostris* is a nocturnally-active vector, greater exposure of men could only be explained as a result of different sleeping habits – but there are no indications for this.

2.2.5.5 *The Islands of the Sibuyan Sea*

In the case of the islands of the Sibuyan Sea, which Wernstedt and Spencer [831] treated as a specific geographical region, data on the frequency of Wuchereriasis are not available, with the exception of all-inclusive figures for the Masbate (3.0%) and Romblon (1.7%) Provinces [290]. Thanks to the prevailing climatic conditions of this region – most of the islands have a 3–4 month dry season early in the year – abaca cultivation does not play a role. Rice and grain cultivation, fishing and stock-breeding dominate there, with coconuts as the only cash crop of importance. It is therefore doubtful whether *Aedes poicilius* has any significance as a vector of *W. bancrofti* in the Sibuyan Sea region. No entomological investigations are available which could assist in the clarification of this problem.

Considering the data on prevalence provided by Estrada and Basio [290] and the number of inhabitants at the time in the early 1960s, about 10,000 persons in the Masbate Province and about 2,300 people in the Romblon Province must have been infected with *W. bancrofti.* This corresponded with a case density of 2.5/km^2 in Masbate and of 1.7/km^2 in Romblon. But the summary data which are available neither take into account the potential frequency variations between the larger islands of the region (Masbate, Burias, Ticao, as well as Romblon, Tablas and Sibuyan), nor do they indicate at all which islands have been included in the investigations. A regional differentiation is therefore not possible.

2.2.5.6 *The Visayan Islands*

The group of the Visayan Islands represents the central part of the Philippines. They show a characteristic climatic division: large parts of Samar and Leyte, the eastern Visayas, are marked by high precipitation (2000–4500 mm) throughout the year, with a distinct maximum around the turn of the year. In western Leyte, the northern tip of Cebu and Bohol, rain falls throughout the year (1500–2500 mm) too, but there is no such maximum. Further to the west this is followed by a region in which the all-year-round rains are interrupted by a fairly short, one to three month, dry season. The main part of Cebu, Negros Oriental (including Siquijor) and the north-western triangle of Panay belong to this part. On the western fringe of the Visayas (Negros Occidental, the south-western half of Panay including Guimaras) finally the dry season lasts even 4 to 6 months. Closely allied to this climatic division are the possibilities of abaca cultivation, which is widely practiced, especially on the eastern Visayas, and together with coconuts represents the most important cash crop. In the early 1970s about 31 per cent of the area cultivated under manila hemp and 28 per cent of the production fall to the share of this area, that of the western Visayas being no more than 2 and 3 per cent each [676]. Centres of abaca cultivation were the northern coast of Samar and the southern half of Leyte [cf map in 831].

Thanks to the paucity of data in the literature, it is hardly possible to establish a direct relationship between abaca cultivation and wuchereriasis for the Visayas though *Aedes poicilius* has been shown to be a vector of *W. bancrofti* in south-eastern Samar [830]. The prevalence rates are distinctly lower than in the Bicol Region, for example, which is at present the most important abaca-growing area (cf 2.2.5.2), and show an east-west gradient. On the basis of the mass examinations carried out at a national level between 1959 and 1963, Estrada and Basio [290] calculated an average microfilaria rate of 4 per cent for Samar. This average value was later confirmed for 39 barrios and sitios in south-east Samar [830], and for 8 barrios in northern Samar [245]. With prevalences of 1.7 per cent (North Leyte) and 1.9 per cent (South Leyte), Leyte already presents a noticeably lower frequency of *Wuchereria* infections. Even Bohol still counts as being endemic, but here the rate is barely 0.3 per cent. The situation in the other Visayas further to the west is beyond final judgement. Cebu and Negros Occidental only produced one case of microfilariae from *W. bancrofti* each out of 3,766 and 1,958 blood samples taken, whilst Negros Occidental and Panay had not been included in the mass examinations at the time [290]. A *Wuchereria* focus is said to exist in Negros Oriental (S. M. Wale 1977, personal communication), though it has not so far been investigated in detail. With regard to Negros Occidental it must be pointed out that Phalen and Nichols [618] had already diagnosed cases of *Wuchereria* there. Since the infected Filipinos originated from Misamis and Cotabato, it was unlikely even then that they were autochthonous infections. A similar thing applies to isolated cases in Panay [618, 790], though with 16.3 per cent inmates of the Bilibid Prison who came from this island had shown a high prevalence in 1913 [120], so that it is not out of the question that wuchereriasis continues to be endemic there.

Although the findings published so far do not allow a regionally more differentiated analysis, they do permit the east-west gradient mentioned above to be recognised. This comes out particularly clearly when the data from the mass examinations carried out by Estrada and Basio [290] are brought to bear upon the population as a whole. In the Visayas, according to this, about 55,000 people had to be assumed to be infected with *W. bancrofti* in the early Sixties, with their distribution showing a distinct decrease from east to west. This is also expressed in the case density (cases per km^2) (cf Table 15). In this way conditions here again show a connection between the climatic situation and abaca cultivation on the one side and *Wuchereria* frequency on the other.

Though such a correlation does not immediately follow from the investigations carried out by Cross et al. [245] in northern Samar, they, too, draw attention to the

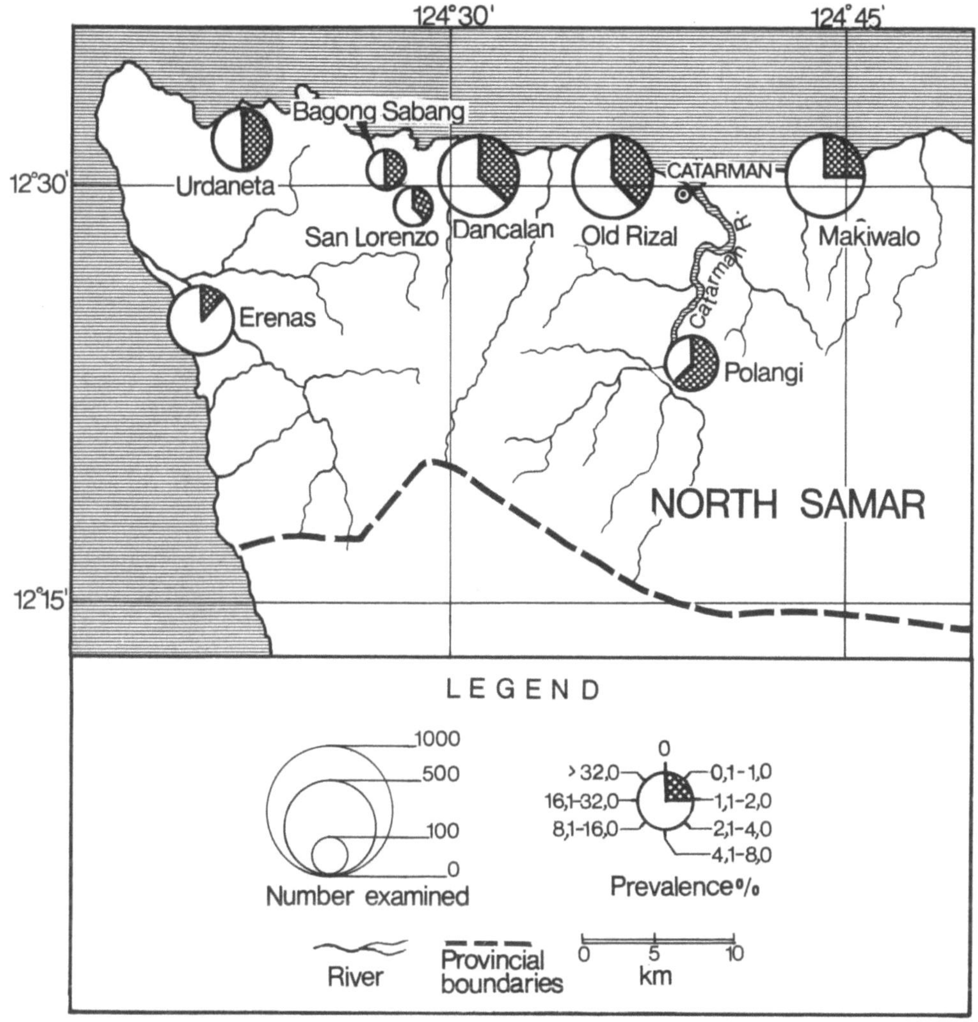

Fig. 18. Wuchereriasis in Northern Samar [Source: 245]

local significance of abaca cultivation; they observed fluctuations in the prevalence rates of the barrios under investigation, ranging between 1 per cent and 11 per cent (cf Fig. 18). Another important point was made by Wenceslao et al. [830] through their discovery in south-eastern Samar of the fourth *Brugia* focus in the Philippines (cf 2.3.1.5), for *W. bancrofti* was also endemic in the region of the Salcedo Municipality (1–17%; cf Fig. 19). Depending on the method of trapping, *Aedes poicilius* contributed a significant proportion of the entire mosquito fauna there (3–20%) and played host to infective larvae (L_3) of *W. bancrofti* in 1.5 per cent of all cases, the rate of infection being 6.7 per cent. The high relative density of this important vector species of course assumes the presence of a fairly large number of plants offering suitable breeding places in their leaf axils. But it has not been unequivocally established so far whether *A. poicilius* functions as the sole vector in the endemic areas of the Vi-

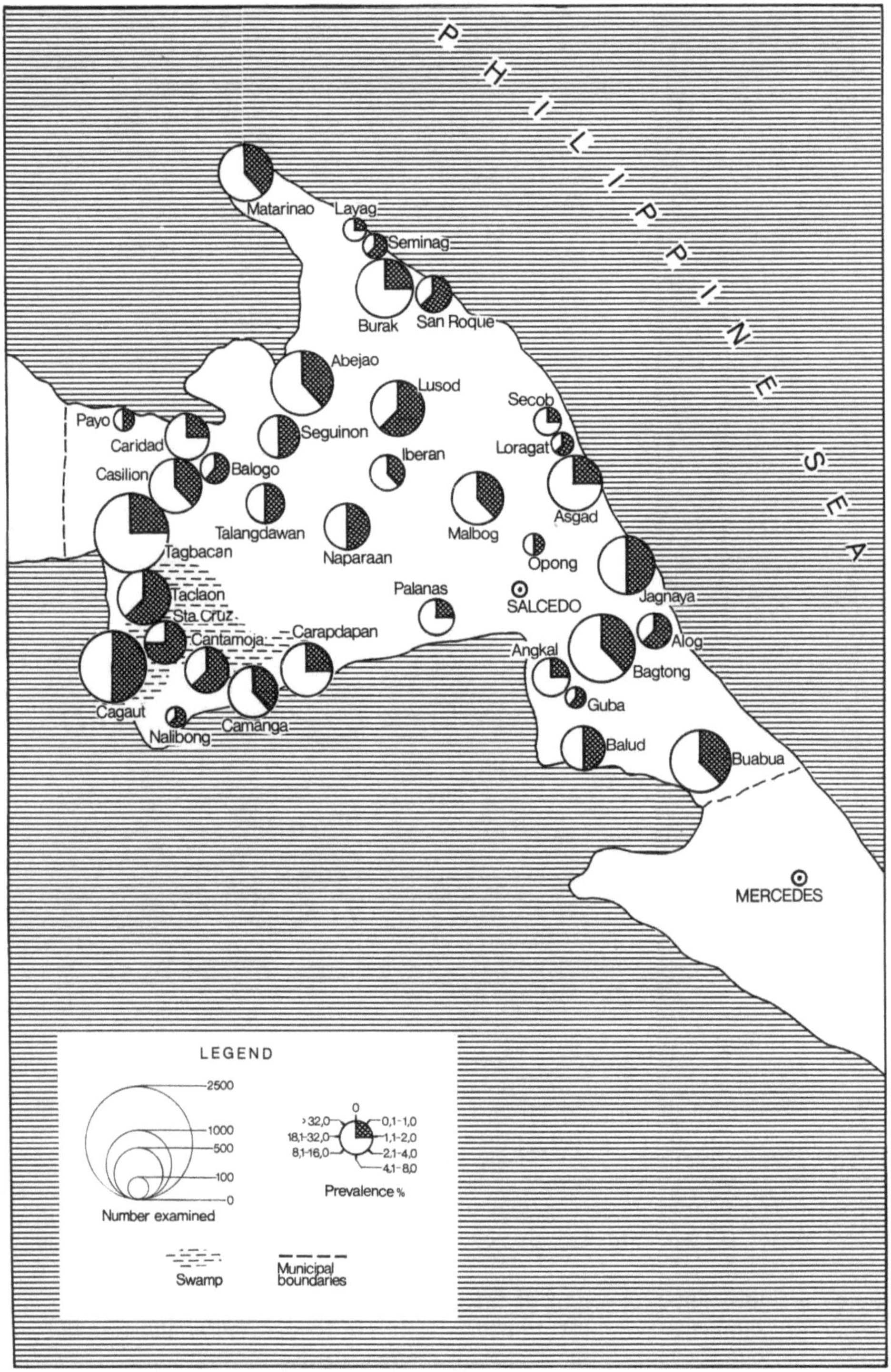

Fig. 19. Wuchereriasis in Eastern Samar [Source: 830]

sayas. Though found to have an infection rate of 5.2 per cent in the Salcedo Municipality, none of the specimens of *Culex quinquefasciatus* dissected contained infective larvae (L_3). *Anopheles flavirostris,* the other potential vector species, did not crop up in the Salcedo catches; this can probably be attributed to the lack of suitable breeding water in the flat and partly swampy terrain there. These isolated findings, however, do not allow generalising conclusions to be drawn for the whole Visayas area.

2.2.5.7 Mindanao

Extending over 97,984 km², Mindanao, the second largest of the islands in the Philippines, is composed of five major regions of varying character: the Pacific Cordilleras (including Dinagat and Siargao), the Agusan-Davao Plain, the Central Highland, the Bukidnon-Lanao Plateau, the Cotabato Basin with the Tiruray Highland and the Zamboanga Peninsula. The eastern part of the island, which extends as far as the Central Highland, belongs to the climatic type having precipitation throughout the year, with high maxima in November/February. A precipitation distribution, which is interrupted by a fairly brief dry season, occurs in the far north, following the western escarpment of the Central Highland, and in the western part of the Zamboanga Peninsula. In the remaining parts of the islands precipitation tends to be evenly distributed throughout the year (cf Map Plate 1). This implies predominantly favourable climatic conditions for abaca cultivation; the mountainous character of large parts of the island, however, restricts the area of potential cultivation to the large plains and coastal areas at lower altitudes. It follows that the highest *Wuchereria* prevalences are found there in north-eastern Mindanao (Surigao and Agusan Provinces), in the region of the Gulf of Davao, in South Cotabato, in the centre-north (Misamis Oriental and Bukidnon), and to a lesser degree also in the north and in the centre-south of the Zamboanga Peninsula [171; cf Fig. 20 a–e and Table 15].

A statistically significant correlation between abaca cultivation and microfilaria rates was established by B. D. Cabrera and Tamondong [171] for the Davao and Agusan Provinces and Bukidnon severally, as well as for Mindanao collectively. Although corresponding entomological investigations are still outstanding, these findings compel us to assume that on Mindanao, too, *Aedes poicilius* must be the chief vector of *W. bancrofti.* An important exception to this, though, may be the conditions in Surigao del Norte. In 1960 a mere 0.32 per cent of the total agriculturally-useful area of 73,754 ha was given over to the cultivation of abaca (= c. 250 ha); nevertheless, a very high average prevalence of 2.73 per cent was observed (cf Table 15). The conclusion that a fundamentally different epidemiological situation might exist in Surigao del Norte is supported by the examination of the indigenous negrito population, who are hunters and gatherers in this area, for members of the negrito tribe of Mamanwas in Surigao del Norte and Agusan del Norte proved to be infected with *W. bancrofti* at a rate of 19.3 to 26.7 per cent [752]. The absence of abaca in the settlement area and the way of life of the Mamanwas therefore hardly argues for a transmission by *Aedes poicilius.* A definitive assessment of the situation there, however, must remain subject to the results of future entomological investigations.

Whilst a relatively high prevalence could be established in Surigao del Norte in spite of a small area under abaca cultivation, the situation was the reverse in Davao del Norte. In the case of the average value of 0.64 per cent given for this province, it ought to be mentioned that 63 per cent of the total of 27,102 examinations had been carried out in the Tagum Municipality. Tagum happens to be the creation of the only

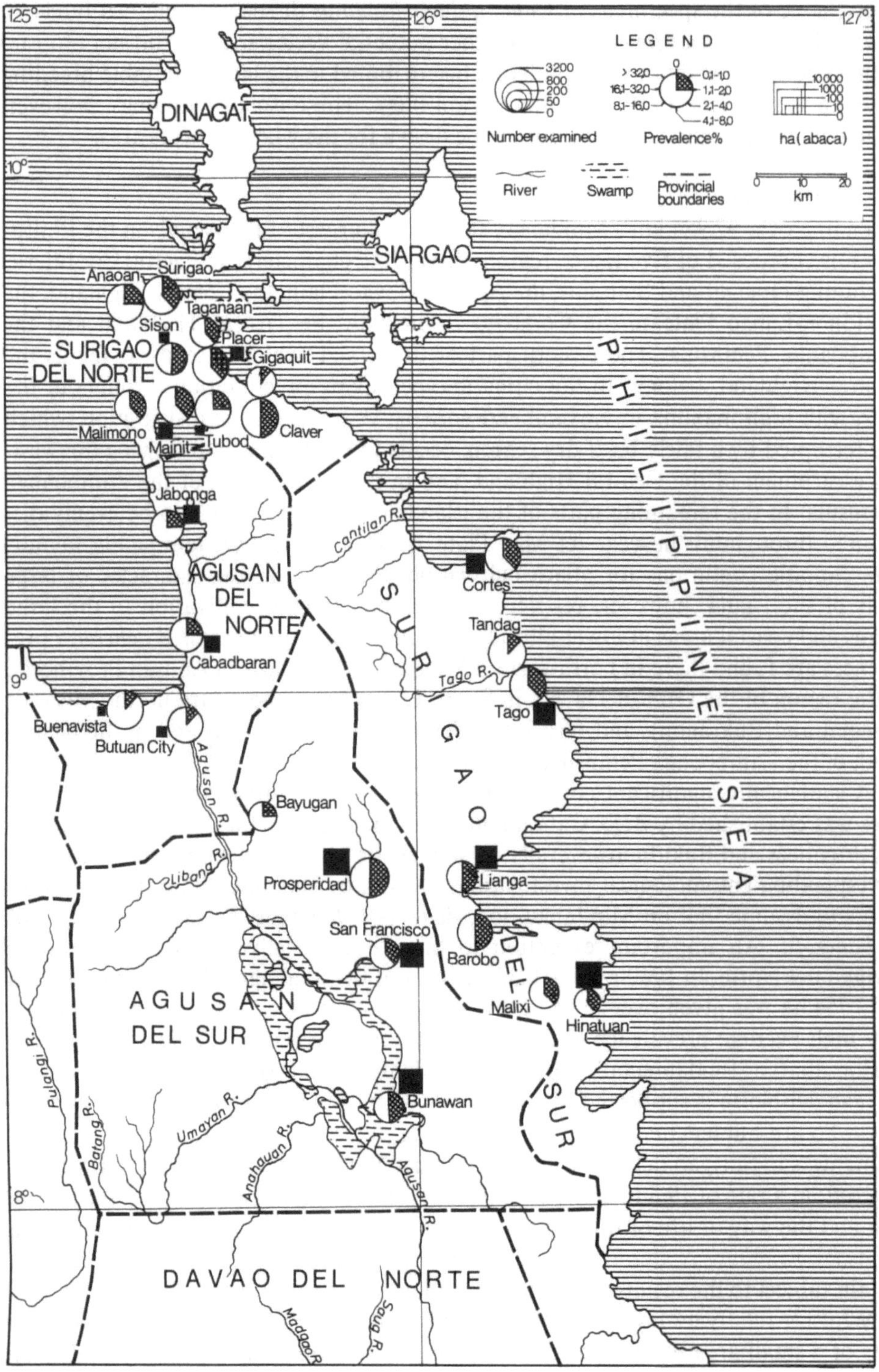

Fig. 20 a. Wuchereriasis and abaca cultivation in north-eastern Mindanao [Source: 171]

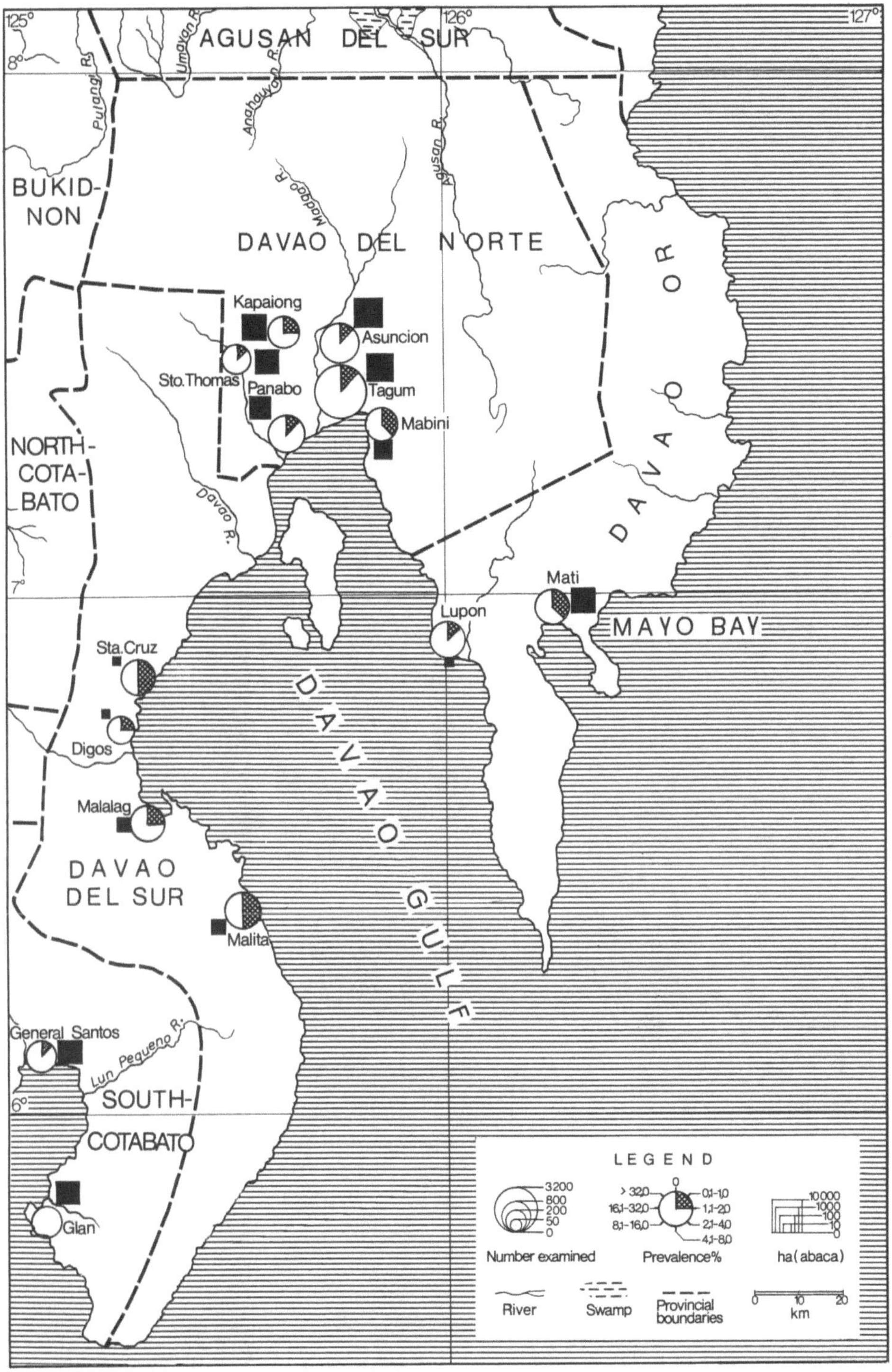

Fig. 20b. Wuchereriasis and abaca cultivation in south-eastern Mindanao [Source: 171]

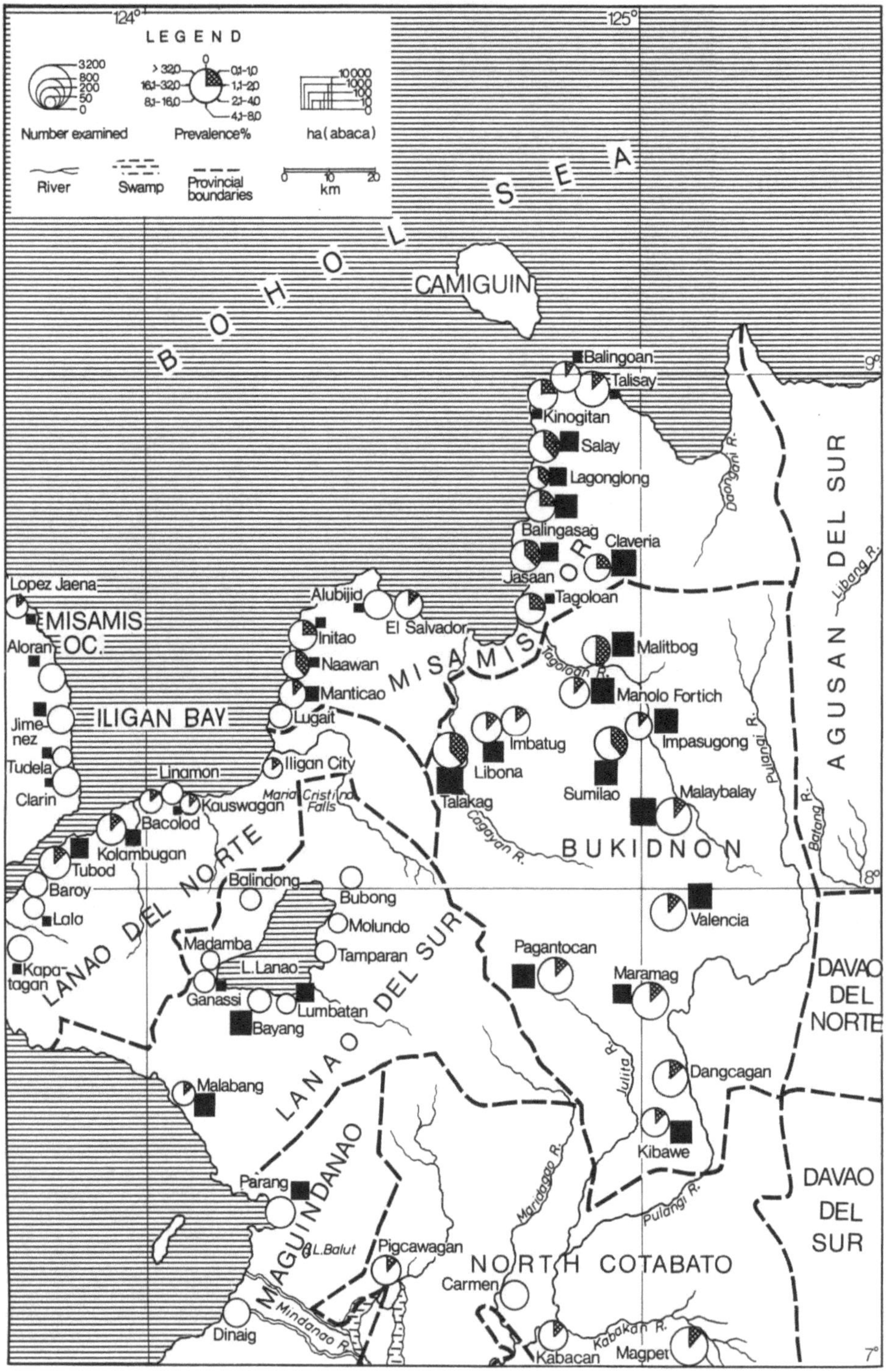

Fig. 20c. Wuchereriasis and abaca cultivation in north-central Mindanao [Source: 171]

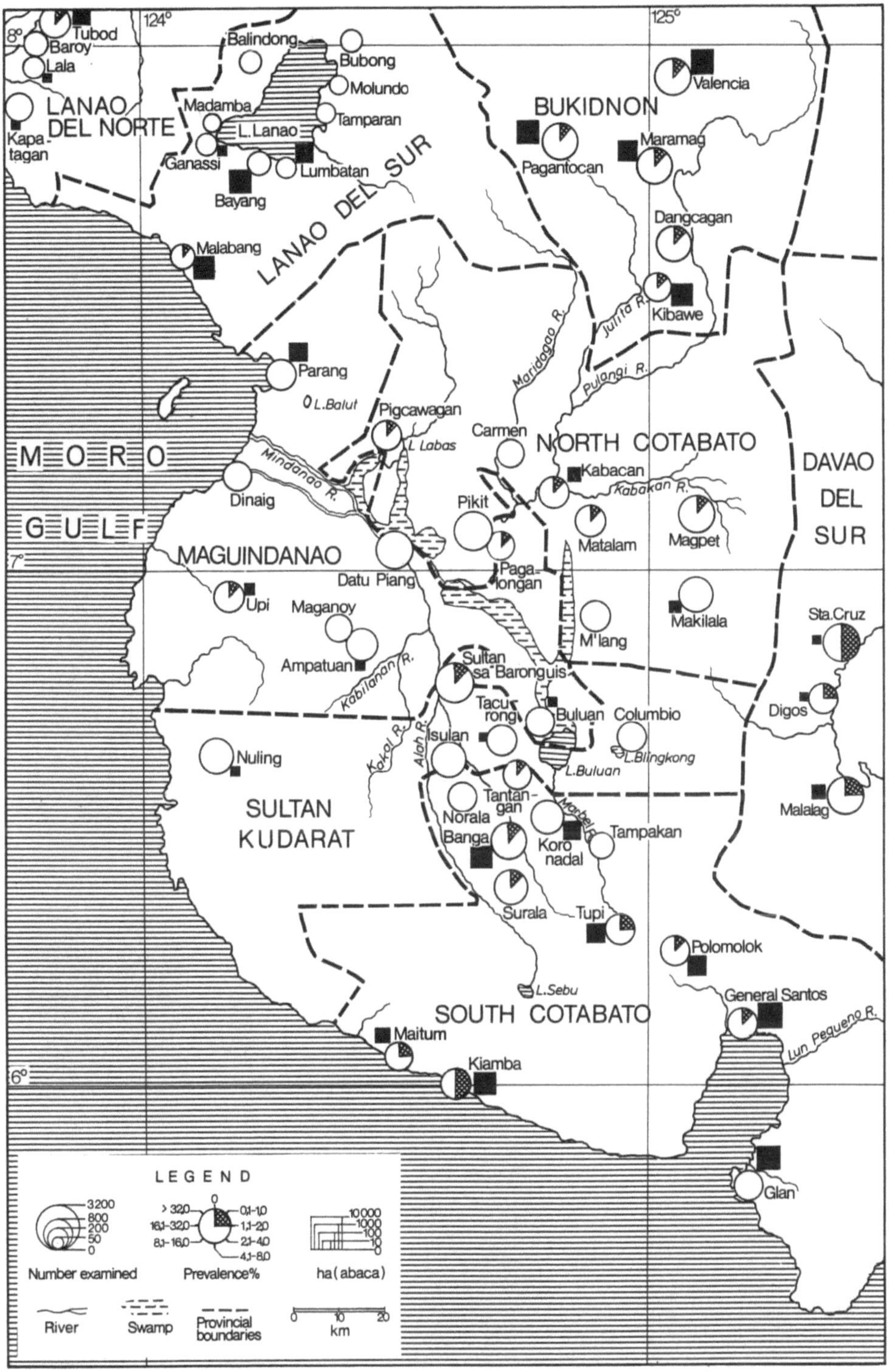

Fig. 20d. Wuchereriasis and abaca cultivation in south-western Mindanao [Source: 171]

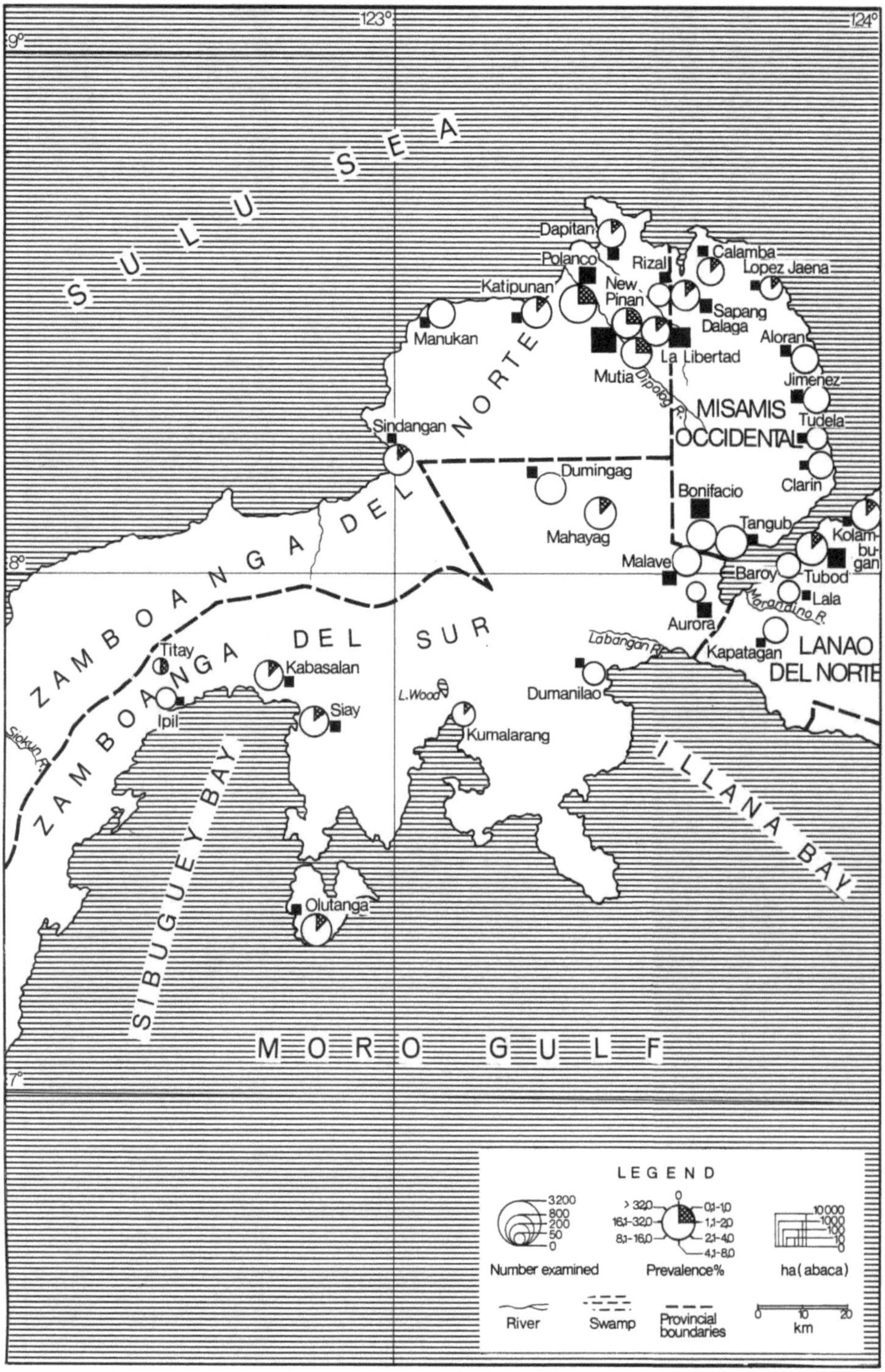

Fig. 20e. Wuchereriasis and abaca cultivation on the Zamboanga Peninsula, Mindanao [Source: 171]

stationary filariasis field unit on Mindanao; quite obviously concentrated more or less on the immediate neighbourhood, it did not fulfill its task of collecting representative data (for criticism of the work of this field unit see 2.4). It must therefore be assumed that Davao del Norte has a higher prevalence than has been stated.

On the other hand, prevalence rates alone are an insufficient measure of evaluation for the problem of wuchereriasis, even on Mindanao. Assuming that the results of the Filaria Control Service which B. D. Cabrera and Tamondong [171] published in 1970 are based upon approximately representative random samples, the conversion of the population afflicted (Table 15) permits a better-grounded evaluation of the situation. Only then does it turn out that in 1960 a very high number of wuchereriasis cases, i.e. about 18,700, were concentrated in Davao del Sur, a province once leading in abaca cultivation (cf 2.2.3.3). The Surigao provinces and Misamis Occidental in the north-east and north of Mindanao evidently follow at a distance (with a total of around 16,500 cases). As far as the density of cases (per km^2) is concerned, Davao del Sur is leading once again (cf Table 15). In 1960 Mindanao accordingly had to assume a total of 55,000 *Wuchereria*-infected persons. As control measures got stuck in their early stages, whilst prevalence rates presumably remained unchanged (not to mention the considerable population growth that must be taken into account), an approximate doubling of these case numbers to 113,000 must have taken place by 1980.

In conclusion it remains to be said that differences in the prevalence rates and the microfilarial density in the peripheral blood between the sexes and the various age groups (cf Tables 10–13) on Mindanao correspond to the conditions in the remaining endemic areas of the Philippines. The higher values among the male population are reflected in a relative transmission index of 78.5 per cent, 33 per cent of which are attributed to the 16–25 year old men alone. This is an age group which is responsible for a mere 10 per cent of the population as a whole.

2.2.5.8 The Sulu Archipelago

The Sulu Archipelago is divided into two provinces: Sulu and Taw-Tawi. A string of islands connecting the southern tip of the Zamboanga Peninsula and Borneo, it belongs to the permanently humid tropical zone, though with frequently recurring dry periods during the months of February and March. Besides coconuts, rice and root crops, abaca is one of the important agricultural products there. Information provided by B. D. Cabrera and I. Cruz [158] indicates that on the Sulu Archipelago abaca cultivation and *Wuchereria* prevalence are correlated once again. They noted that with a proportion of 0.1 per cent under cultivation with abaca out of the agriculturally used total, the *Wuchereria* prevalence amounted to 0.65 per cent. Where the proportion rose to 1 per cent, the prevalence rates were between 3.78 and 10.26 per cent. Where abaca cultivation formed an even greater proportion 3.72–11.45 per cent of the persons examined proved to be infected.

The great frequency of wuchereriasis on the islands of the Sulu Archipelago had first been recognised after mass examinations were carried out throughout the country during the period 1959–1963 [290]. With a prevalence rate of 10.2 per cent they were top in the list of the endemic provinces. Only three of the 17 municipalities so far investigated were shown to be free from *Wuchereria* infections [158, 169]. Moreover, it emerged that on some islands *Brugia malayi* was endemic as well (cf 2.2.1.3). In general high prevalences tend to occur on the large islands, whilst smaller ones contain relatively few infected persons, or are even free from *Wuchereria* (cf Fig. 21).

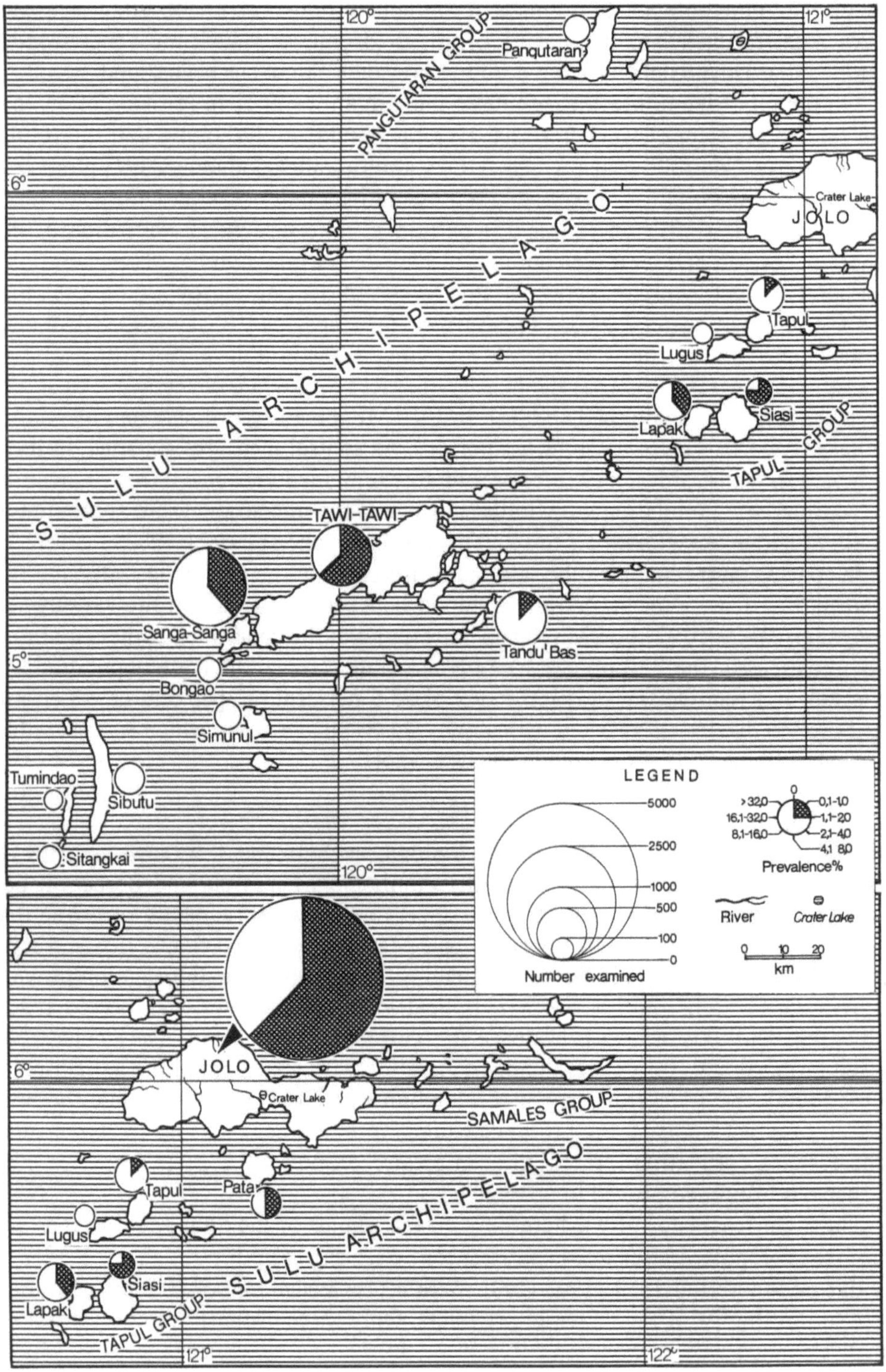

Fig. 21. Wuchereriasis on the Sulu Archipelago [Sources: 158, 169]

Even these findings indicate an obvious correlation with the extent of abaca cultivation, 97.8 per cent of which – i.e. the total of 5,650 ha – is concentrated in the eight municipalities of the island of Jolo, where 11.4 per cent of the population is infected with *W. bancrofti* [169]. Even the municipality of Siasi with 105 ha cultivated with abaca has a higher prevalence rate (3.7%), though this applies to the Balimbing (10.3%) and Pata (5.1%) municipalities as well, despite their abaca areas totalling a mere 5 ha each. Nonetheless, wherever abaca is not grown wuchereriasis occurs rarely or not at all.

On Jolo, the most important area of abaca cultivation, *Aedes poicilius* proved to have a share of 44 per cent, and thus to be altogether the most frequent mosquito species. Among those mosquitoes approaching man, 90 per cent even belonged to this particular species, so that it may be regarded as the chief vector of *W. bancrofti* in that endemic area as well. This is especially so as the total of 5.2 per cent proved to be infected with the corresponding filaria larvae; 0.4 per cent contained L_3 [169].

In the Bongao Municipality, Tawi-Tawi Province on the other hand – where abaca cultivation is insignificant – conditions were different. *A. poicilius* was rare (a 0.8% share); the 35 specimens that were trapped and dissected were uninfected. *Anopheles flavirostris,* by contrast, found suitable breeding places in the running waters of the hill country there, so that after *Anopheles franciscoi* (53.2%) this mosquito was the second most frequent species (11.1%) to be represented in the adult catches [127]. On the other hand these investigations also confirmed the obvious zoophilia of *A. flavirostris.* It occurred very frequently (11.3%) among the carabao-baited catches, but distinctly less often in "off human" ones (1.1%). Nonetheless, their infection and infectivity rates were relatively high, with 7.2 and 1.2 per cent respectively. On the evidence of these findings *A. flavirostris* may be regarded as the main vector of *W. bancrofti* in the southern Sulu Archipelago. In spite of a high population density, their transmitting frequency is restricted; their marked zoophilia results in relatively low rates of prevalence among man. In areas without abaca cultivation it is not the small numbers of *A. poicilius* alone which contributes to the reduction in transmissions.

The separate data presented here (cf Table I) – 9.9 per cent for Sulu (main vector: *A. poicilius*) and 3.5 per cent for Tawi-Tawi (main vector: *A. flavirostris*) result in a difference in average prevalences in the two provinces. A computer prediction for the overall population therefore arrived at approximately 27,000 *Wuchereria* infections in the entire archipelago in 1960 (cf Table 15). Estimates of the present situation are different because the population statistics, though citing a marked increase for Sulu till 1970 (from 248,000 to 316,000), go back to the rather lower figure for inhabitants in 1975 (240,000) [1980 Philippines Statistical Yearbook]. It is impossible to make out whether this a matter of genuine reduction in numbers or merely a correction of previous data.

2.2.5.9 Summary Treatment of the Endemic Areas of Wuchereria bancrofti in Its Correlation with Population Density, Case Density and Abaca Cultivation

Besides paying attention to the prevalence rate, a discussion of the geographical distribution of wuchereriasis in the Philippines also requires the regional comparison of case distribution as the only way of weighting the endemic areas. However, such a regional comparison requires that the mass examinations which have been carried out can be ranked as representative samples. Things being what they are, this condition tends only to be fulfilled where it is a matter of data in geographically closely-limited areas: the material compiled for the Philippines (cf Table I), which is primarily concerned with the results of countrywide mass examinations, might differ from these demands in many cases. Not infrequently the location of places indicates that the accessibility which has been achieved by traffic engineering has also had some say in their selection; remote, inaccessible areas are less well represented. On the other hand, it must be borne in mind that such inaccessible areas tend also to be the less densely populated ones, and they are therefore more likely to play a subordinate role in the statistical survey of total case numbers.

In spite of the statistical short-comings which could underlie the existing data, the attempt will be made to achieve a geographical differentiation of the distribution of cases. The basis of this is provided by the prevalence rates resulting from mass investigations on the one hand [158, 169–171, 290] as well as from the statistical data for the provinces concerned on the other [1980 Philippines Statistical Yearbook]. Since the population development during the past twenty years shows distinct differences in the various provinces (even caused, for instance, by migration processes within the country), the calculations presented here were based on the figures available for 1960 (cf Table 15).

Calculated from the number of inhabitants and the prevalence rates, the analysis of the overall frequency reveals four focal points of wuchereriasis: the Bicol provinces, the eastern Visayas, Davao del Sur, and the islands of the Sulu Province. Approximately 80 per cent of all *Wuchereria* infections were concentrated in these areas, though only about one third of the population of all the endemic provinces lived there. Since they occupied little more than 20 per cent of the area, this meant a significantly higher population density than in those provinces where there was a lower case frequency of wuchereriasis. When calculating the case density (per km^2) for every province for 1960 (cf Table 15) it appears that in the provinces with the highest population density the greatest case density can be observed as well. Although the data concerning the distribution of cases are only based on rough estimates, an attempt has been made to work them up statistically and to examine them for correlation between the two parameters. Initially provinces with a low estimated number of cases (< 1000) were also included in the analysis, although the computer predictions for them were based upon individual diagnoses, and thus did not allow for an estimation of the real extent of wuchereriasis. A second calculation then included those provinces which showed a prevalence of at least 0.25 per cent and more than an estimated 1000 cases.

The result of the two calculations can be summarized as follows: when taking into account all the provinces mentioned in Table 15, as well as when limiting them to those 27 provinces which can be regarded as definitively wuchereriasis-endemic, the analysis reveals a statistically assured correlation ($p < 0.05$) in the form of a parabolic function ($\hat{Y} = 0.6886 - 0.172x + 0.0423x^2$; x = population density per km^2; y = case density per 0.1 km^2); i.e. an increase in the population density is coupled with a rise in the density of cases, though not in a linear relationship, but as a disproportionate increase in case density dependent upon population density. Whether this formal correlation can be matched with a causal analogy therefore needs to be discussed.

Besides observing that disproportionately high numbers of wuchereriasis cases occur in the most densely populated areas, it should be noted that these are the very areas in which abaca cultivation is at its most intense. It is quite obviously a multifactorial problem, the particular factors of which correlate with each other and in a number of ways. In certain areas the agricultural development, for instance, in the course of which the cultivation of manila hemp played an important role for decades, only set the population going or had a significant part in it. A high population density in conjunction with a high population density of the transmitter, however, guarantees frequent contacts between mosquito and man, a precondition for a high transmission frequency of *Wuchereria bancrofti.* To that extent it must be assumed that the population density constitutes an important factor in the epidemiology of this helminthiasis as well, and that it is correlated with the case density. But the case density on its part also presents an important dimension if the significance of a disease is to be weighted and regional differentiation is to be attempted. Such a differentiation at provincial level is presented on the map shown in Fig. 22.

2.2.6 Geomedical Consideration of the Package of Ecological Conditions in the Nexus of Effects of *Wuchereria bancrofti* Distribution in the Philippines

As has been demonstrated in the preceding sections, the distribution of wuchereriasis in the Philippines is concentrated in the areas of abaca cultivation, where the anthropophilic *Aedes poicilius,* the vector, finds a large number of breeding places. In this abaca cultivation, vector density, the development of *Wuchereria* in the vector, as well as the percutaneous penetration of the larvae of filariasis depend upon geo-factors which singly or in interaction determine the epidemiological process.

Among the various geo-factors, climatic conditions are the most significant. Abaca flourishes sufficiently only in those places where precipitation is high and falls as evenly distributed as possible throughout the year. Such a pattern of precipitation, however, also forms the condition for development of the vectors, being as it were the sole guarantee of sufficient water in the breeding places in leaf axils of abaca over a longer period of time. So too, evenly high temperatures are an important condition which must be met when abaca is to be cultivated profitably; on the other hand, they, too, have a favourable effect upon the development of the larvae of *A. poicilius,* and admit a quick succession of generations, besides playing an additional important role for development of *Wuchereria* larvae in the vector. Again, precipitation and temperature are co-determinants of the atmospheric humidity required for abaca cultivation. On the other hand, a high atmospheric humidity also favours the development of *Wuchereria* larvae in the vectors, and promotes the process of penetration by infective larvae, since it keeps the surface of the skin moist – not least by maintaining the secretion of sweat – and protects the larvae from drying out. Such climatic conditions prevail only in two of the four climatic zones of the Philippines (cf Map Plate 1) in which wuchereriasis is accordingly concentrated.

But within these two climatic zones the distribution of abaca cultivation, and thereby of wuchereriasis, is checked by another climatic factor. Since abaca is a relatively shallow-rooting bush, and very susceptible to wind-breakage, it can only be grown outside the main typhoon zones. In spite of otherwise favourable climatic conditions, a relative frequency of typhoons of 32 per cent does not permit abaca to be cultivated on the east coast of Luzon (north of 14°N). Admittedly, typhoons often cross the Bicol Peninsula, Samar and North Leyte as well (16–19% relative frequency), but in some areas attempts have been made to reduce the destructive effects of the typhoons by planting protective trees.

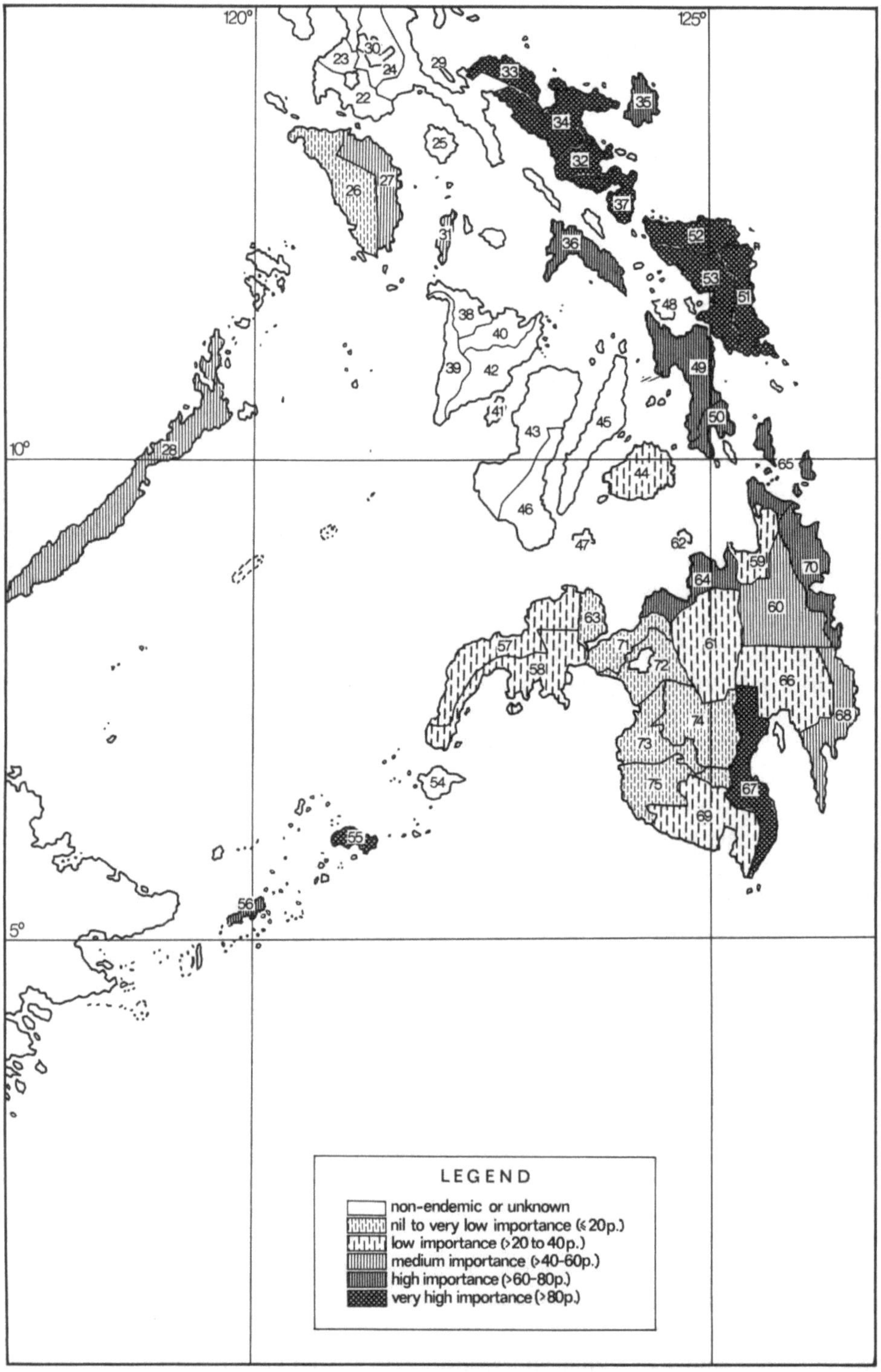

Fig. 22. Wuchereriasis in the Philippines: the significance in the endemic provinces with respect to prevalence, number of cases and case density (names of provinces in map plate 1)

As mentioned previously (cf 2.2.3), soil conditions and ground-water levels also play a role, in so far as abaca makes certain demands on the soil structure, which determines the anchorage of the plant and the permeability of the soil – abaca reacts to too high a level of static ground-water by root-rot. Philipson [626], analysing the growing of useful plants on the Philippines in their dependence upon soil conditions and climatic factors, commented on this as follows:

"Although abaca is planted on the level and gently sloping soils of coastal and mixed alluvium, it is generally more prevalent on the less intensively cropped, moderate and steeper sloping upland soils, of a variety of sedimentary, igneous and metamorphic parent materials. Concentrations are relatively high on the well to excessively drained mollisols, ultisols and oxisols, of clay and clay loam surface textures, and they are relatively low on the poorer drained soil units, being especially low on vertisols."

Thus the factors listed above determine whether abaca can be cultivated, whether *Aedes poicilius* achieves a high density, whether the development of the *Wuchereria* larvae can take place in the vector, and whether conditions are suitable for the penetration of infective larvae. A quantitative evaluation of the actual extent of the transmission has, however, only been attempted by Valeza and Grove [803]. Thanks to the results they arrived at in their investigations in Catanduanes, they drew the conclusion that per year every inhabitant of the endemic area under review is exposed to an average of 80.3 infectious bites and a transmission of 271 infective larvae. Considering the infectivity rate (L_3) of *A. poicilius* given in Table 4, the number of larvae harboured by them, and the biting density in the course of a year (Fig. 5), the risk of infection might well reach similar dimensions in Sorsogon Province, on East Samar and on Jolo. The situation on Mindanao and in the other endemic areas, where *A. poicilius* is the vector, can, however, not yet be estimated due to the lack of data.

Again, geofactors determine the frequency and distribution of *Anopheles flavirostris*, which has been shown to be the vector of *W. bancrofti* in Palawan, Bongao and Calaccad. Here flow velocity and chemical composition are the main factors which determine whether surface water constitutes a suitable breeding place or not. Both factors can be influenced by the annual precipitation regime. Heavy precipitation can increase the flow velocity and wash away larvae and pupae if it exceeds the critical level of 30 cm/sec, beyond that having the effect of diluting some water-soluble substances or of enriching others (such as oxygen). Above all, however, flow velocity depends on the gradient of the terrain as a geomorphological element, together with the extent of tree-cover on the slopes as one for regulation. These geofactors are responsible for *A. flavirostris'* predilection for breeding in the slow-moving water of foothills and mountain slopes, and for its chance of performing as vector of *W. bancrofti* in areas where these conditions are met. But thanks to its exophilia and zoophilia, the transmission intensity achieved by *A. flavirostris* is outstripped by that attained by *Aedes poicilius.*

2.3 Brugiasis

Wuchereriasis having been identified in the Philippines already at the beginning of this century (see above), the occurrence of infections with *Brugia malayi* in an endemic area in the south-west coast of Palawan was first reported in 1964 by Cabrera and Rozeboom [165]. Since then further foci were discovered in the Sulu Archipelago [158], in the Agusan Province in Mindanao [170, 171] and in the south-western tip of Samar [830]. *Brugia* never occurs by itself, but always with *Wuchereria.*

Apart from Palawan, where further investigations [163] had shown *Brugia* infections as occurring in the Roxas and Puerto Princesa municipalities as well, it is a matter of relatively limited areas, all of which are characterized by the presence of fresh-water swamps with a rich variety of aquatic plants. As elsewhere in South and Southeast Asia, so in the Philippines, too, the close correlation between the bionomics of vectors and the occurrence of *Brugia* infections is expressed in this. As a result of the peculiarities in oxygen intake of their larvae and pupae, species of the mosquito genus *Mansonia,* which are the chief vectors of this filariasis, present particularly high population densities in the range of breeding waters with ample vegetation. Thus *Mansonia bonneae* and *M. uniformis* as representatives of this genus have turned out to be the most important, if not the sole, vectors in the Philippines as well.

Although locally of great significance, on the whole the frequency of brugiasis by no means equals that of wuchereriasis. The number of *Brugia* infections registered so far and established by proof of microfilariae only totals 305 cases (cf Table II), but a further distribution through infected persons and/or the existing animal reservoir cannot be excluded for the future.

2.3.1 The Endemic Areas

On the evidence of the present state of knowledge the following endemic areas in the Philippines will be particularized and briefly characterized. Isolated cases of brugiasis have, however, also been diagnosed at greater distances from the focus centre, though mostly without providing a clear answer to the question of whether these infections were autochthonous or picked up elsewhere.

2.3.1.1 Quezon (Palawan)

The first focus to be discovered was on the south-west coast of Palawan in the municipality of Quezon with a centre in Panitian, the Gungnan, Marerong, Kambing, Tagbanaba, Tagpisa, Malatgao and Taganilao sitios of which were the most severely affected with their prevalence rates ranging from 27.7 to 56.6 per cent [168, 666]. All these sitios are in the vicinity of the so-called crocodile swamps ("Kakrokan"; cf Fig. 23), a wooded, swamp water district, the characteristic tree of which is the giant pandanus.

The Palawanos who inhabit this district tend to devote themselves to rice cultivation, in which the entire family is engaged. In cases where the fields are some distance away from the sitios, the inhabitants frequently stay out over-night, in other cases they return to the sitios for the night. Thanks to the close proximity of sitios, fields and lines of communication to the fresh-water swamps the exposure to bites from the transmitting mosquitoes (*Mansonia bonneae*) is extraordinarily high. With 12 per cent (carabao bait) and 95 per cent (human bait) out of the total mosquito population, this species came second and first in the frequency scale there, and was the only species to be infected by the *Brugia* larvae (0.98%).

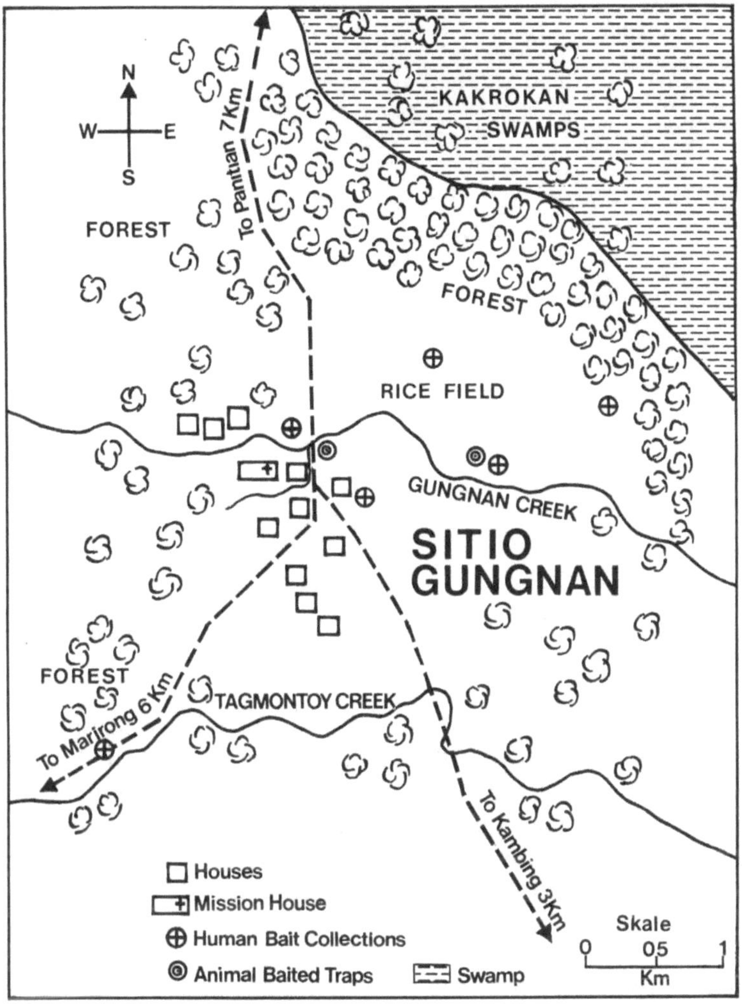

Fig. 23. Sketch map of Sitio Gungnan, Barrio Panitian, Palawan [Source: 666]

According to Rozeboom & Cabrera [668] the brugiasis focus in south-western Palawan has been there since prehistoric times. Its origin might be traced back to immigrations from the south in times when there was still a land-bridge between Borneo and Palawan.

2.3.1.2 Puerto Princesa – Roxas (Palawan)

In 1971 investigations in the municipalities of Puerto Princesa and Roxas led to the discovery of further occurrences of *Brugia* on Palawan [163]. In five barrios, which are situated between 10°N. and 10°25′N. (cf Fig. 24), 17 of 248 persons undergoing the examination (= 6.8%) proved to be infected (for details see data in Table II). There is no evidence in the information supplied by B. D. Cabrera and Jueco [163] as to the presence of fresh-water swamps or other kinds of breeding places of *Mansonia*

Fig. 24. Sketch map of the Municipalities of Puerto Princesa and Roxas, Palawan [Source: 163]

in this area, but the proportion of the two potential vector species in the overall mosquito population is distinctly smaller than in Quezon, *M. bonneae* totalling 0.78 per cent and *M. uniformis* 2.81 per cent. This is also reflected in the difference in the rates of prevalence among men. At present it is not possible to determine whether the foci established in Quezon and Puerto Princesa – Roxas are parts of one uniform area of endemicity. This requires further investigation.

2.3.1.3 Sulu Archipelago

In 1967 some islands of the Sulu Archipelago adjacent to Quezon (Palawan) were found to be endemic for *Brugia malayi* [158]. The total of 29 cases was shared by the islands of Lapak (Tapul Group), Tandu-Bas, Sanga-Sanga and Tawi-Tawi (all of which belong to the Tawi-Tawi Group; cf Table II). Totalling 17.5 and 6.4 per cent each, Masantong (Tawi-Tawi) and Tingol-Tingol (Sanga-Sanga), the two barrios in the vicinity of fresh-water swamps, showed the higher prevalence rates (cf Fig. 25, too). Mosquitoes of the genus *Mansonia* had a very high share in the overall mosquito density, with 41.2 per cent netted in traps with carabao as bait, and 47.5 per cent with those approaching humans [134]. *M. uniformis* (21.9 and 29.6 per cent) was the dominant species, followed by *M. bonneae* (9.9 and 17.3 per cent) and *M. ochracea* (9.4 and 0.5% respectively), whereas a mere 22 specimens (0.1%) of *M. crassipes* were caught, all of them in carabao traps. The infection rates with *Brugia malayi* larvae, however, proved to be very low indeed (1 out of 5,737 *M. uniformis,* 2 out of 2,631 *M. bonneae).*

2.3.1.4 Mindanao

Extensive mass examinations carried out in 1968 on Mindanao [170, 171] resulted in information about another *Brugia* focus in the upper reaches of the Agusan and its

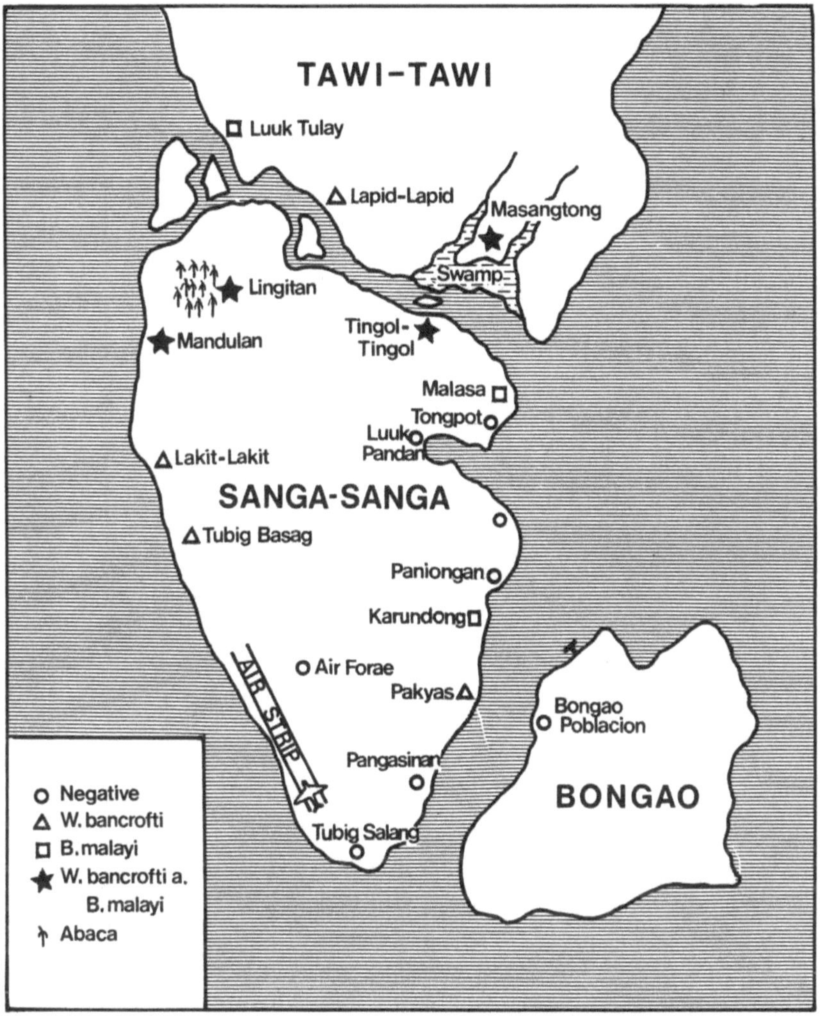

Fig. 25. Distribution of wuchereriasis and brugiasis in the Municipality of Bongao, Tawi-Tawi [Source: 158]

catchment area (Agusan del Sur Province; cf Fig. 26). In the area of the Libertad and San Andres (Bunawan Municipality) barrios, which manifested the highest prevalence rates with 2.2 and 5.1 per cent respectively, data on the composition of the mosquito fauna showed a distinct dominance of *Mansonia uniformis* (64.8% and 69.7%), followed at quite an interval by *M. bonneae* with 4.3 and 13.9 per cent of the catches [131]. The breeding places there are located in flat, swampy areas cultivated with rice. As part of a rift valley which runs in a north-south direction and of the Agusan-Davao Plain, the endemic area is amply endowed with swamps, which afford the conditions for a high population density of the vector mosquitoes in many places. Notwithstanding this, an examination of altogether 2,034 persons in the Agusan provinces yielded only 25 *Brugia* infections.

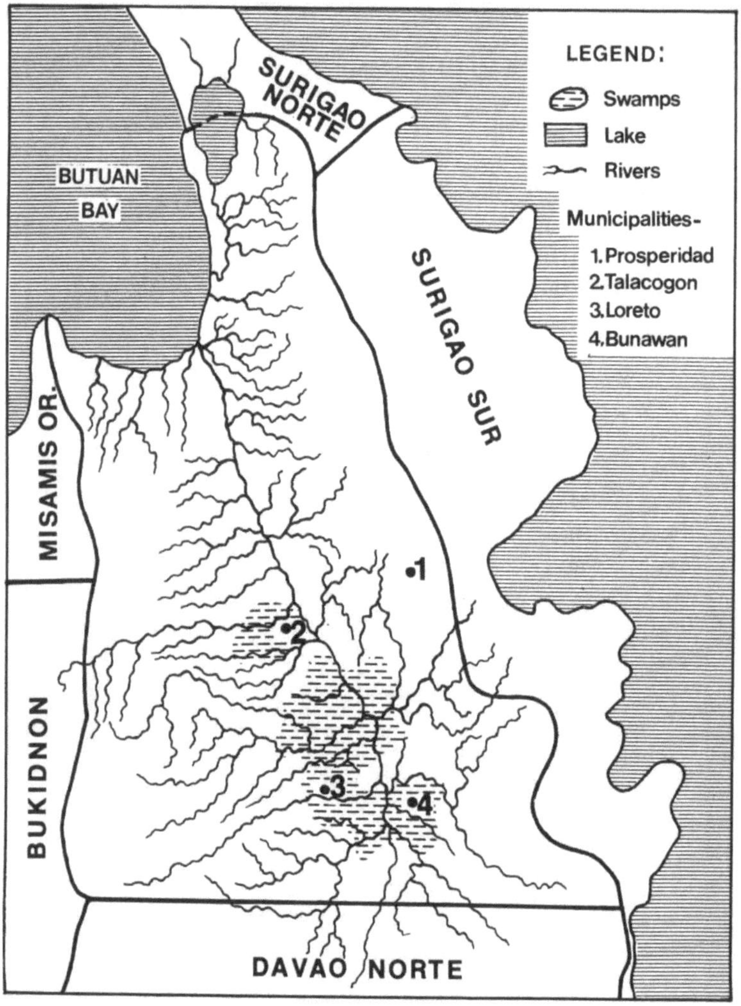

Fig. 26. Distribution of brugiasis in Agusan del Sur, Mindanao [Source: 171]

2.3.1.5 Samar

The last *Brugia* focus so far described for the Philippines was a swampy area in the south-east of Samar [830]. The centre of this endemic area is situated in Santa Cruz (4.4%) and Taclaon (2.3%), two sitios in the south-west of the Salcedo Municipality, which are among fresh-water swamps near the Taclaon River (cf Figs. 27 and 28). According to data supplied by Wenceslao et al. [830], some of the cases diagnosed as such appeared to be autochthonous infections, whilst others might have become infected when on Palawan. Since isolated cases have also been reported from Quinapundan, which borders on Salcedo in the west, the overall distribution pattern of *B. malayi* cannot yet be assessed in full.

In terms of the results of entomological investigations, the proportion of the genus *Mansonia* in the overall mosquito fauna varies between 8.3 and 24.7 per cent accord-

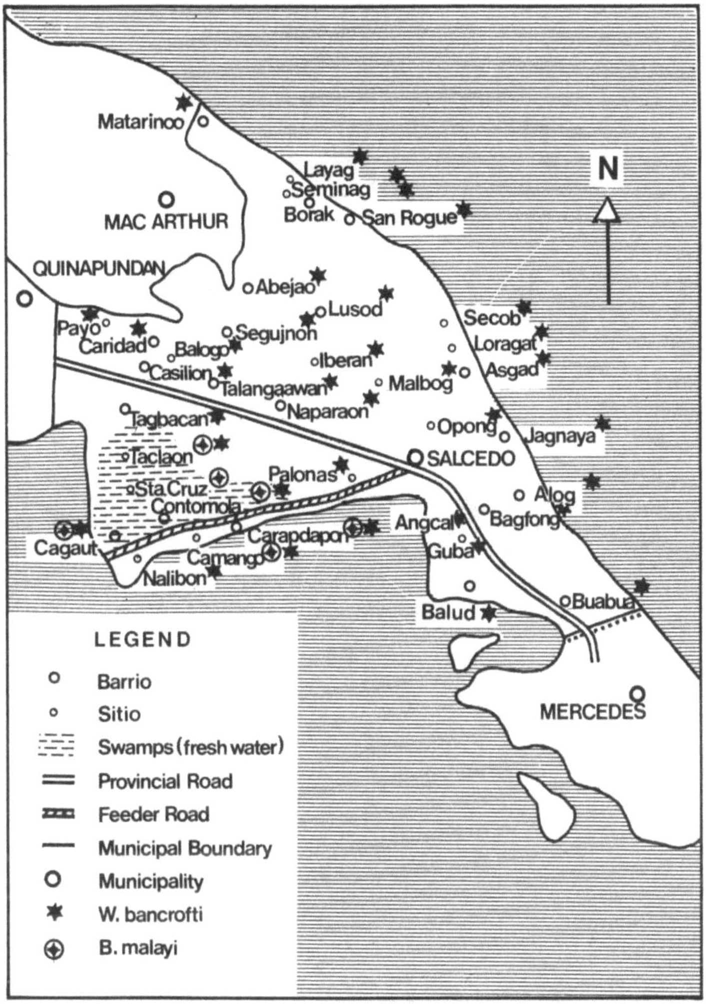

Fig. 27. Distribution of wuchereriasis and brugiasis in the Municipality of Salcedo, Eastern Samar [Source: 830]

ing to the trapping method. *M. uniformis* (8.1 to 23.9%) clearly came before *M. dives* (0.1–2.9%) and *M. crassipes* (0.1–0.2%). The dissection of 479 *M. uniformis* resulted in a rate of infection of 2.9 per cent; the infectivity rate was 0.62 per cent.

2.3.2 The Vectors and Their Bionomics

Among the various transmitting species of the genera *Mansonia, Aedes* and *Anopheles,* which are so far known in South and South-east Asia, in the Philippines only specimens of *Mansonia bonneae* and of *M. uniformis* were found to be naturally infected by *B. malayi;* though the two species differ in their proportions of the overall mosquito fauna in the endemic areas of the Philippines, their population density is so high everywhere that they may be regarded as the sole vectors of *Brugia malayi.*

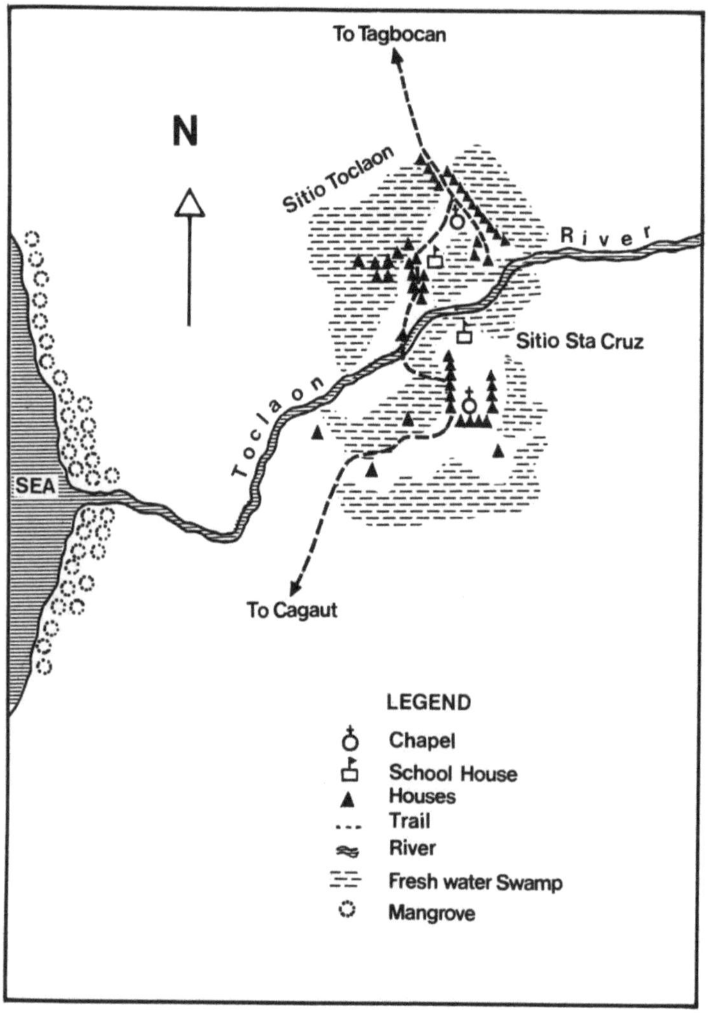

Fig. 28. Sketch map of the Sitios Taclaon and Santa Cruz, Salcedo, Eastern Samar [Source: 830]

Apart from the endemic area of Puerto Princesa-Roxas (Palawan) all the foci were found to be in the immediate vicinity of fresh-water swamps. The ample plant cover of these swamps is the condition for massive increases among the *Mansonia* species, as the larvae and pupae of these mosquitoes are dependent upon the oxygen produced by the aquatic plants, which they obtain from their leaves and root-systems by scratching with a cutting apparatus on the siphon specifically equipped for this purpose.

Investigations carried out in the Philippines into host preferences as well, as into the biting density during the course of the day, are exceedingly scarce for both species of *Mansonia. M. uniformis* is known from India and Malaysia to be a predominantly zoophilic species; its index of anthropophilia there reached only 2.8 per cent [586] and 2 per cent [836]. Neither is *M. bonneae* a predominantly anthropophilic species: it approaches man as well as animals for blood. It bites outside as well as inside houses without resting there. Admittedly *M. uniformis*, an exophilic species, also enters houses in order to bite.

In turns of diurnal activities *M. bonneae* can be included in the group of those mosquitoes whose chief phase of activity falls between the onset of dusk and 10 to 11 p.m. [666]. On Palawan the population density is subject to considerable seasonal fluctuations; a high density of bites must be expected mainly in the months from September to February (i.e. beginning about four months after the start of the rainy season, and persisting until the end of the second month of the dry season [cf 666]). This implies that the transmission of brugiasis is very much related to the seasons, at least in those endemic areas which experience a more marked division between rainy season and dry season.

2.3.3 The Periodicity of Microfilariae

Investigations of the periodicity of the microfilariae of *Brugia malayi* in the Philippines are available for the three endemic areas on Palawan, the Sulu Archipelago and Samar [158, 166, 830]. According to the authors' view, the periodicity is that of the nocturnal sub-periodic type in every case, but a comparative study appears to give rise to doubt on this view, for the fact that there are differences in the periodicity of the three endemic areas can be proved in several ways.

This applies even when following the method introduced by Turner and Edeson [797], which was later adopted for the Philippines calculating the periodicity graph from the "average per cent of peak count" (cf 2.2.1.2). Even the application of this manner of presentation shows different courses for the brugiasis graph in Quezon (Palawan), Bongao (Tawi-Tawi) and Salcedo (Samar) (Fig. 29), which is due to differences in the amplitude. With approximately equal mean maxima the values do not dip below 25 per cent in Quezon during the time of day with the lowest density of microfilariae in the peripheral blood, whereas they drop to 9.9 and 5.8 per cent respectively in Bongao and Salcedo. The decisive cause of this are the differences in the variability of the temporal occurrence of the individual microfilaria maxima in the three endemic areas, for in the counts carried out at two-hourly intervals only Quezon does not produce a maximum (peak value) in an infected person at 12 a.m.; in Bongao (between 8 a.m. and 6 p.m.) and in Salcedo (between 5 a.m. and 5 p.m.), on the other hand, maximum values are never observed during approximately half the days (Fig. 30).

Even the method of constructing the periodicity graph from the mean enumerated values suggests the supposition that the type of microfilariae met in Salcedo might not be of the nocturnal sub-periodic variety (Fig. 31). Due to the varying absolute density of microfilariae in the endemic areas this manner of presentation does, however, give the impression that the periodicity is least marked in Bongao. This is not the case, though, considering that the average minimum (1.9 MF/20 µl) here amounts to only 21 per cent of the mean maximum (9.0 MF/20 µl); i.e. from the point of the greatest microfilarial density (1 a.m.) until 3 p.m. seventy nine per cent of the microfilariae disappear from the peripheral blood. In the Brugia cases from Quezon this drop from an average of 71.5 to 27.2 MF/20 µl amounted to only 62 per cent. Among the patients in Salcedo on the other hand, it is very pronounced at 97 per cent (from 37.8 to 1.1 MF/20 µl).

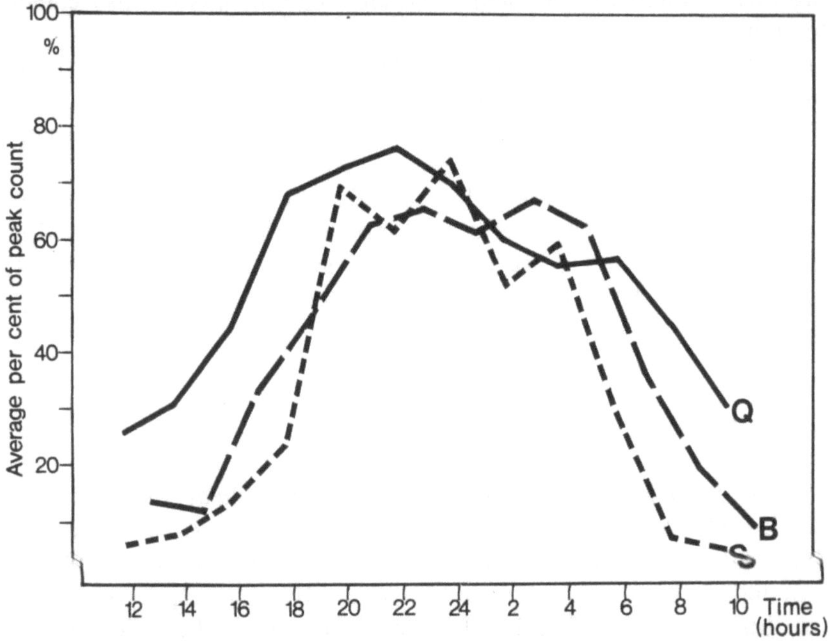

Fig. 29. Microfilarial periodicity of *Brugia malayi* in Quezon, Palawan (*Q*), Bongao, Tawi-Tawi (*B*) and Salcedo, Eastern Samar (*S*) [Sources: 158, 166, 830]

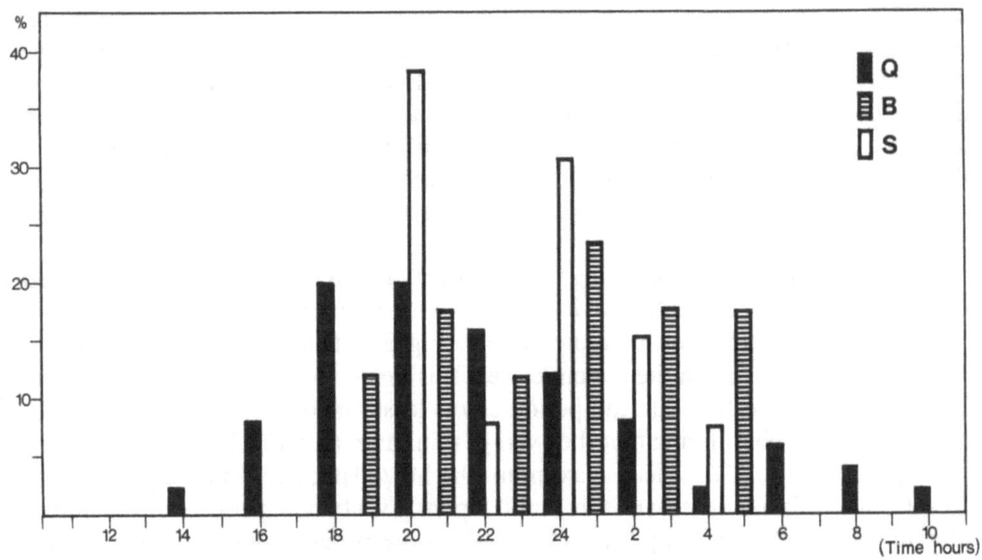

Fig. 30. Percentage distribution of maximum microfilarial densities in brugiasis cases during the course of the day in Quezon, Palawan (*Q*), Bongao, Tawi-Tawi (*B*) and Salcedo, Eastern Samar (*S*) [Sources: 158, 166, 830]

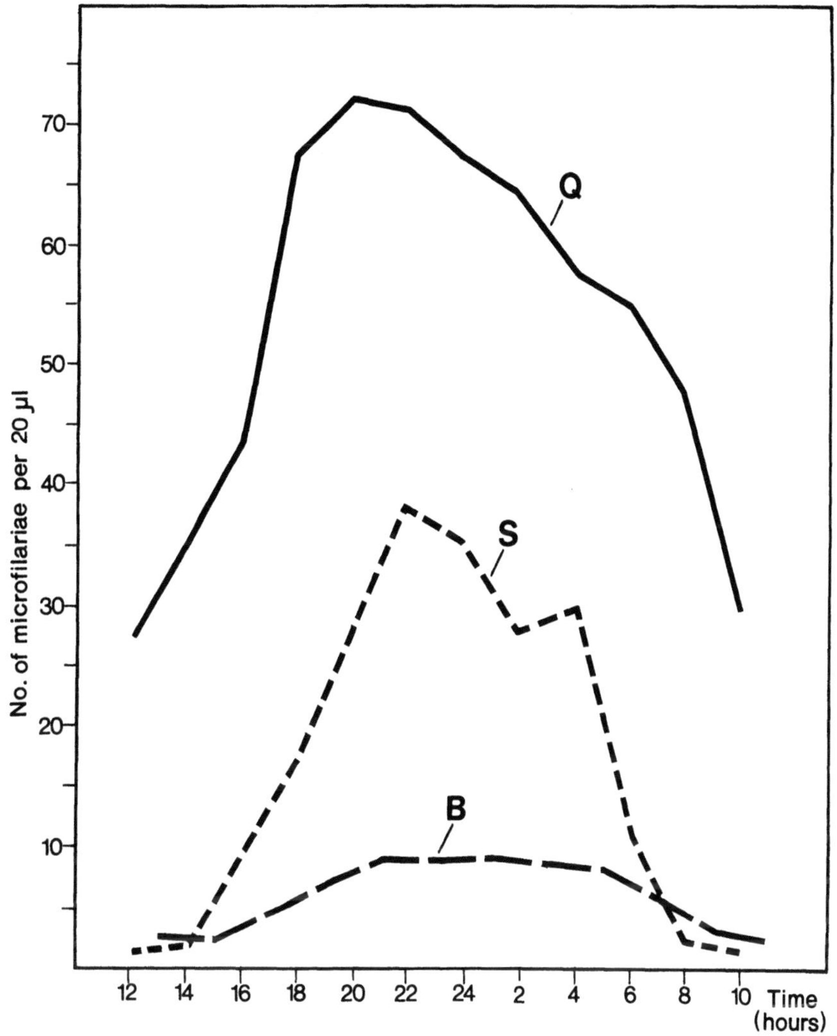

Fig. 31. Microfilarial periodicity in *Brugia malayi* cases: average number of microfilariae per positive per 20 µl blood in Quezon, Palawan (*Q*, 32 patients), Bongao, Tawi-Tawi (*B*, 13 pat.) and Salcedo, Eastern Samar (*S*, 13 pat.) [Sources: 158, 166, 830]

These relative conditions also produce a better comparability of the periodicity in the three endemic areas of the Philippines. A method of presentation which might profitably be used for this is the one which Habs [355] applied for the classification of seasonal disease frequencies in the World Atlas of Epidemic Diseases (Welt-Seuchen-Atlas), with the only difference that instead of the monthly or four-weekly mean value the mean number of microfilariae counted at two-hourly intervals throughout the day was set down as 100 per cent. As the sum of these mean values is equally large (at 1200) for each endemic area, this method permits a direct comparison and provides a simple way of recognizing differences (cf Fig. 32).

At the same time mean values converted in this way act as a starting point for finding a so-called periodicity index, as developed by Sasa and Tanaka [699–700], as

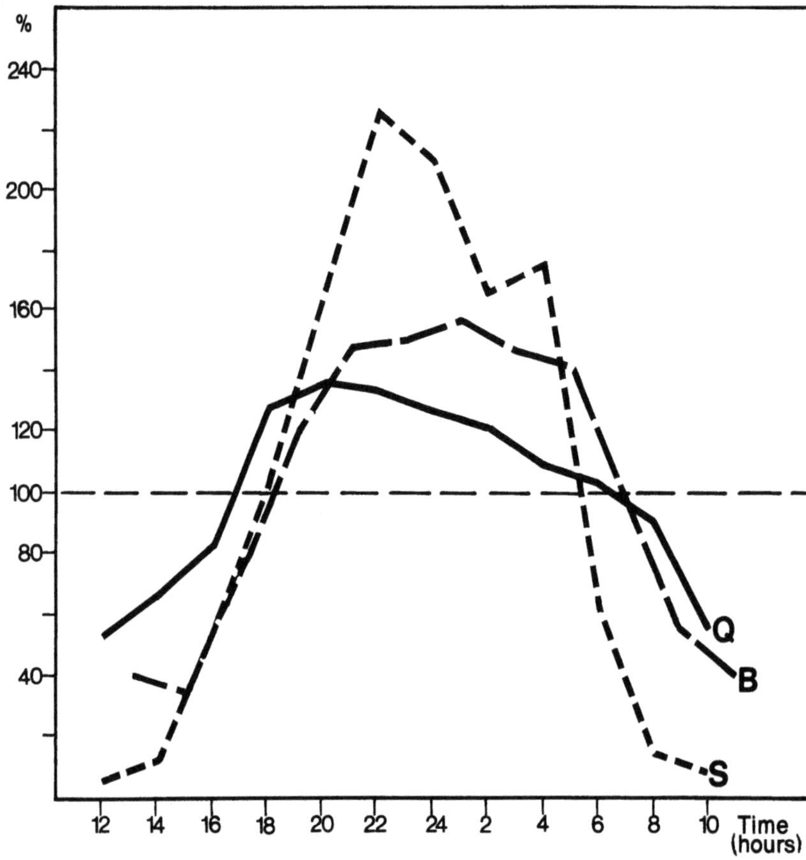

Fig. 32. Microfilarial periodicity in *Brugia malayi* cases: percentage distribution according to the daily mean count (= 100%) in Quezon, Palawan (*Q*), Bongao, Tawi-Tawi (*B*) and Salcedo, Eastern Samar (*S*) [Sources: 158, 166, 830]

well as by Aikat and Das [37], as an evaluative measure for the type of periodicity [cf 750]. At the same time these methods allow the calculation of that moment at which the microfilaraemia is to reach its "peak hour". (The Aikat-Das method shows the Quezon peak hour to be at 11.08 p.m., the Bongao peak as occurring at 0.37 a.m., and Salcedo as reaching its peak hour at 11.25 p.m.)

In the examples calculated by Sasa and Tanaka [699–700], the index for the typical nocturnal-periodic form fluctuated between 83.60 and 110.16, for the nocturnal sub-periodic type between 22.79 and 34.36. The Aikat-Das method, arriving at values from 107.96 to 147.97 and from 26.94 to 45.91 for the same endemic areas, presents a somewhat higher level without complicating the allocation. On the other hand, both the methods of calculation require that the periodicity of the microfilariae of *Brugia malayi* in Salcedo be allocated to the nocturnal-periodic type[1].

---

[1] Calculations carried out by Gajudo et al. [308] in accordance with the Sasa-Tanaka process resulted in a periodicity index of 86.80 for naturally-infected dogs in Quezon, Palawan, which is also high but differs markedly from the values found there for man

But before a final agreement can be reached, it remains to be seen whether the absolute microfilarial density determines the periodicity index or not. Since the mean maximum for Quezon is approximately twice as high as for Salcedo, it should be considered whether this might account for the differences in the degree of periodicity. In the context it is significant to observe that decreasing peak values indicate an increase in the probability that microfilariae are only traceable in the peripheral blood for a limited period. This problem had already been investigated by Sasa and Tanaka [700] without leading to discovery of such a relationship. The analysis of the available data for Quezon leads to similar results; the grouping of patients according to their peak values does not produce different microfilarial density-related periodicity indexes:

MF/20 µl (peak count)	periodicity index
>200	34.99
101–200	26.94
51–100	44.54
26– 50	34.80
≤ 25	37.20

These findings can be illustrated by a comparison of progress charts of patients with peak counts of >200 and those of peak counts of ≤25 MF/20 µl (Fig. 33). It admits the conclusion that the differences in periodicity between Quezon and Salcedo are not due to differences in the absolute density of microfilariae, but that the type occurring in Salcedo is to be described as nocturnal-periodic. Admittedly Wenceslao et al. [830], who discovered the Salcedo focus, point out that the periodicity there, though

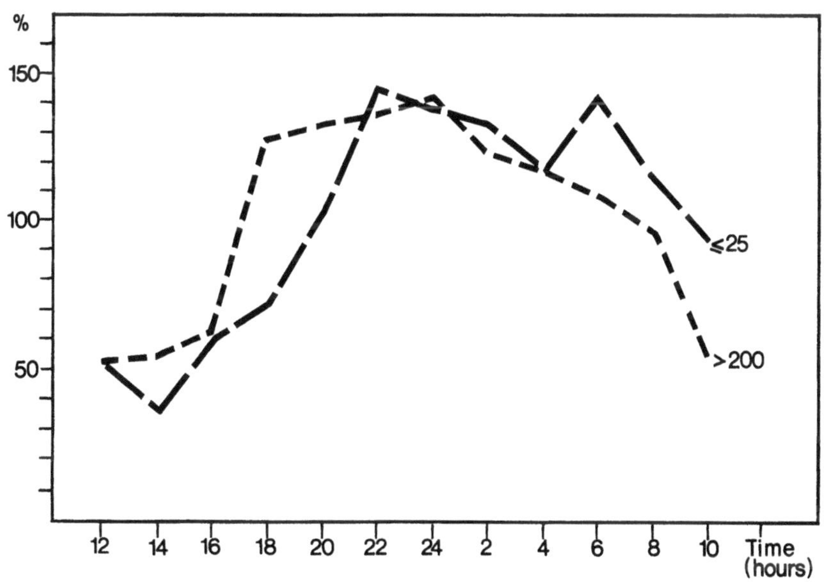

Fig. 33. Microfilarial periodicity in *Brugia malayi* cases in Quezon, Palawan: percentage distribution according to the daily mean count (= 100%) in patients with different peak counts (25 or less and more than 200 microfilariae/20 µl blood respectively) [Source: 166]

close to the nocturnal-periodic type, cannot be classified accordingly, because all the microfilariae that were observed had been found inside their sheath. This is indeed a characteristic of the sub-periodic type [cf 698]. (No further data are available on Salcedo so far as to other features of this physiological race.) Accordingly it might be a matter of an originally sub-periodic variant, which has converted to genuine nocturnal periodicity. It is not certain whether it has existed in Salcedo for some time in an autochthonous way, or whether the focus there owes its origin to an introduction from West Palawan [830]. The distinctly isolated occurrence might be an important reason for the presence of a *B. malayi* variant with an altered microfilarial periodicity, since geographical isolation favours formative processes of races and species.

2.3.4 The Transmission Index

As earlier mentioned, in the case of wuchereriasis (cf section 2.2.4) the transmission index is an important measure for recognizing epidemiological correlations. What is more, it forms a partial basis for the devising of control programmes. Whilst in the case of wuchereriasis man alone presents the point of departure for the transmission of the causal organism by the insect, in the case of brugiasis a number of mammals join man in his role as final host. Appropriate data are so far available only for Palawan, where endemicity was discovered first. In the sitios of the Barrio Panitian Rozeboom and Cabrera [666] found prevalences of 33 per cent among cats and of 9 per cent among dogs. Investigations in the region of the Quezon muncipality [308] on the other hand, which were carried out at a later date, returned considerably lower values. As nothing is known about this or the other *Brugia* foci, nor about the total number of potential animal final hosts, nothing can be said about this transmission potential either. Evaluation must therefore be confined to the human populations and their grouping according to age and sex.

On account of the low number of cases in the foci on Mindanao, Samar and the Sulu Archipelago only the data for Quezon, Palawan [168] can be regarded as adequate and be employed in an evaluation (without reference to the <1 year age group).

Just as was observed in the case of wuchereriasis, infections with *B. malayi* attain much higher rates of microfilaraemia among males (28.5%) than among females (17.8%). This difference is highly significant ($p<0.001$). A comparison of the age-related results in differing graphs for ♂♂ and ♀♀ (Fig. 34, Table 16) shows that there is a noticeable regression among females from the 1–5 year olds to the 11–15 year olds; after that microfilaraemia rates remain largely constant until old age (~11%). Matters are different among the male sex: here the graph shows a double-peaked course with maxima among the 6–10 year olds and the 34–45 year olds, though there are no statistically proven differences between the age groups.

Since information on the way of life of the affected population in south-western Palawan is scarce, the causal interpretation of the findings quoted above meets with certain difficulties. The differences among the sexes indicate differences in exposure. At the onset of puberty exposure among girls appears to be greatly reduced, whilst remaining constant among boys and men. In contrast to wuchereriasis (cf 2.2.4) there is no rise in the rate of microfilaraemia in the male sex with increasing age; this might be attributed to a rapidly-developing relative immunity. The same applies to

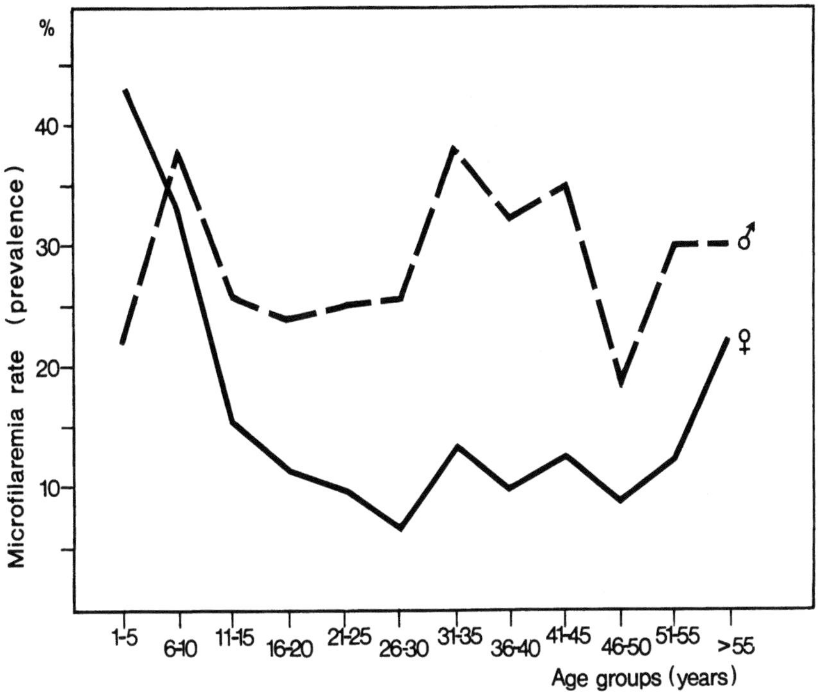

Fig. 34. *Brugia malayi* microfilaremia rates in Quezon, Palawan, by age and sex [Source: 168]

the considerable decrease among girls when further exposure suddenly stops. But whether the differences are in fact conditioned by exposure remains questionable in so far as it is not discernible why women should be less exposed to a vector which is active from the onset of dawn until around 11 p.m., or man more so. This would only be the case if men (cf 2.3.1.1) (as well as the children) more frequently remained in fields at a greater distance from home where they acquire the infection, whilst the women tended to remain behind in the villages or to return there after their daily work in the fields unless their activities were in any case concentrated on the fields nearer home.

The microfilaraemia rates are frequently coupled with the values for the density of microfilariae, i.e. greater exposure results not only in an increase in the number of infected people within a population, but also in the intensity of their infection. This rule is not fully confirmed by the data (cf Tab. 10) available for the endemic area of south-western Palawan. Although there are recognisable overall differences between the sexes (♂♂: 28.7; ♀♀ 21.1 MF/20 µl per infected person), no relationship between rates of microfilaraemia and density of microfilariae can be established for the various age groups. This is possibly due to the considerable individual fluctuations, together with the low absolute number of infected persons in the higher age groups.

2.4 Control of Filariases

When mass examinations in 18 provinces had resulted in a filariasis prevalence of 8.8 per cent, this outcome acted as a signal for the health authorities to set up a Filariasis

Committee on February 13th, 1957. From this derived the development of a Filariasis Control Pilot Project in 1959; at first intended as a five year venture, it later became a permanent institution within the Bureau of Disease Control in the Department of Health, and was known as the Division of Filariasis [129, 135]. The pilot project included one mobile and two stationary teams with in part different tasks.

The chief task of the mobile team was to carry out mass examinations in all the country's provinces. It was intended to observe a strictly random selection of municipalities in a province and of the barrios of these muncipalities. Blood samples of representative random samples of the population were finally to provide a survey of the proportion of microfilariae carriers in the overall population. Based on 282,712 blood samples taken in 49 of 56 provinces, the inventory made in 1963 recorded a microfilaraemia rate of 3.7 per cent [290]. Besides mass examinations, entomological investigations (collection of mosquito imagoes and larvae etc.) were also part of the programme of the mobile teams.

The tasks of the stationary teams working in the Sorsogon and Palawan Provinces were rather wider. It fell to them to examine the inhabitants of all the barrios of the municipality for microfilariae in the blood, to carry out the clinical examination of microfilariae carriers, and to divide the area of competence into test areas with different control measures; in the case of the Sorsogon Province, B. D. Cabrera [135] specifies the following arrangement:

Area I–A: weekly diethylcarbamazine (DEC, Hetrazan) treatment
I–B: monthly DEC treatment
II: weekly DEC treatment and DDT spraying
III–A: annual DDT spraying
III–B: semi-annual DDT spraying
IV: no control measures
V: DEC mass-treatment

The spectrum of entomological investigations by the stationary teams included the collection of adult mosquitoes in 20 selected houses, human bait trapping, search for and collection of mosquito imagoes in an abaca plantation, collection of *Aedes poicilius* from the leaf axils of *Musa textilis*, as well as determination of the material collected, dissection of adult mosquitoes and their examination for filariae larvae.

The stationary team working in the muncipalities of Brooke's Point and Quezon on Palawan does not seem to have existed for long, as it is only mentioned by Estrada and Basio [290] and no longer in later publications. But in addition to the field unit operating at Gubat, Sorsogon, another two field units were established in Guiuan, Eastern Samar and in Tagum, Davao del Norte. Recurring political unrest at Samar led to the Guiuan Unit being moved to Leyte later on and being called the Tacloban Unit.

On the basis of experience gained in Sorsogon, Camacho [190] produced cost-benefit analysis of filariasis control for this province as a model for a national control programme. His calculation is based upon 40,908 cases of filariasis in 1970, 13,497 of which affected persons in employment. On the basis of the observation that 34 per cent of the infected people suffered from lymphangitis and experienced an average of 3.5 attacks of fever a year, each lasting three days, Camacho [190] estimated the loss of working days caused by wuchereriasis to be 141,718 for the Sorsogon Province alone. At a minimum wage of 3.50 pesos this equalled a loss of more than 496,013 pesos. These indirect costs are to be added to the direct costs of 62 pesos per case (= 2,536,296 pesos for 40,908 infected persons) for medical examinations, treatment, medicine and laboratory work. The total costs would thus be 3,032,309 pesos. Expenditure specified for a five-year control programme (staff, materials, maintenance) on the other hand would amount to a mere 653,744 pesos for the first year. (Camacho contrasted this sum with a total net gain of 19,229,229 pesos for the 5-year period.)

But this cost-benefit analysis is guilty of one noticeable omission in so far as it assumes that all filariasis cases are already known, and do not cause expenditure through "case finding". But on Sorsogon alone the Gubat Unit had to spend 162.67 pesos per case for this during the period 1975–1979. In areas with a lower prevalence the proportion of expenditure was even less favourable: 353.29 pesos/case in the Tacloban Unit and even 2,735.87 pesos/case in the Davao Unit. Moreover, it remains doubtful, considering the high unemployment rate in the Philippines, whether the indirect costs calculated for the loss of labour can really be entered in, as this implies continued payment of wages in the case of disease.

The fact is that no control programme has been carried out; neither the one limited to Sorsogon Province nor the other which was to include the entire country. B. D. Cabrera [135] comments on this: "I regret to say that the Filaria Control Project has not been able to extend the program to the entire country as anticipated. The reasons for this may be as follows:
(a) lack of funds; (b) lack of well-trained personnel, both in the central office as well as in the field units; (c) red tape(s) in the release of funds and insecticides; (d) lack of interest, coordination and teamwork; (e) low moral among the personnel in the field; (f) lack of transport, field equipment and supplies."

Even in 1980 the Division of Filariasis still only consisted of the Central Office and the three field units at Gubat (Sorsogon), Tacloban (Leyte) and Tagum (Davao del Norte). Their tasks include:

- case finding surveys
- case management/treatment
- vector surveillance/control
- health education
- rehabilitation

The staff of the three field units consists of the following:

2 Medical Specialists I
2 Entomologists
6 Senior Medical Technicians
8 Medical Technicians
1 Driver
19 Others

In the period from 1965 to 1979, these field units examined 729,974 persons in the 19 provinces of the Health Regions V, VIII and XI, diagnosing 17,613 cases of filariasis, but treating a mere 7,861 because of the lack of chemotherapeutics. In their achievements all the field units thus remained considerably below the prevalence known to exist in their project areas. This is to be attributed to the inadequate availability of funds for travel as well as to the lack of motivation among the staff after years of work, which led to examination of the same, easily reached (and in parts already repeatedly treated) population groups.

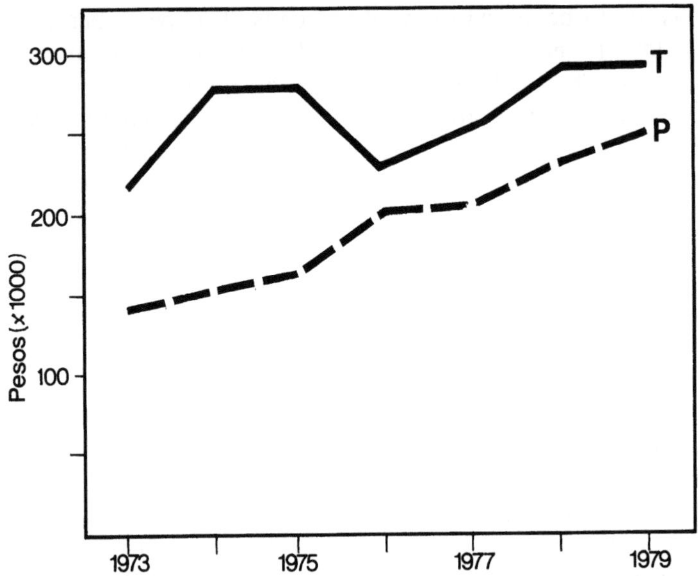

Fig. 35. Expenditures on filariasis control, 1973–1979. *T* total expenditures, *P* expenditures for personnel

On the evidence of the expenditure for the period 1973–1979 it can also be proved that the intended programme could not be maintained, let alone expanded. There is a continuous rise in payments of staff salaries (though staffing levels remained almost unchanged), but the overall sum available was not increased significantly; the rise in staff salaries was therefore largely achieved at the expense of consumables (cf Fig. 35).

Between 1973 and 1975 there was a distinct drop in the numbers of persons examined (by 32%, from 71,238 to 48,580), and an even greater one in the number of cases diagnosed (by 62%, from 1,657 to 633). Considering the means available and actually spent, this implies a drastic increase in the costs of examinations: from 3.06 to 5.78 pesos per person examined and from 132 to 444 pesos per infected person. Only in the following years did the situation as regards price become more favourable again (cf Fig. 36), although there were marked differences between the three field units in respect of the cost-benefit relationship: at the Bicol Unit the costs per case amounted to 162.67 pesos, at the Tacloban Unit they were 353.29 pesos, whilst the Davao Unit spent an average of 2,735.87 pesos per case. Though in part these differences are attributable to variations in prevalence, other causes cannot be excluded. The Davao Unit, for instance, only traced 143 cases among 82,961 persons examined (0.17%), although the known prevalence amounts to 3.28 per cent in this region, whilst distinctly more favourable relations were attained for the Tacloban Unit (0.91% compared to 2.55%), and still better ones for the Bicol Unit (3.11% compared to 5.85%).

Even when taking into account the inadequate financial endowment of the filariasis field units, it must be stated that the achievements remained behind the expectations. From 1973 until May, 1980, only 7,039 filariasis infections were diagnosed, and a mere 3,150 cases treated. In face of the fact that in 1975 there were about 8.5 million people living in the three regions for which the field units were responsible, and

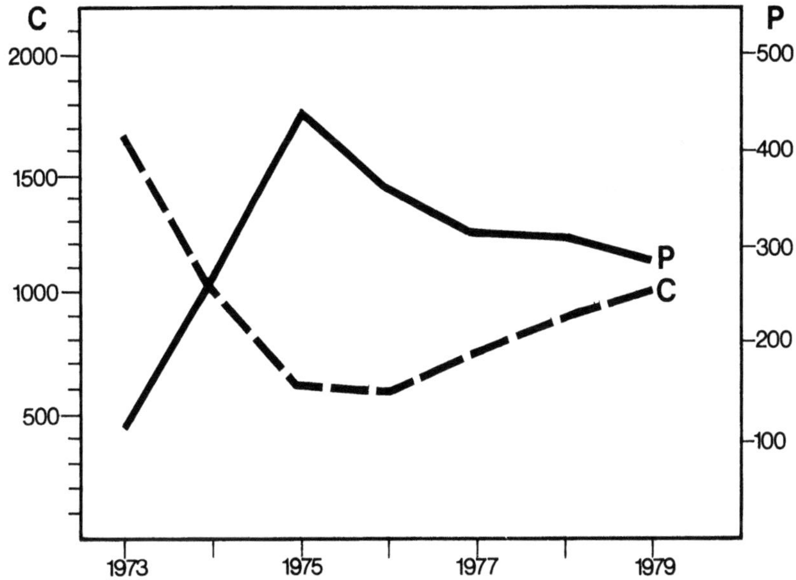

Fig. 36. Number of filariasis cases diagnosed (*C*) and costs per case in pesos (*P*), 1973–1979

342,200 are estimated to be infected with filariae, it is admittedly scarcely possible to expect significant progress in filariasis control so long as the necessary means cannot be found. This not only applies to the three regions where field units are already operating, but to the remaining endemic areas as well. At present 18 million Filipinos in 42 provinces of the country are exposed to the risk of infection; altogether more than 600,000 people are likely to be infected by filariae. The control measures taken so far have had little impact upon the situation, especially since no specific control measures aimed at wuchereriasis vectors have been taken since 1970.

3 Water-Borne Helminthiases: Schistosomiasis (Bilharziasis)

A special feature of the geomedical situation of the Philippines archipelago is the presence of a disease which occurs only in some of the islands and in connection with an amphibian species of snail, which was in turn found early in this century to depend upon a parasitic worm, the oriental blood fluke *Schistosoma japonicum.* The clinical picture of the Philippine schistosomiasis is classified as one of the hepatolienal forms, and is characterised by a chronic progress which, when undetected, very often leads to prolonged ill health and loss of the capacity to work.

Schistosoma japonicum is a parasite of the mesenteric veins. In order to develop, the eggs – having been deposited in the blood vessels by the female – have to proceed to the intestinal lumen and to be excreted with the stools. The intermediate host is the amphibian snail *Oncomelania hupensis quadrasi,* in which the miracidia having entered then develop via the sporocysts into furcocercariae; on leaving the snails, the cercariae swarm into the surrounding water and are able to force their way percutaneously into a suitable final host when one happens to enter the infested waters. After a phase of migration and growth and mating in the liver of the final host, the cycle recommences with the settling of adult pairs in the mesenteric veins. It is of epidemiological significance that, apart from man, a number of mammal species can function as a final host: Schistosomiasis japonica is an anthropozoonosis.

Although some authors attribute the presence of *S. japonicum* in the Philippines to importations by immigrants [e.g. 642], Hsü and Hsü [380] are of the opinion that all the currently known geographical races (cf 3.2) stem from one common ancestor, a native of southern China. The diffusion from there to Japan, Formosa and the Philippines is thought to have taken place either before or during the Pleistocene, i.e. before the separation of the islands from the mainland of Asia. This view is supported by the fact that in each endemic area the different geographical races (or strains or sub-species respectively) of *S. japonicum* are each adapted to a particular sub-species of *Oncomelania hupensis* as intermediate host. Evidently it is a matter of parallel developments which were furthered by geographical isolation and are to be regarded as co-evolution.

According to Hubendick [393], the snail which acts as intermediate host (*O. h. quadrasi*) and occurs in the Philippines was originally a native of the forest, the *S. japonicum* occurring there as a primary agent of a rodent zoonosis, which went on to establish itself as a parasite of man wherever he cleared forest in order to create new agricultural areas. In this regard the cultivation of rice plays a significant role.

3.1 Historical Survey

In the Philippines *Schistosoma japonicum* has been known as parasite of man since the beginning of this century. Shortly after the discovery of this trematode by Katsurada [437] and Catto [204], Stiles [733] reported the findings of an autopsy carried out by Wooley and subsequently described by him in detail as the first case of a *S. japonicum* infection of a Filipino [844]; this was published in 1906. As well as later diagnoses of further cases [114, 252, 337, 365, 616, 743, 837], this information attracted so little attention that the annual report of the Bureau of Health for the year 1912/13 reported another and similar finding of an autopsy as being the first case involving a Filipino [119, p. 257].

Until that date, and indeed in the years that followed [372, 534–536, 639, 842], the results had predominantly been obtained from examinations of the inmates of the Bilibid Prison or through autopsies in Manila. Nevertheless, the origin of these infected people permitted the drawing of conclusions as to the place of infection as the majority of them came from Samar, Leyte and Mindanao. This localization finally made Tubangui [781] choose Palo on Leyte as the area for his investigations when searching for the intermediate host of *S. japonicum*, which was then not known to be operating in the Philippines. On the basis of the amphibian *'Blanfordia quadrasi'* which he had collected there, and now known as a sub-species of *Oncomelania hupensis* he was successful in establishing that this specific snail was the intermediate host of *S. japonicum*. In addition he was able to produce adult worms in mammals through experimental infection for the first time, something which had also been previously unknown in the Philippines[1], and to establish that they were identical with the flukes occurring in Japan, China and Formosa.

A survey carried out by Africa and Garcia [16] made it appear likely that besides Leyte, Samar and Surigao, Sorsogon and Mindoro were also to be included in the list of endemic areas. Higher prevalences having been observed in China and Japan, Africa and Garcia were the first researchers to attribute greater importance to schistosomiasis in the Philippines than had previously been assumed: "Schistosomiasis japonicum deserves the same consideration if not more than that given to malaria, typhoid fever, the other forms of dysentery, and intestinal parasitism in provinces where it is endemic." Presenting a four-point programme, they therefore recommended the carrying out of mass examinations of people and their domestic animals throughout the country, together with the investigation of the geographical distribution of the intermediate host. This had already taken place by 1938/39, when with the support of the National Research Council of the Philippines, Tubangui and Pasco [794, 795] began to investigate schistosomiasis in various parts of the country and established prevalences of 17.1–45.5% in Leyte, Samar, Mindoro and Mindanao. Investigations carried out shortly afterwards (from April, 1940) showed 6.0 and 21.6% of the population in the Lake Mainit area (Agusan del Norte/Surigao del Norte) respectively to be infected with *S. japonicum* [592].

---

[1] All the previous diagnoses in man had been based exclusively on proof of eggs in the stool or in tissue sections

As early as July, 1940, three schistosomiasis units were set up and commenced work in the established endemic areas (Surigao, Leyte, Mindoro). Each of them included:

1 epidemiologist,
1 clinician,
1 bacteriologist,
3 nurses, and
3 assistant sanitary inspectors

Their tasks were wide ranging, and included the following activities [cf 702]:

(1) the collection of specimens of the snail intermediary host known as *Blanfordia (Oncomelania) quadrasi;*
(2) the examination of the snails for larval stages (cercariae) of the parasite;
(3) general physical examination of school children, as well as the persons in the localities surveyed, to detect evidences of the disease;
(4) collection and laboratory examination of stool and blood specimens for specific identification of the infestation;
(5) free treatment of cases and carriers of schistosomiasis with fuadin and emetine hydrochloride, as well as all other patients consulting the dispensaries;
(6) house-to-house canvass for case finding;
(7) extensive and intensive campaign for establishment of sanitary toilets;
(8) educational lectures for the enlightenment of the people on the prevention and control of the disease; and
(9) study by laboratory and field tests if it is possible to find an effective, economical and practical method or methods of controlling the snail.

A list of tasks had thus been set up which already included a large part of what constitutes the current programme of control measures (cf 3.7). Candido M. Africa and Luis Guerrero are to be regarded as the initiators responsible for promoting research into and control of schistosomiasis in the Philippines at this time. In an editorial (on the menace of schistosomiasis in the Philippines) in the Journal of the Philippine Islands Medical Association [12] Africa had once again forcefully drawn attention to the problem. Considerably earlier, namely already in 1927, Guerrero, the then head of the Department of Tropical Medicine and Parasitology at the College of Medicine, University of the Philippines, had issued a memorandum to the president of the university recommending mass examinations on Leyte and Samar, as well as the search for the then still unknown intermediate host. In 1940 he commented on the then current schistosomiasis campaign in a letter to the chief of the Division of Epidemiology, Bureau of Health. Pesigan [599], reporting on this, closed with the remark: "due to the outbreak of World War II, the campaign outlined by Dr. Guerrero could not be carried out." All research came to an abrupt end at this point in time.

The war in the Pacific, however, not only merely interrupted research into schistosomiasis in the Philippines but was also the cause of the even greater interest with which it was taken up again at a later date, for when American troops landed on Leyte on October 20th, 1944 and began the reconquest of the country they ran into a highly endemic area, the dangers of which were unknown to them (a detailed description of the events may be found in Ferguson and Bang [296]). Within a short time, about 1700 members of the American forces [846] as well as 144 of the 560 members of the Australian Air Force [259] were infected. This raised schistosomiasis to the rank of a research project of military importance; on April 25th, 1945 the members of the Sub-Commission on Schistosomiasis, Commission on Tropical Diseases, Army Epidemiological Board, arrived on Leyte, and established a laboratory at Tacloban. The investigations which were brought into being by the infection of members of the armed forces are, lastly, reflected in a multitude of publications [e.g. 96, 124, 202, 222, 257, 366, 397, 399, 451, 463, 514, 515, 745, 829, 843]. Besides these results, which predominantly concerned infected Americans, prevalence rates among the native population on Leyte, Mindoro and Mindanao are taken into consideration as well [76, 398, 849], together with investigation of the infection of rats and of the ecology of intermediate hosts [471, 472].

The Philippine side returned to schistosomiasis research in the field in 1948. From that time their investigations have been associated with the name of Trinidad P. Pesigan for twenty years. He and his colleagues investigated further endemic areas on Mindanao [594, 610], and recognized the province of Sorsogon as being clearly endemic [593]. So too, the legislature recognised the significance of schistosomiasis for the Philippines, as well as the need to take measures for its control: in 1948 the Republic Act No. 299 (cf Appendix A) was passed, the effective date of which made available 200,000 pesos for the prevention and control of schistosomiasis. Pesigan, Senior Parasitologist and Acting Chief of the Schistosomiasis Control Program, Division of Laboratories, Department of Health, took charge of the organisation, co-ordination and implementation of the entire programme, which was inaugurated on Mindoro on May 4th, 1949, following preliminary work on the possibilities of the control of intermediate hosts. Members of that team formed a field unit which was stationed at Naujan, Oriental Mindoro. On September 22nd, 1949, some of the staff were sent to Leyte and commissioned with the task of setting up another field unit. A third one was formed on October 2nd for the purpose of commencing work in the Panguil Bay area (with responsibility for Misamis Occidental as well as the Lanao and Zamboanga Provinces); three months later these were to be followed by a further trio of field units for Sorsogon, Samar and Northeast Mindanao (Surigao and Agusan Provinces). Each of these units was composed of one or two medical officers, one or two technicians and two auxillary staff [595, 597]. Their tasks consisted of the following:

1. Case-finding by mass stool examination,
2. Treatment of the cases found positive for schistosoma ova,
3. Health education and propaganda work, and
4. Study of the snail host with particular reference to its control [597].

Supported by prolific documentation, the results were published as a special number of the Journal of the Philippine Medical Association (Vol. XXVII, No. 4, 1951), and summarized again by Pesigan [597] after the completion of the project. In the period from May 4th, 1949 to December 31st, 1951, a total of 123,291 persons had been examined in 132 towns in 16 provinces. In the endemic areas the average prevalence was 12.2%. There were more than 11,000 infected persons. Approximately half these people were treated with Fuadin (or Nilodin). Almost 250 km of the total of 1200 km of water courses investigated proved to be *Oncomelania*-positive; 122,000 m^2 of snail-breeding areas had been treated with chemicals or flame throwers. In the very year – 1951 – the Philippines government took the significance of the results into account by setting up a Schistosomiasis Division within the Department of Health. A petition addressed to the World Health Organization led to the signing of a formal agreement on March 17th, 1952, in accordance with which the WHO sent three short-term experts to the Philippines; the Philippines Council for United States Aid (PHILCUSA) had made available additional means for this. The terms of reference of the team of experts were [cf 763]:

1. to study the schistosomiasis problem in the Philippine Islands;
2. to examine the control work that was being done; and
3. to make recommendations for a national programme.

The detailed report of the experts concluded with the recommendation that a beginning should be made by concentrating all means and efforts on one pilot project. This was to be a concerted action carried out by the WHO (by sending three advisers) and the Division of Schistosomiasis (by seconding some of its staff). After building a project centre at Palo on Leyte, the pilot project was able to commence work as early as June, 1953, having been assured of the support of the WHO, PHILCUSA, the Foreign Operations Administration of the USA, and of other organizations such as UNICEF, and the Japanese International Co-operation Agency from 1972, in later years. Its tasks were defined as follows [606]:

1. To determine the most effective and economical means of controlling bilharziasis in the Philippines.
2. To train local professional and auxiliary personnel in the various phases of work envisaged in this project.
3. To conduct studies and make observations on the human, other vertebrate and snail hosts of *Schistosoma japonicum* as well as on the parasite itself; to obtain data for the rational control of the disease.
4. To make a thorough epidemiological study of the disease in a highly endemic area.
5. To design and test control measures on the results of data obtained by basic studies and observations.
6. To plan and prepare for an expanded bilharziasis control programme.

On June 22nd 1957 the Republic Act No. 2011 (Appendix B) became effective under which the government made available a further 34,500 pesos for "emergency labourers and foremen to be employed in the prevention or control of schistosomiasis". In 1958 the Pilot Project presented the first detailed results of the investigation of epidemiology, of the intermediate host and preliminary experiments on control [606, 607, 609]. Research work carried out by the Pilot Project eventually led to the setting up of a National Schistosomiasis Control Commission, which was created by Republic Act No. 4359 (Appendix C) on June 19th, 1965. Because of insufficient financial means this control commission was able to fulfill its task to only a limited extent. The necessary reorganization which, among other things, put the multi-disciplinary aspect to the fore, was made by Presidential Decree No. 893 (Appendix D) dated February 16th, 1976. The Control Commission was changed into a Schistosomiasis Control Council with a board of directors, which is chaired by the Minister of Health and has six further members. The board of directors is associated with an executive committee with eleven members, which in turn has four sub-committees, whilst the technical staff are directly responsible to the executive committee (see Appendix E).

Besides the Schistosomiasis Control Council, which is responsible for the co-ordination of all the control measures, there is another institution: the Schistosomiasis Control and Research Service (SCRS), Ministry of Health, which is concerned with the problem of schistosomiasis in the Philippines. Attached to the SCRS are the research centre at Palo on Leyte, as well as the schistosomiasis control teams (SCTs). In 1980 there were 19 such SCTs, stationed in the following towns [705]:

Victoria, Oriental Mindoro
Irosin, Sorsogon
Trinidad, Bohol
Mayorga, Leyte
Sta. Fe., Leyte
Catarman, Northern Samar
Catbalogan, Western Samar
Dolores, Eastern Samar
Kapatagan, Lanao del Norte
Molave, Zamboanga del Sur
Bonifacio, Misamis Occidental
Cabadbaran, Agusan del Norte
San Francisco, Agusan del Norte
Malaybalay, Bukidnon
Tandag, Surigao del Sur
Kidapawan, North Cotabato
Tagum, Davao del Norte
Digos, Davao del Sur

This setting up and perfecting over several decades of institutions which had to concern themselves with research into and control of schistosomiasis (cf 3.5) led to an almost complete picture of the distribution of this worm infection in the Philippines. Whilst the earliest estimates remained within the range of 25,000 to 33,000 infected persons [795], corrections had to be applied regularly at later dates. The number of cases was then estimated to be between 100,000 and 300,000 [525, 796, 846]. On the basis of a report for the year 1979 the Schistosomiasis Control Council [705] finally arrived at an estimated total of 626,980 cases distributed throughout 134 towns or municipalities in 22 provinces, with prevalences fluctuating between 0.01 and 52.19%.

3.2 *Schistosoma japonicum* and Its Biology

The distribution of *Schistosoma japonicum* is limited to East and South-east Asia where, in co-evolution with its intermediate host (*Oncomelania hupensis*) it has split up into several distinct strains which G. M. Davis [264] raised to the rank of "subspecies". These strains differed in a number of characteristics, such as morphology, size and form of the eggs, the pre-patent period, the protein-electrophoretic pattern, virulence, sensitivity towards schistosomicides and infectivity for the intermediate host [cf in this connection 225, 228–230, 241, 249, 263, 270, 271, 292, 299, 377–391, 444, 452, 550–552, 669, 823, 824].

In his detailed surveys G. M. Davis [265, 266] has presented a summary of the present state of knowledge of the evolution of *O. hupensis*, its presently known subspecies and the co-evolution of *S. japonicum*. on the basis of it the following orderings can now be set out:

Schistosoma japonicum strains:	*Oncomelania hupensis* subspecies:
Formosan	– *O. h. formosana*
Chinese	– *O. h. hupensis*
Indonesian	– *O. h. lindoensis*
Japanese	– *O. h. nosophora*
Philippine	– *O. h. quadrasi*

Accordingly it is solely *S. japonicum*, the sub-species which develops in *O. h. quadrasi* as its intermediate host, that occurs as a parasite of man in the Philippines[1]. All further data refer to this sub-species unless otherwise indicated.

3.2.1 Pre-patent Period, Egg Production and Transmission Index

Neither pre-patent period nor egg production of *S. japonicum* are fixed dimensions, but are significantly co-determined by the infection dose and the worm burden [357, 379, 386]. An experiment, for instance, in which mice had been infected with two cercariae (1♂, 1♀) showed the average pre-patent period to be 39.4 days (32–52) in contrast to only 34.6 days (33–38) after an infection with 100 cercariae (50♂♂, 50♀♀). However, this is not to be attributed to a difference in the onset of egg laying, as Moloney and Webbe [548] observed in investigations of the Chinese sub-species of *S. japonicum*. It seems rather that a threshold value of the number of eggs in the intestinal wall needs to be exceeded before a passing of eggs in the faeces takes place. Such a threshold value is reached sooner when the intensity of infection (worm burden) is higher than in cases where only a few pairs are present in the final host.

Apart from the species of the host and its age, the intensity of infection is also an important parameter in determining the daily as well as the overall egg production of a female worm [cf 357]. As far as man is concerned it has been found that the passing out of eggs decreases as the intensity of infection increases in consequence of the crowding effect, and that among 10 to 14 year-olds distinctly more eggs are passed in the stool per female per day than among other age groups (Fig. 37). However, it is not the parasites' reproduction rate in an individual host that is epidemiologically decisive but that of the host population. Hairston [357] gives the following relevant factors in this connection:

Vertebrate hosts
  Population of each species in the area
  Density of species in the area

---

[1] *Schistosoma philippinensis* as described by E. Y. Garcia [319, 320, 323] was separated from *S. japonicum* mainly because of the structure of its seven testes in clusters (cf the controversy between Chanco and Pesigan on the one side and E. Y. Garcia on the other as published in the Journal of the Philippine Medical Association, 1960, pp. 57–59 and 131–132). Such a structure, however, can also be observed among about 25% of the Philippine sub-species of *S. japonicum* [cf. 313]. *S. philippinensis* was neither recognized as a new species nor ordered as a new sub-species of *S. japonicum*. Even the search for strains with varying adaptation to different final hosts did not yield any information as to the occurrence of anthropophilic and zoophilic *S. japonicum* in the Philippines [606, 609]. Nevertheless it is not possible to rule out that parasite-intermediate host complexes with differences in compatability do exist in some endemic areas [cf 636]

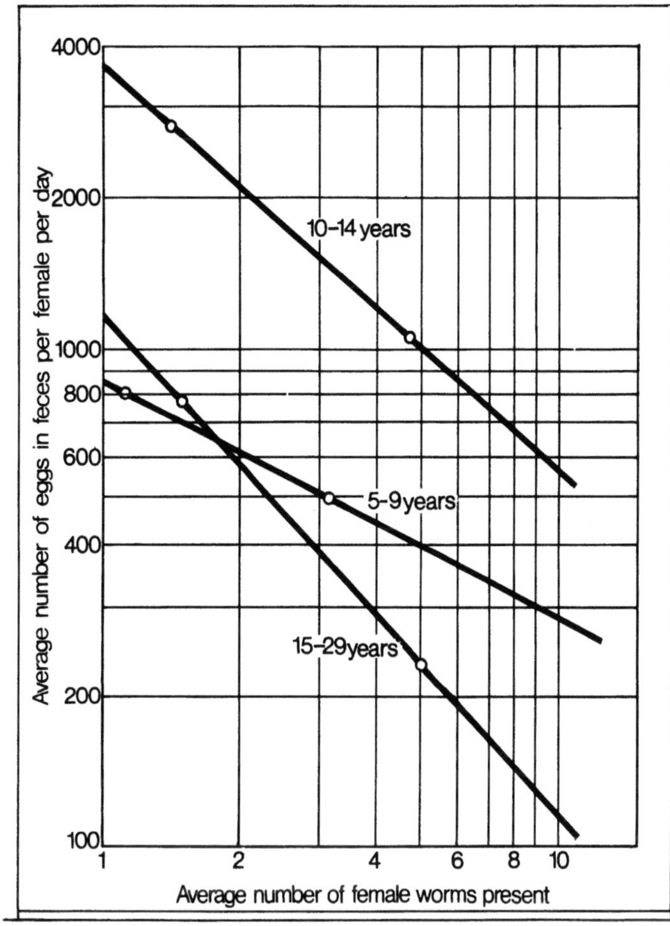

Fig. 37. Average egg output per female *Schistosoma japonicum* as a function of infection intensity (worm burden) [Source: 357]

Mortality rate of each species
Population fluctuation of each species

Parasites in vertebrates
Prevalence for each species of host
Prevalence of each age group of long-lived hosts
Rate at which hosts become infected
Mortality rate in vertebrate hosts
Total potential miracidia produced by each prevalence category

The total of hatchable eggs produced by all the final hosts – that is to say of the hatched miracidia – permits the calculation of relative transmission indices for every prevalence category, which establish their potential significance for the further distribution of schistosomiasis within certain endemic areas. Calculations of this sort have so far been published for Leyte by Pesigan et al. [606] and Hairston [357]. The significant difference between these two publications is the assessment of rats with indices of 1.36% and 52.49% respectively, which has been especially emphasized by Hairston.

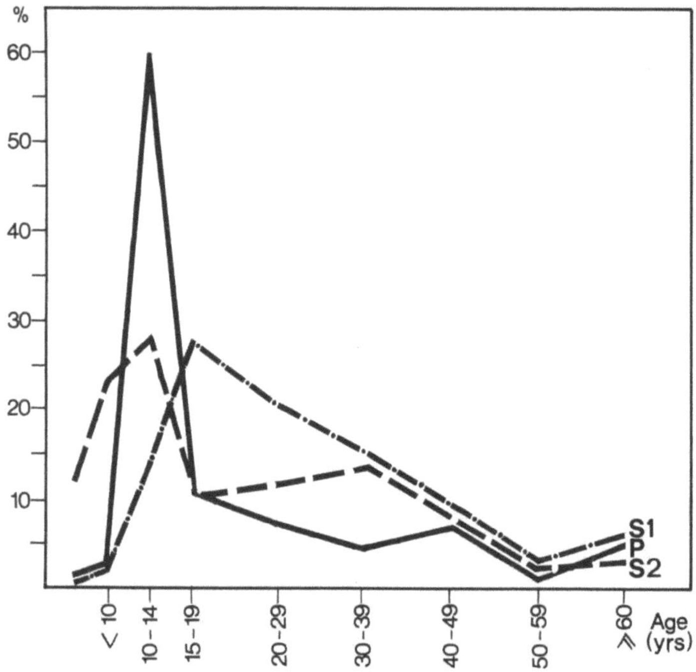

Fig. 38. Relative schistosomiasis transmission indices among humans in Palo (*P*) and Santa Rosa (*S 1, S 2*), Leyte [Sources: 277, 577, 606]

This is also the reason for the noticeable differences between the two investigations in respect of the relative transmission indices of man and other mammalian final hosts (cf Table 17).

If this inconclusive situation of the animal final host is disregarded in favour of an analysis of indices among selected human populations only, a direct comparison of different investigations can be attempted. The findings from two surveys at Santa Rosa, Barugo (Leyte) [277, 577], were drawn upon, and compared with the frequently cited data of Pesigan et al. [606] in order to test the stability, and thus the value of the evidence contained in the transmission index as well. Given that there were no population statistics for Santa Rosa, the proportional consideration of the various age groups followed the Philippines' overall mean 1975 values [1980 Philippine Statistical Yearbook], which presented no serious age differences from the age structure for Palo as quoted by Pesigan et al. [606]. Additional values (mean daily stool weight, hatchability) required for the calculation of the daily production of miracidia were also taken from the data for Palo. Since the results for Santa Rosa were based on examinations of 92% (1980) and 82% (1983) of the village population, a direct comparison is permissable although it had not been a case of a longitudinal study.

This comparison of age-dependent transmission indices (Fig. 38) shows distinct differences between Palo and Santa Rosa on the one hand and both the inquiries made at Santa Rosa on the other. The high value of almost 60% in the 10–14 year-old age group at Palo is striking, with the maxima for Santa Rosa reaching a mere c. 27–28%. (An even lower peak value of 21.6% results from calculations for Irosin, Sorsogon, which were made on the basis of data from the geometric mean egg count; [838].) Beyond that a precession of the maximum can be determined, moving from the 15–19 year-olds (1980) to the 14–15 year-olds (1983), and already attaining a very high index of 23.3% in the youngest age group (< 10 years).

The completed analysis shows that the relative transmission index does not present a uniform figure, nor does it remain stable within the same population over a longer time period. The question therefore arises in the first instance whether variations in the overall prevalence are responsible for this: for a particularly high degree of environmental contamination could lead to high prevalence rates even among small children, so that their very high proportion among the population as a whole results in an extremely high relative transmission index. To examine this problem recourse was made to the data provided by Olveda et al. [577] for Santol (San Miguel), Santa Rosa (Barugo) and Macanip (Jaro). Less than 10–20 km distant from each other, none of these three places had previously undergone a control programme; practically the entire population is engaged in rice cultivation; climate and topography are the same. Distinct differences occur in the prevalence rates (26%, 39%, 44%), however, so that the transmission indices of these three villages are particularly well suited to the comparison intended here. The calculations (Table 18) do not indicate that a higher average prevalence exerts a definable influence upon the course of the age-related transmission graphs, however. As examinations of the same population carried out at intervals of only a few years show marked differences in the indices (Fig. 38), they are of limited value as evidence. Nevertheless, the comparison of data available so far permits the assertion to be made that it is above all children and youths who – due to the high indices in conjunction with their defaecation habits – contribute to the contamination of the environment with schistosome eggs.

3.2.2 Passing of Eggs and Hatching of Miracidia

With the passing of eggs in the stool the miracidium is directly exposed to the influences of the environment. Though enclosed by the egg shell, the miracidium is fully developed at this moment and represents the first larval stage. The duration of the miracidia's survival and the process of hatching both depend on the humidity of the environment together with temperature, salinity, oxygen content and the pH value of water, whereas light conditions are said to play no role.

Optimal conditions for the hatching of miracidia only occur when the stool is discharged straight into water. If it is deposited on the ground the majority of the miracidia will die within forty hours [402], though on Leyte a survival time of at least one week was recorded in stool deposited in the shade [313]. Thus even in this phase in the development of *S. japonicum* frequency and intensity of precipitation are of epidemiological significance. In this regard Leyte, for instance, belongs largely to that climatic zone which is characterized by precipitation throughout the year, with high maxima around the turn of the year and great typhoon frequency. This accounts for the great probability that stool containing eggs is washed into water.

In this context the habits of the population are of decisive importance. Tiglao [769] noted on Leyte that 82.2% of the inhabitants of the area under investigation owned toilets, though only 41.5% of them were regularly used. Among those for whom no toilet is available, 10% regularly defecate into water. When questioned on the reasons for contact with water, however, only 42 replies (= 1.6%) out of a total of 2603 answers admitted defecation (in rivers, streams, sea/beach, canals and rice fields). This does not seem to account for the true extent of contamination of water, particularly so as the investigation was carried out among adults (≥25 years). The main transmission potential is found in the 10–14 year-old age group (cf 3.2.1), so that it is primarily their behaviour which decides upon the number of eggs entering the water. The habit of discharging stool into water is probably more often found among children and young people than among adults.

Optimal conditions for hatching are given at water temperatures ranging from 25° to 31 °C, as they permit, under experimental conditions, the miracidia to survive for approximately 24 hours. Higher (37 °C) or lower (18 °C) temperatures, though not killing the miracidia, prevent hatching [473]. This phase of the development of schistosomes is thus, once again, best guaranteed in the two climatic zones of the Philippines presenting relatively constant temperature conditions throughout the year (cf Map Plate 3).

Besides a sufficiently high oxygen content, hatching also requires a pH value of the water which is provided by the slightly alkaline environment. At a pH value of 6.4–6.8, the hatching rate remains at 15.9%, rising to 44.4% at a pH value of 6.9–7.1% and to an average of 70.5% at a pH value of 7.2–7.6, the latter pH 7.6 being considered to be the optimum [402]. The salinity of the water is also of significance: a salt content of 0.7% stops hatching altogether. Whilst a salt content of 0.2% allows ready-hatched miracidia to maintain their normal level of activity, 0.66% leads to death within 70 minutes, and 1.3% to the same within even two minutes. In waters or coastal flats where tides or floods (in the wake of storms or typhoons) cause major salination, the development cycle of *S. japonicum* may therefore permanently or periodically be prevented.

Since miracidia do not take up nutrients, the duration of their survival is primarily determined by the extent of the existing reserve supplies. Following the depletion of the reserve substances the chances of survival depend entirely upon the environmental factors; after that the miracidia waste away within a short time. According to data supplied by Hairston [359] about 80% of the *S. japonicum* miracidia survive for a period of ten hours after hatching. After a further five hours the survival rate is no more than 20%; 18 hours after hatching all the miracidia have died merely as a result of exhaustion of their reserve substances.

3.2.3 Larval Development in the Intermediate Host and Shedding of Cercariae

Mateo [516] specifies the larval development of *S. japonicum* in *O. h. quadrasi* as taking place over a period of altogether at least 42 days. After penetration the miracidium loses its cilia within 24 hours and turns into a mother sporocyst within a further three days. About 21 days p.i. daughter sporocysts appear and hatch after another three days. In snails which have been penetrated by one miracidium only the release of cercariae into water begins 42 days after this event at the earliest, and no sooner than 45 days after the process has been started in those which have been infected by 2 to 5 miracidia. The mean duration from infection and first release of cercariae, however, is 62 or 64 days [609].

These results from Palo on Leyte are evidently based on investigations carried out at room temperature. The information supplied by Pesigan et al. [609] and by Mateo

[516] therefore does not afford an insight into a connection between temperature and the speed of development. Such a correlation was defined by Pflüger [613, 614] and Pflüger et al. [615] for the development of *Schistosoma mansoni* and of *S. haematobium* in their specific intermediate hosts at constant temperatures between 16° and 32 °C in the form of a hyperbolic relationship. Diurnal, periodically alternating temperatures led to the exact addition of temperature totals. It is therefore to be assumed that the temperature for the development of *S. japonicum* in *O. h. quadrasi* is of similar significance. Since the kind of relationship is not known, nothing can be said about the manner in which the speed of development is influenced by temperature conditions in the different climatic zones of the Philippines.

With an average of 232 (1 miracidium) or as the case may be 279 (2–5 miracidia), the total number of cercariae released by one *O. h. quadrasi* is very low [609]. There is no doubt that this represents an adaptation of the parasite to the very small intermediate host (maximum shell length: 4.8 mm), the average life duration of which must not be decisively shortened if the schistosome cycle is to be maintained. Consequently there is a positive correlation in all schistosome intermediate host systems between the size of the intermediate host and the extent of the production of cercariae [465].

For a single snail under laboratory conditions the average duration of releasing cercariae is 32 (1 miracidum) or 62 days (2–5 miracidia), but for populations of snails which have undergone single or multiple simultaneous infections respectively, it is 150 or 200 days [609]. Thus even a single infection in a population of snails leads to a long-lasting contamination of waters with cercariae. This fact ensures the continuity of the parasite cycle, although the individual cercaria tends to lose its infectivity after 36 hours and its own life cycle after 2–3 days. The number of cercariae released by a snail population, however, decreases markedly after only 80–90 days p.i. (Fig. 39).

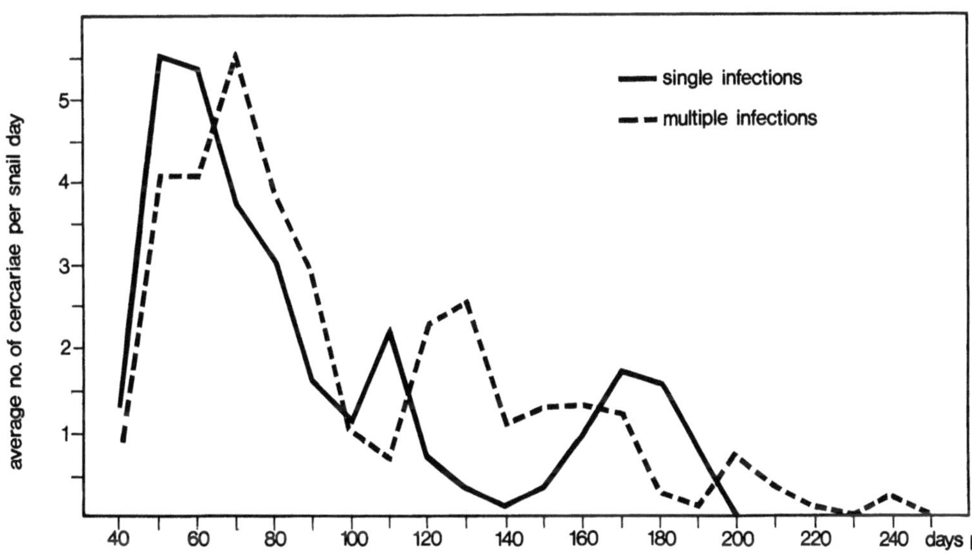

Fig. 39. Number of *S. japonicum* cercariae emerging from singly or multiply infected *O. h. quadrasi* [Source: 609]

However, the dynamics of transmission are not only assessed by the production of cercariae per infected surviving snail. In this the natural quota of deaths among the intermediate hosts plays a decisive role as well. Assuming death rates per diem of 0.019 for infected ♀♀, and of 0.027 for infected ♂♂ of *O. h. quadrasi* as suggested by Pesigan et al. [609] and by Hairston [358], almost 60% of females and 70% of the males in a snail population that have been subjected to a single infection only will have died until the end of the prepatent period (45 days); in this way the potential number of cercariae which could proceed from one miracidium is reduced by a corresponding percentage. Thanks to the continuing death rate of the snails during the entire period of cercariae release (45 to 195 days post infectionem), the net productive rate of the miracidia decreases drastically once again [cf 359]. Instead of a total of 307 cercariae, one miracidium only produces an average of 63 in female snails or 38 in male snails (cf Fig. 40).

Epidemiological importance is attached to the question of whether the release of the cercariae from the snails is influenced or even steered by outside factors, i.e. whether there are possibly any diurnal and/or seasonal cycles. The findings on diurnally periodic swarming presented in this context by Baumann et al. [87], Pesigan et al. [609] and E. G. Garcia [313] had to be revised once again because of their short observation period of only 1–2 days, for investigations carried out by Nojima et al. [563, 564] over a period of 12 days and at room temperature showed that the introduction of snails into the water is followed by an initial, non-periodic release of cercariae. Only then does a diurnally determined rhythm evolve: the swarming of *S. japonicum* cercariae from *O. h. quadrasi* is subject to a diurnal periodicity with a mean maximum between 2 and 4 p.m. (Fig. 41). In this duration of exposure to light and light intensity are evidently of great significance. Cloudy skies, for instance, result in a distinctly lower rate of releases, with the maximum occurring two hours later between 4 and 6 p.m. The greatest danger of infection for man thus arises during the late afternoon, since about 80% of the cercariae have swarmed at dusk.

Besides the diurnal periodicity there is a short rhythm of 3–4 days [563, 564] which is primarily based upon the fact that a release phase must be followed by another one during which more cercariae mature in the intermediate host. The number of de facto released cercariae is, however, subsequently modified by the weather conditions, i.e. by the duration of exposure to light at a specific intensity (cf Fig. 42). There is a positive correlation with the duration of sunshine; i.e. sunny weather results in a distinctly higher risk of infection.

In addition to light conditions, temperature and pH value of water bring influence to bear on the release of cercariae. Temperature, however, remains of secondary importance, since significant quantitative change does not occur with an interval spanning 19° to 30 °C (within which the natural conditions largely fall in the Philippines) (cf the investigations by Mao et al. [510] on *O. h. formosana*). As far as the concentration of hydrogen ions is concerned, a value of 7.6 has a beneficial effect and a slightly acid environment (rain water of pH 6.4 to 7.0) a retarding effect on the release [87].

A seasonal cycle in the release of cercariae has not been observed so far. That there are, nevertheless, seasonal differences in the contamination of waters with cercariae has other reasons; they depend primarily upon the extent of the snail population and upon the seasonal fluctuation of prevalence rates of final and intermediate hosts.

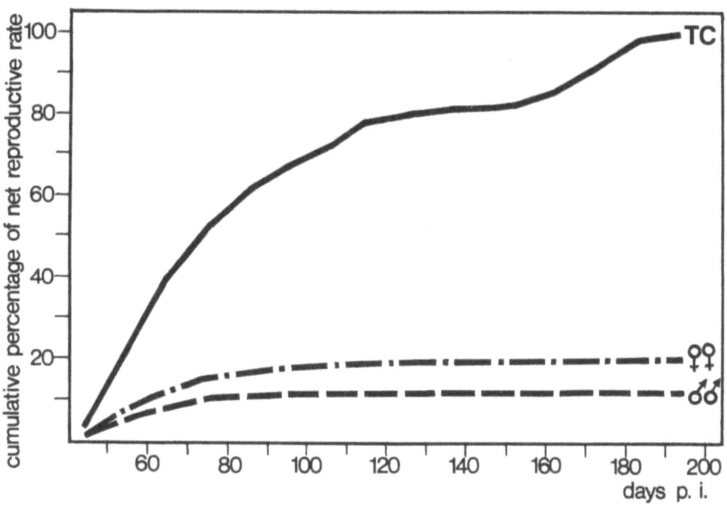

Fig. 40. Net reproduction of *S. japonicum* miracidia: cumulative percentage of cercariae emerging from *O. h. quadrasi* with regard to a daily death rate of 0.019 and 0.027 of female (♀♀) and male (♂♂) snails respectively. *TC* theoretical curve where all snails survive the whole period of cercarial emergence [Source: 359]

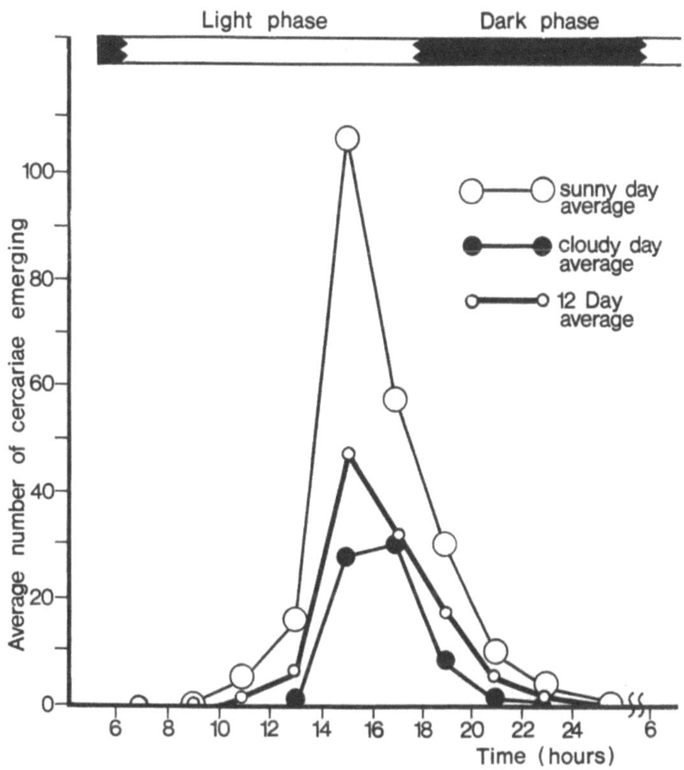

Fig. 41. Emergence periodicity of *S. japonicum* cercariae [Sources: 563, 564]

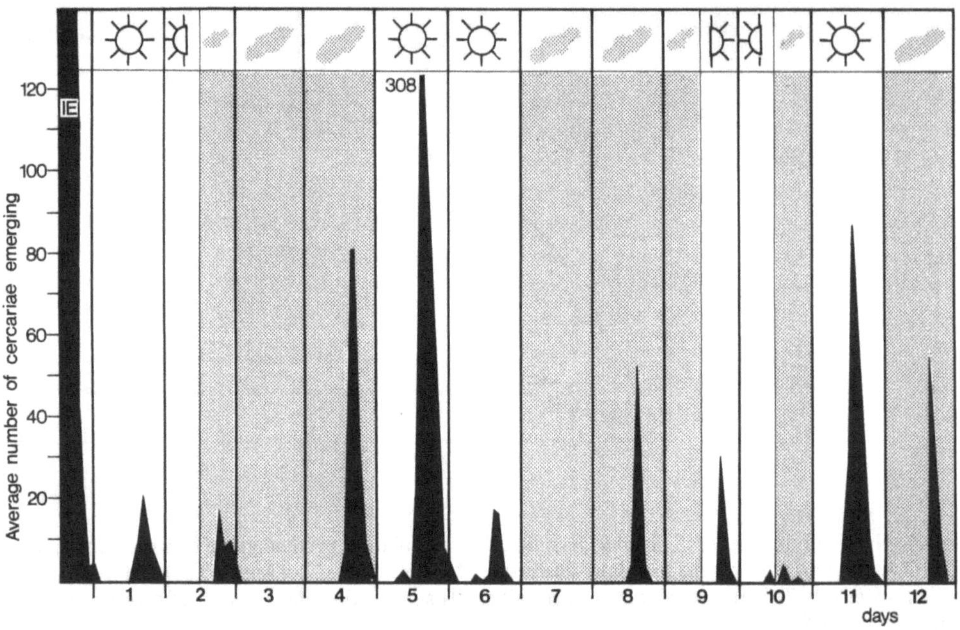

Fig. 42. Weather conditions and emergence of *S. japonicum* cercariae from *O. h. quadrasi*. *IE* initial emergence after submerging [Sources: 563, 564]

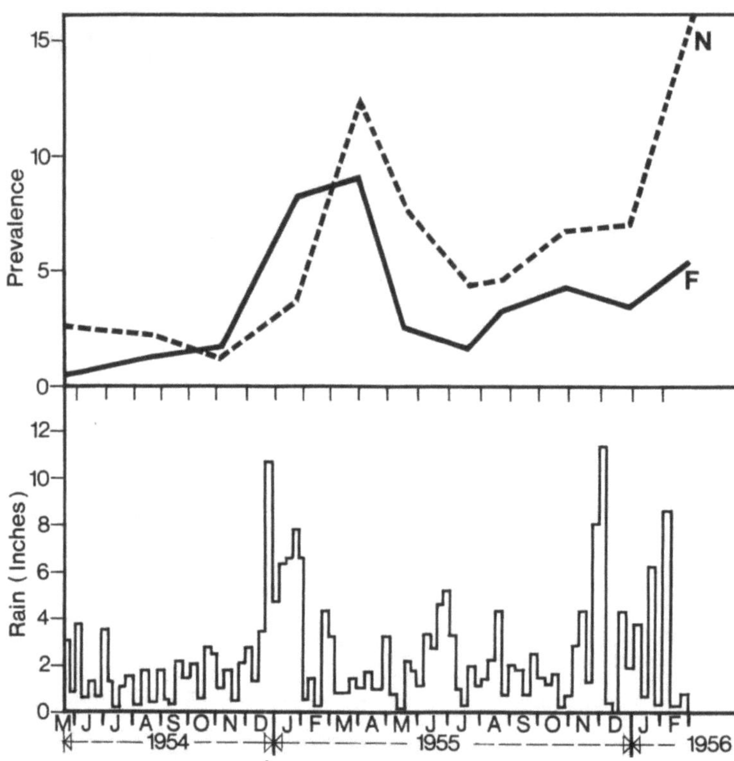

Fig. 43. Seasonal prevalence of *O. h. quadrasi* and rainfall in Malirong, Leyte. *N* five snail colonies near houses, *F* five snail colonies far from houses [Source: 609]

3.2.4 Prevalence in the Intermediate Host

The prevalence rates of *O. h. quadrasi* not only show considerable differences between endemic areas (cf Table 19), but also among the snail colonies of one endemic area, and can be up to 55%. Beyond this a clearly seasonal dependency has been established. Differences in the contamination of snail habitats through host faeces, as well as climatic factors, are responsible for this. Pesigan et al. [609] were able to establish a distinct correlation between the amount of precipitation and prevalence in the course of the year (Fig. 43). Time and time again a marked increase in prevalence among the snails could be observed eight weeks after the rains at the turn of the year, which are particularly heavy, i.e. after the end of the pre-patent period. The influence of precipitation is attributed to the fact that heavy rains wash the faeces – together with the eggs they contain – into the waters before the latter are able to dry out; at the same time more snails are covered by water and thus exposed to penetration by miracidia. (Similar season-related differences in prevalence are reported by Kamiya et al. [433–435] on the basis of their investigations at Dagami on Leyte.)

Besides the differences described above, it also emerged that larger snails are more often found to be infected than small ones, females more frequently than males [398, 408, 609]. As size is related to age, the increase in the prevalence rates with increasing size may well be attributed to the duration of exposure increasing with age. The reason for the difference between the sexes has so far failed to come to light. It could be a matter of differences in resistance, but it cannot be ruled out that, under natural conditions, miracidia might be better at penetrating large female snails, or that differences in exposure occur thanks to sex-specific behavioural patterns.

3.2.5 Duration of Cercaria Survival

As in the case of miracidia, the survival time of cercariae released from snails is primarily limited by the fact that they do not take up nutrients and waste away as soon as their store of reserves is exhausted. This occurs after 48 hours at the latest [359], although on the other hand the survival time is also determined by environmental factors. It is markedly reduced in higher water temperatures, whereas lower temperatures are endured for longer periods (Fig. 44). The pH value plays no role in this. Jones and Bradley [422] discovered that a temperature of 50 °C caused the death of all cercariae within three minutes; at 45 °C death occurred after 20 minutes at the latest, at 40 °C after four hours at the latest, whereas they survived temperatures of 20°–35 °C for one to two and a half days. This implies that cercariae living in shallow waters, or sections thereof which are exposed to sunlight and thus to very high temperatures when there are no clouds, have only a very short period within which to infect a final host. This is the case in rice fields prior to planting, for instance; at later stages the growing rice shoots increasingly offer shade and thus prevent a greater temperature rise through exposure to the sun. For the farming population this can mean seasonal differences in the infection risk, which also happen to be correlated with the cultivation of the fields. At winter temperatures of 20°–25 °C conditions are relatively favourable, ensuring a uniform survival duration of about two days (cf Fig. 44). It is true that none of these data provide information on whether the surviving cercariae have maintained their infectivity or not.

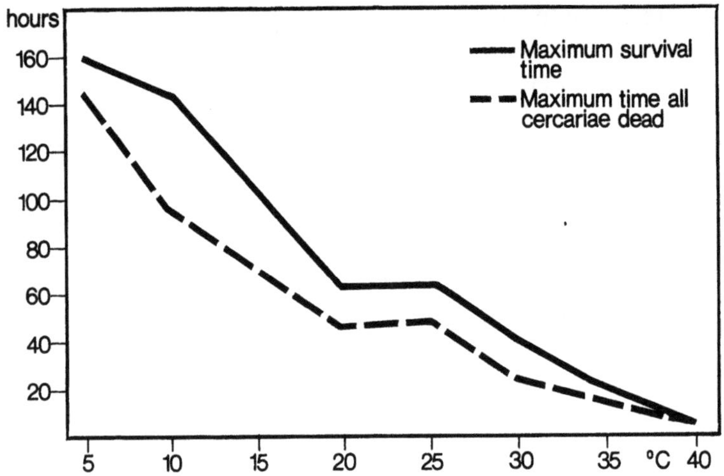

Fig. 44. Survival of *S. japonicum* cercariae as a function of temperature [Source: 422]

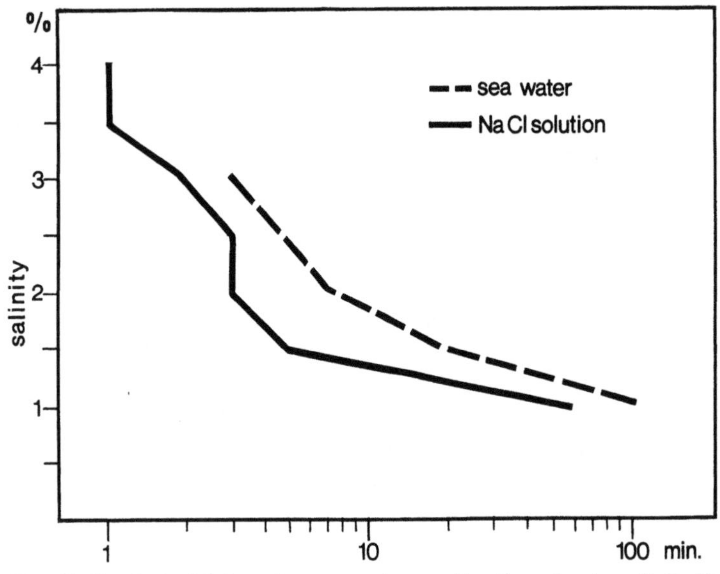

Fig. 45. Survival of *S. japonicum* cercariae as a function of water salinity [Source: 401]

Similar considerations hold good for the survival duration of cercariae in salt water. Relevant investigations by Ingalls [401] showed that NaCl solutions prepared with river water produce a greater cercaricidal effect than sea water of the same salt concentration (cf Fig. 45). In both cases a salt content of ≥ 3 per cent led to the dying off of all the cercariae within three minutes. Decreasing salinity results in increasing lifespan which may be as much as two hours or more at 1 per cent. A risk of infection that eludes precise calculation therefore continues to exist in the coastal area in the immediate vicinity of river mouths, as well as in flats that are flooded by sea water.

3.2.6 Man's Contact with Water

As the infection of man by percutaneous penetration of the cercariae can only take place when man comes into contact with water containing them, the question arises of the when and where of such contacts. Tiglao [769] has carried out an extensive health interview survey on this in four barangays of the municipalities of Mayorga, Sta. Fe, Tolosa and Javier in the north-east of Leyte (cf Table 20). Based on the replies from 1,935 household respondents it emerged that contact with water occurred through rivers in 42.7 per cent of all cases, through streams in 18.6 per cent, through rice fields in 13.5 per cent, and through canals in 5.5 per cent. The most frequently occurring single causes were described as washing clothes (34.1%), bathing and swimming (31.4%), and farming in rice fields (12.7%). Since the people thus asked were free to supply several answers, the information does not lend itself to providing any detail on the proportion of those in the total population who had come into contact with water. The ability to inform is further limited by the fact that 78.7 per cent of those interviewed were women, and only 21.2 per cent farmers. A survey of a representative sample might therefore produce different results, though it must be assumed that washing clothes, bathing and swimming, and farming in rice fields are indeed the most frequent causes of water contact and thus of an infection by *S. japonicum*. (Tiglao and Camacho [770] were unable to establish direct links between the duration of water contact and the intensity of infection for Sta. Fe on Leyte). On the other hand, the risk of infection for 24.7 per cent of those interviewed who are in contact with sea-water when bathing, swimming or fishing cannot be calculated as the unknown salinity of the water with which they are in contact plays a role which determines the survival time of the cercariae (see above).

Apart from the number of people in contact with water, it is the frequency and duration of their exposure which is decisive in the prevalence and intensity of infection within a population. Hairston [356] presents the results of his observations in Leyte on this matter. He recorded a total of 653 contacts with water made among 752 inhabitants in the course of a single day (5 a.m. to 7 p.m.). The total amounted to 104.4 hours, the average time to 9.6 minutes (with variations ranging from less than a minute to approximately one hour). There were marked differences in the frequency and duration of contact as between the sexes: a total of 59.6 per cent of the contacts and 71.3 per cent of the exposure time affected the female sex (cf Table 21). This did not, however, imply, a higher risk of infection for women and girls, since their contacts with water were largely limited to a river which had been graded as being far less hazardous. A very high risk of infection did, however, exist for those men and boys who most often stayed in the very habitats of the snails, remaining there altogether twice as long as the women and girls. Hairston [359] also draws attention to the fact that the probability of a penetration by cercariae is positively correlated with the exposed area of the skin. This is small when – as in the washing of clothes – only the hands and arms are in contact with water; it is large when, during swimming and bathing, almost the entire body surface is available for cercariae to enter.

An estimation of the risk of infection also requires the quantitative distribution of cercariae within the waters to be taken into consideration. This is primarily determined by the density of the intermediate hosts and the prevalence among the snail population. Due to the high locational constancy of the snail colonies (cf 3.3.3) it implies a relative stability in the distribution of cercariae in slow-moving waters and thereby in the infection risk for the final host as well, though increased flow velocity may well result in cercariae being washed downstream, where they cause infections even at considerable distances from the snails' habitats [609]. Sections of such waters situated downstream from infected snail populations can therefore harbour greater risks for people entering them. Though containing fewer cercariae per unit of water volume than those of the snail habitats proper, these water sections can compensate for that through the larger quantity of water flowing past a bather when the velocity of flow is increased, thereby also enabling a great number of cercariae to come into contact with his or her skin.

In summary it can be stated that though dependent upon the degree of contamination of waters by cercariae, it is the behaviour of man in respect of his contact with water which decides whether an infection occurs or not. Frequency, duration and kind of exposure are finally decisive in the incidence, prevalence and intensity of infection.

3.3 The Intermediate Host and Its Bionomics

Since Schistosomiasis japonica can only occur in those places in the Philippines where intermediate snail hosts are present, knowledge of the development and ecology of *Oncomelania hupensis quadrasi* (von Moellendorff, 1895) is of decisive geomedical significance. (Questions of nomenclature and morphology are not be taken up here [cf 1, 3, 42, 82, 83, 90, 112, 264–266, 517, 518, 634, 650, 651].)

3.3.1 Development, Maturity and Life-span

Being about 0.5–0.7 mm high (with 1½ shell-whorls), newly hatched *O. h. quadrasi* start by passing through a fortnight's aquatic phase before embarking on the amphibian way of life [2]. At the beginning their weekly increase in size is about 0.25 mm [524] and after attaining maturity only 0.05 mm [656]. By the time they have attained the age of 7 weeks, male and female snails have grown to about 2 mm. After that the males grow at a slower rate than the females, reaching a final size of 4.5 mm – by contrast with 5 mm in females – at an age of 180 days. Sexual maturity is, however, attained at a much earlier stage; copulation was first observed in snails of a length of 2.5 mm (♂) and 3 mm (♀) respectively [609].

The mean life-span of *O. h. quadrasi* is said to be 25–35 weeks [853]. Under experimental conditions females continued to live for 65.8 days and males for 47.6 days after reaching maturity [525]. According to investigations undertaken by Pesigan et al. [609] daily mortality was 0.76 per cent (♀) and 1.05 per cent (♂). Considering the average time span required for the development of schistosomes in snails to be 62–64 days as well (cf 3.2.3), *O. h. quadrasi* can only be an effective intermediate host if the infection with miracidia happens at a very early age in the young snails, but it is not known whether this can take place in the fortnight's aquatic phase immediately after hatching – that is during a phase in the development of the snails which affords particularly favourable conditions for the penetration of miracidia.

The schistosome infection affects the lifespan, growth and reproduction of the intermediate hosts, the snails, as well. As Pesigan et al. [609] were able to demonstrate, the lifespan of *O. h. quadrasi* is considered to be reduced as a consequence of the infection – the more so the younger the snails had been at the moment of infection. The lethality of snails which had been infected when 2.7 mm long 20 weeks after infection was 83 per cent as compared with 41 per cent in non-infected controls. If the infection takes place when the snails are only 2 mm long, lethality rises to 79 per cent, compared to only 10 per cent in the control group after a period of only 11 weeks. Hairston [358] has calculated the overall death rate to be 0.010 (♀) and 0.017 (♂) among non-infected *O. h. quadrasi*, and 0.019 (♀) and 0.027 (♂) among infected ones.

The growth of infected snails is also noticeably affected by schistosome infection. In this a retardation can be observed which is negatively correlated with age [609]. In addition the infection has a damaging effect on the number of eggs laid and on their hatchability. The egg production of infected snails is reduced by 80 per cent, the hatchability of these eggs by 24 per cent – an overall reduction in the reproduction rate of 83.3 per cent. Snail colonies which show a high prevalence can therefore soon arrive at a superannuated population and thus experience a decline in their population density. As a rule, however, *Oncomelania* populations are relatively stable and settle in all areas satisfying their environmental demands.

3.3.2 Geofactors as a Pre-condition for the Distribution of *Oncomelania hupensis quadrasi*

The distribution of *O. h. quadrasi* as an amphibiously-living snail is primarily determined by the climatic conditions. Its sensitivity to aridity restricts its distribution range to those two climatic zones which receive precipitation throughout the year, and this in turn defines the potential area of schistosomiasis distribution in the Philippines (cf Map Plate 3). According to various estimates, between 46,300 and 162,800 ha of these areas between 6° and 14° N. are inhabited by this snail species [684, 703–706]. Eighty one per cent of these are on Mindanao, 18 per cent in the Visayas, and only 1 per cent in the southern part of the Bicol region and on Mindoro Oriental. The total extent of the areas within the two climatic zones being many times larger, it remains to be seen which factors are instrumental in limiting the area of potential distribution. This is a matter of the details of hydrological, edaphic, orographic, atmospheric, climatic and biological factors which may condition one another or depend upon each other, although represented in a network the components of which cannot be considered in isolation. The analysis which follows will therefore have to take this fact into account when proceeding with its classification.

3.3.2.1 Flow Velocity of the Waters, Depth of Water, and Changes in Water Levels

As the maximum flow velocity of waters that *Oncomelania* tolerates is about 20 m per minute, this factor is of considerable significance for the restriction of its distribution. Such velocity of flow is determined by the orography, the distribution and quantity of precipitation, the land and water vegetation and the course followed by the waters themselves.

Orography plays an important role in so far as a very sharply falling terrain causes the flow velocity to exceed the tolerance limit of *Oncomelania*, especially during the aquatic phase of the young snails. For this reason neither snails nor schistosomiasis are to be found in the hills and mountains of the Philippines. It follows that favour-

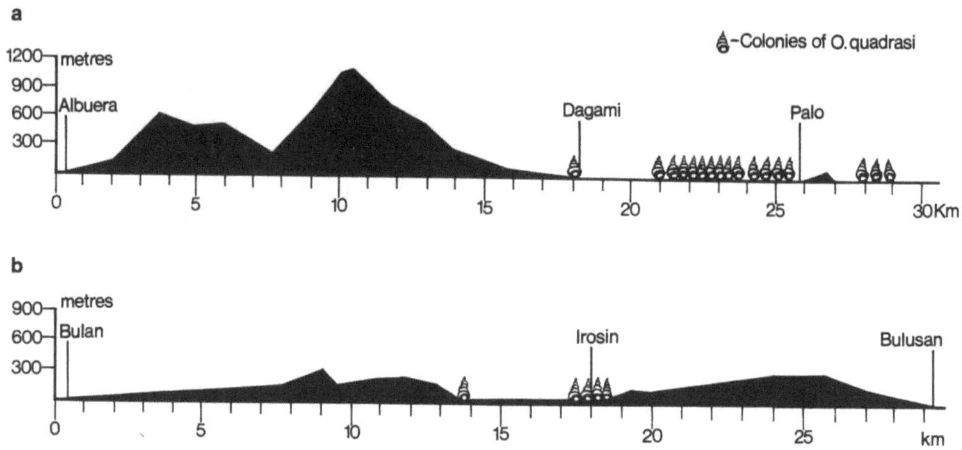

Fig. 46 a, b. Distribution of *O. h. quadrasi* and orography of Leyte (**a**) and Sorsogon Province (**b**) [Source: 609]

able conditions exist above all in the large riverain and coastal flats, as well as on high plateau (such as the Bukidnon Plateau) which have slight gradients. Thus *Oncomelania* habitats are characterized by their flatness. Typical settlement areas of this sort are the large plain in the north-east of Leyte and the valley of the Caloscos River in the Sorsogon Province, for instance (Fig. 46). In this part of Leyte the altitudinal range over a distance of more than 19 km is less than 30 m. Included in this is the Manat River Swamp in Davao Province (Mindanao), together with a 2 km tributary. Extending over a length of 20 km and with a breadth of 2–3 km, it is the largest continuous snail area hitherto known in the Philippines [609].

Apart from the regular wetting of the subsurface and the high atmospheric humidity, both of which are needed by snails when not actually in the water, relatively constant water-levels – and thus small fluctuations in flow velocity – are secured by a precipitation which is evenly distributed throughout the year. By contrast, if a high level of precipitation is concentrated in a distinct rainy season, the threshold value of 20 m/minute is very soon exceeded, so that the *Oncomelania* are flooded out of their habitats. This is the reason why colonization by snails cannot occur in areas with markedly alternating rainy and dry seasons, since an *Oncomelania* colony which has suffered an 85 per cent loss of its inhabitants takes about 240 days to repopulate itself. To a certain degree, nonetheless, *Oncomelania* can compensate for changes in the flow velocity by their movement pattern; they leave the water when its speed increases [756].

Another important regulator of discharge and flow velocity is found in the vegetation of the land. Trees on the hills and mountain slopes store the water, ensure a relatively continous run-off even during periods with little precipitation, and prevent the direct discharge at times of ample precipitation, thereby ensuring that the habitats of the snails persist. Human interference in this structure brought about by felling timber without reafforestation or by extensive clearance for the plantation economy – which also occurs in the Philippines – can therefore bring about permanent changes in water-levels and flow velocities of waters in the lowland so that *Oncomelania* can no longer survive in their traditional habitats. No instance of this kind has so far been known to occur in the Philippines, however.

The water vegetation also performs a function in regulating the flow velocity. Ample plant growth and/or weed growth in waters can reduce their flow velocity so much that they are the very cause by which a habitat suitable for snails is created. The same effects – namely reduction of flow velocity and, arising from it, the formation of snail habitats – have been observed in more vigorously meandering river courses. Removal of aquatic vegetation and the straightening of running waters therefore present important components of measures aimed at controlling snails, and thus in the control of schistosomiasis (cf 3.7).

On the other hand, it is not only the natural conditions mentioned above which have an effect upon the velocity of flow. Interference by man (such as the construction of dams) can also contribute to the creation of new snail habitats. On Leyte fishing in streams and the use of streams as places where the water buffaloes (carabaos) can wallow seem to favour the formation of snail habitats. Pesigan et al. [609] have described this situation as follows: "Several species of fish, particularly the mudfish (*Ophicephalus* sp.), inhabit these streams and are widely sought for food. The method used in catching them is to dike off or dam a part of the stream, and then to wade in and capture the fish by hand. The repeated temporary damming up of the stream prevents the water from making anything like a permanent channel. In addition, the streams are very widely used as wallowing sites for the water buffalo (carabao), the principal work-animal of the Philippines. The wallows are temporary, and many different sites may be used by the same animal. This wallowing has the same effect as the fishing – it forces the water away from any defined channel, softens the whole stream bed, and by making the flow more sluggish promotes the growth of semi-aquatic grasses and other plants, which in turn impede the water still further."

The depth of water was found to be in inverse proportion to the density of *Oncomelania:* the deeper the water, the fewer were the snails observed in the water itself, as well as in its immediate environs [609]. Deep waters, especially those with steep banks, are not suited to as *Oncomelania* habitats. Fluctuations of the water-level up to 50 cm are, on the other hand, tolerated by the snails. They react to changing water-levels by directional migration. When the water rises, they move away, returning as it recedes [753].

3.3.2.2 Nature of Water and Soil

The available publications do not show clearly in how far the nature of water and soil limits the distribution of *O. h. quadrasi* within the Philippines. Although the soil and water analyses carried out on Leyte show certain differences in the mean values of the mineral content of areas inhabited by snails, and other which are not, the range of values was so large that they overlapped in every case (cf Table 22). The question of why biotopes of the same water and soil qualities are in some cases inhabited by *O. h. quadrasi* and not in others, must therefore remain open for the time being.

The degree of turbidity of the water is, however, significant for the occurrence of *Oncomelania.* Preference is given to clear waters, the average turbidity of which amounts to 5–10 p.p.m. SiO_2 equivalent, but snails have been found where turbidity had reached values of up to 30 p.p.m. SiO_2 equivalent. The material suspended in the water tended to be of organic origin. On the whole it may be assumed that the less turbidity there is, the more *Oncomelania* will be found in the water [306].

As far as the pH value of the water is concerned, the optimum for *O. h. quadrasi* lies in the slightly alkaline range at pH 7.4–7.6 [398], but snails have also been observed in waters with lower (6.2) or higher (up to 7.8) pH values [609]. The pH value of the soil in snail habitats is classified as being closer to the acid milieu (4.6–7.2). In general it may be stated that the suitability of an area as a snail habitat depends on the fluctuations of the pH value: the more stable the values, the better suited is the area for the settlement of *Oncomelania.*

A clearly limiting factor in the occurrence of *O. h. quadrasi* is the salt content of the water. Only fluctuations of 10–430 p.p.m. are tolerated. More heavily salinated areas in the coastal lowlands are therefore unsuited as snail habitats [609].

3.3.2.3 Temperature, Light Intensity, Atmospheric Humidity and Atmospheric Pressure

Among the factors of the abiotic environment which contribute to the conditions for *O. h. quadrasi's* occurrence, temperature plays an important role. A water temperature of 23–30 °C is considered optimal [3]. Values remaining above or below this over a long period of time have the effect of rendering an area unsuitable as a snail habitat. The same applies to rice fields within *O. h. quadrasi's* area of distribution where the waters concerned may reach temperatures of 40 °C or more. However, temperature also affects the pattern of movement of the snails: a rise from 20 °C to 35 °C is reflected in a marked increase in the intensity of movement. Traversing 3.77 cm (♀) or 3.51 cm (♂) respectively within 10 minutes at a temperature of 20 °C, the snails accelerate their movement so as to cover 8.69 cm (♀) or 6.99 cm (♂) respectively over the same time period at a temperature of 35 °C. This can not be regarded as a positive influence of temperature, however; higher values stimulate the snails to seek more shaded and hence cooler places. Movements dependent upon light intensity are to be interpreted in the same way: the increase in movement activity serves to allow escape from unfavourable conditions.

The activation of locomotion through rising atmospheric humidity and falling air pressure on the other hand presents a positive stimulus. This can also be observed in the copulation behaviour of *O. h. quadrasi.* Copulation frequency increases markedly with rainy weather and great atmospheric humidity. This influence of environmental factors also finds expression in seasonal and diurnal differences, such as noted by Pesigan et al. [609] on Leyte. The number of copulations thus reaches its maximum in September when the precipitation frequency increases; its minima occur in June (lowest atmospheric humidity and precipitation frequency and highest light intensity) and December (extreme precipitation maxima with negative effects). The diurnal periodicity of the copulation frequency is also related to the seasons, on average with higher values during the night hours. A particularly distinct rhythm emerged from investigations carried out in February; they clearly show the relations to air temperature and atmospheric humidity (Fig. 47). So it is these two factors which govern the rate of *O. h. quadrasi's* increase by influencing the copulation frequency, and thus decide whether stable snail populations can grow up in an area and guarantee a permanent settlement.

3.3.2.4 The Inter-action of the Geofactors in O. h. quadrasi's Ecosystem

The inter-action of factors in the abiotic environment determines the distribution of *O. h. quadrasi* in the Philippines. However, its occurrence within the distribution areas depends additionally and predominantly on biological conditions. Pesigan et al.

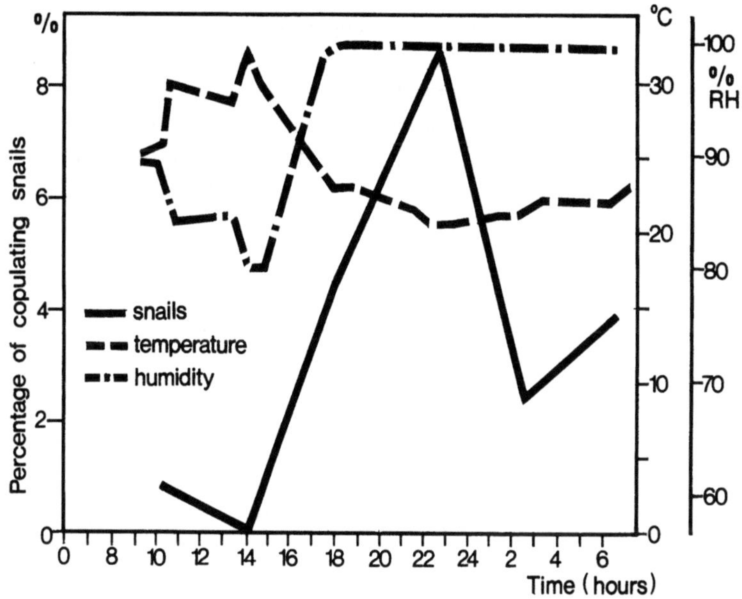

Fig. 47. Temperature, relative humidity and copulation in snails during the course of the day (Source: 609]

[609] characterize the situation as follows: "A species, in order to maintain itself, must be adapted to its whole environment, and it is the components of the whole environment, interacting in innumerable different ways, that determine the survival of the species. Ecological results have indicated that the biological part of the environment is ordinarily more important than the physico-chemical part." In this context the authors draw attention, for example, to a possible competition between *Oncomelania* and other gastropods. However, this snail species can only persist in places which never dry out and maintain an undisturbed situation. Just as everywhere in ecology, so in the case of *O. h. quadrasi* a network of geofactors has to interact for a habitat to come into existence.

Snail colonies investigated in the Philippines showed a very marked locational constancy; *O. h. quadrasi* is able to persist for decades in a suitable habitat. Such habitats nonetheless differ so considerably in their features, that a general characterization cannot yet be attempted. The most important types of those existing in the Philippines have been clearly described by Pesigan et al. [609], and extracts therefrom are therefore reproduced in the appendix (E). They include the following categories:

Flood-plain forests and swamps
Rice-fields
Streams
Small swamps
Road ditches and borrow-pits.

The concepts developed by Pesigan et al. [609] concerning the occurrence of snails in "colonies", the locational constancy, and the "negative" (i.e. snail-free) areas have been expanded or, as the case may be, questioned by Frohne [306]. He was of the opinion that *Oncomelania* can occur not only in colonies but also in pockets along streams (i.e. the "watershed" concept) and, moreover, throughout the entire habitat ("continuum" concept), though their concentration may be rather less. As far as "negative" areas are concerned he disagreed with the term, since he was able to prove the existence of high *Oncomelania* densities even in such places as had previously been held to be free from snails. Whether these findings, which had after all been arrived at during a relatively short period of observation, have been confirmed or refuted by investigations extending over several years, does not emerge from the available literature.

The vertical and horizontal distribution of *O. h. quadrasi* within the habitat is much affected by the depth of water, the light intensity, the flow velocity, the steepness of the bank and the plant cover (3.3.2). In general most of the snails remain in the immediate vicinity of the banks outside the water. On the basis of investigations of a variety of water types, Tanaka et al. [756] arrived at the conclusion that *O. h. quadrasi* "is determined to be not so aquatic in habitual character than was presumed before". [Further details in connection with the distribution of *O. h. quadrasi* may be found in the following publications: 474–480, 722, 753–755, 757–759.]

3.4 Frequency of *Schistosoma japonicum* Infections

Infection with *Schistosoma japonicum* being an anthropozoonosis, not only man but also a number of mammal final hosts contribute to the persistence of this worm infection. Whilst the frequency in man is very well documented for almost all the endemic areas in the Philippines, comprehensive data on prevalence in mammals is lacking.

3.4.1 Frequency in Man

The endemic occurrence of *S. japonicum* in man has so far been established for 22 provinces (Table 23; separate data in Table III). In 1975 these endemic areas were inhabited by approximately 4 million people, a total of almost 630,000 of whom were regarded as infected [706], the average prevalence being 15.9 per cent. For a number of reasons, however, it is difficult to observe at present in how far these values, which are based on computer reckonings, correspond to the situation as it really is. If the computer predictions are based on the results of investigations by the schistosomiasis control teams they tend to arrive at values that are clearly too high, since they are derived from prevalences for a select pool of patients [cf 200 for example]. The same happens when schoolchildren and young people are over-represented in mass examinations in most endemic areas, as this age-group shows markedly higher prevalence rates than other sections of the population (see below). In addition the result of computer calculations also depends upon the diagnostic methods employed in the sample surveys. Thus stool examinations based on only the stool smear method revealed lower prevalence rates than those based on concentration techniques like the formalin-ether concentration technique (FECT), merthiolate-iodine-formaldehyde concentration (MIFC) and cellophane thick smear technique (KATO). The direct concentration techniques which the schistosomiasis control teams tend to use – and which are presumably passed on to the Schistosomiasis Control Council as a basis for their computer reckoning – again register markedly fewer (in parts only 50 per cent) of the cases which prove positive when serological tests are applied [cf 139, 245, 568, 871]. In the case of these serodiagnostic methods, such as circumoval precipitin test (COPT) and enzyme-linked immunosorbent assay (ELISA), however, the question of specificity has not been elucidated with absolute certainty.

It is for these reasons that case numbers contained in Table 15 can only be regarded as rough estimates. On the other hand, even the evaluation of the separate data (Table III) has to take into account the diagnostic technique employed. Moreover, average prevalences, even in the case of age- and sex-adjusted rates, present only an incomplete picture of the distribution of cases, thanks to differences between sections of the population; these differences are primarily to be attributed to the pattern of water contact, and thus to exposure, though in infections which persist over a longer time period the result of exposure may be modified by development of relative immunity. Nothing is known, however, about the difference in disposition among different population groups [cf 874].

As has already been mentioned (3.2.6), girls and women in an endemic area on Leyte were more frequently in contact with water – altogether for twice as long as boys and men, although these stayed more often in the snail habitats proper on the other hand. This contact with the more heavily cercariae-infested water of the snail habitats evidently contributes to the fact that higher prevalence rates have been observed among the male sex than among the female one (Table 24). Though differences between the sexes fluctuate from one endemic area to the next, a general trend can be established and confirmed by a regression analysis based on data from Table 24 (cf Fig. 48). It follows that an increase in prevalence of 1% in the male population is matched by a rise of only 0.73% in that of the female. If, by contrast, a rise in prevalence of 1% is observed among women and girls, the increase among men and boys will be 1.13%. This correlation is statistically highly significant ($r = 0.882$; $p < 0.001$).

Since differences in prevalence rates are primarily due to frequency and duration of exposure, a positive relationship between age and prevalence is to be expected as well. This is confirmed by numerous mass examinations, although it applies to younger age groups only (Fig. 49 and Tables 25 and 26), for very high prevalence rates have as a rule already been attained among the 10–19 year olds; in areas with a high overall prevalence they do not rise further with increasing age, but even manifest a declining tendency in areas with a lesser frequency.

These age and sex differences in schistosome infections are confirmed by a comparison of the case distribution with the population structure of the Philippines (cf Fig. 50). In both sexes the 1–9 year age-group shows markedly fewer infections than their share in the total population would indicate. With increasing age the female sex achieves a balanced proportion, whereas 45% of all cases among men and boys were observed in those between the ages of 10 and 49 years. Constituting about 30% of the population, it is this very group of young males and adult men, who perform the hardest physical work in agriculture, that is affected. Moreover, those predominantly engaged in farming tend to show particularly high rates of prevalence, such as 74.1% in Palo by contrast with 46.1% among the remaining population, 62.3% in Tanauan compared with 43.8%, 7.9% in Carigara compared with 2.2%, and 69.0% in Bagahupi compared with 37.8% [606].

As a rule higher prevalences due to increased exposure are coupled with a greater intensity of infection. There is, however, no suitable criterion for measuring the in-

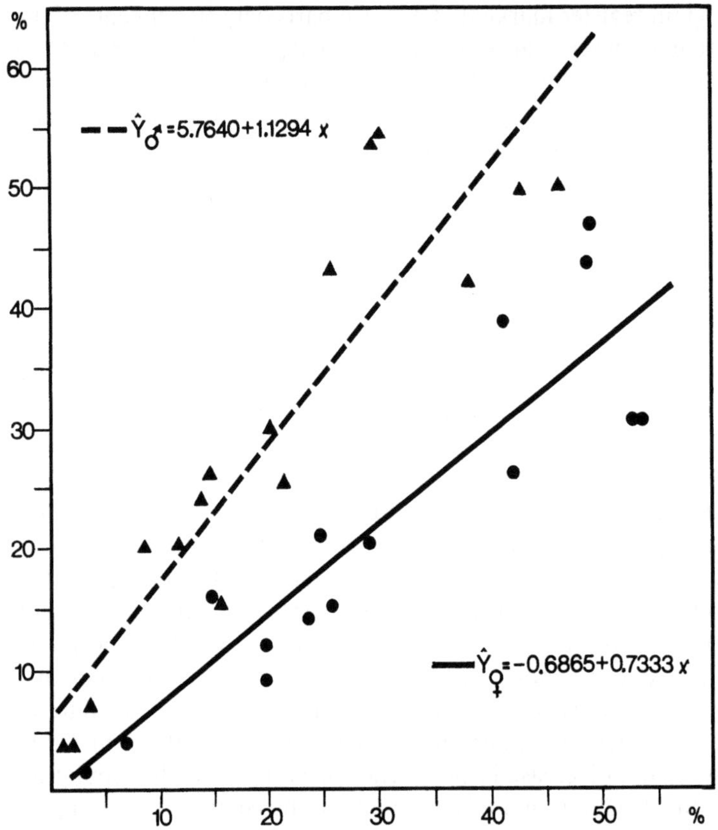

Fig. 48. Relationship of male and female prevalence rates in human *S. japonicum* infections (cf table 16)

tensity of infection since the passing of eggs per schistosome-♀ with the faeces depends on the intensity of the infection itself (crowding effect) and is influenced by the condition of immunity and the age of the host (cf 3.2.1). On the other hand, the rate of egg-passing presents an important yardstick when it comes to evaluating the significance of sections of the population for the dynamics of transmission. Differences in the transmission indices of particular age-groups have already been pointed out (3.2.1); as a result of different prevalences and egg-passing rates, differences of this kind occur between the sexes as well. With an index of 65%, the male sex furnishes on average two-thirds of the total transmission potential (particularly marked among the 10–24 year-olds: ♂♂ 35%, ♀♀ 18%; cf Fig. 51).

To sum up. Allowing for the estimated figure of 630,000 infected people in the Philippines (see above) it can be said that a total of about 380,000 men and boys as well as 250,000 women and girls are affected by schistosomiasis. The 160,000 infected 10–19 year-olds constitute the most important transmission potential (index 48.3%).

3.4.2 The Significance of Mammals for the Persistence of Schistosomiasis

Despite the significance attributed to several species of mammals as reserve hosts for *Schistosoma japonicum*, knowledge of their prevalence rates gained over the past thir-

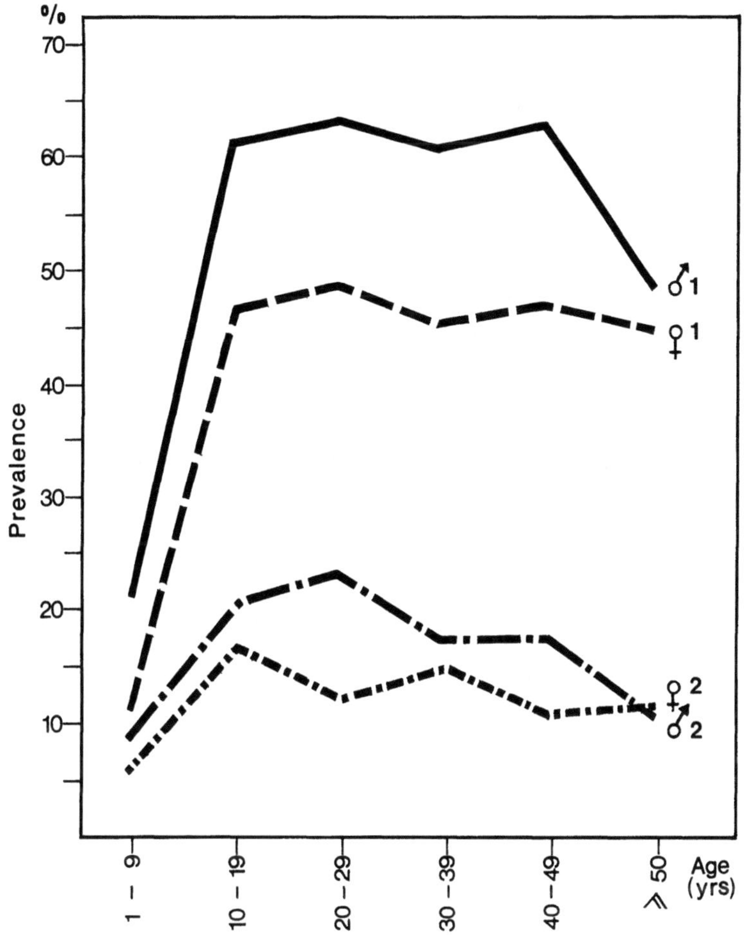

Fig. 49. Schistosomiasis prevalence by age and sex in areas of differing endemicity [Sources: 1: 277, 456, 606; 2: 139, 606, 795]

ty years is based exclusively on investigations carried out on Leyte (cf compilation in Table 28). Conclusions drawn from these results are only conditionally possible in respect of the situation in other endemic areas.

On the basis of results arrived at on Leyte the greatest significance next to man himself, attaches to the field rat (*Rattus rattus mindanensis*) because of the numbers of its individuals, its way of life and its high prevalence rates. The exact evaluation of the role played by rats in the persistence of schistosomiasis in the various endemic areas is beset by great uncertainties, in so far as the population density cannot be established exactly, and the prevalence in the different municipalities on Leyte is subject to substantial fluctuations. In the publications available, calculations for the definition of the transmission index were based on a figure of 60 or 70 rats per hectare [281, 606], and the overall population of the area under investigation was inferred from this. The total number of infected rats was also determined using the computer on the basis of a random sample examination, which amounted to about 0.01–0.1% of the estimated total number only. This results in a further source of errors, even if

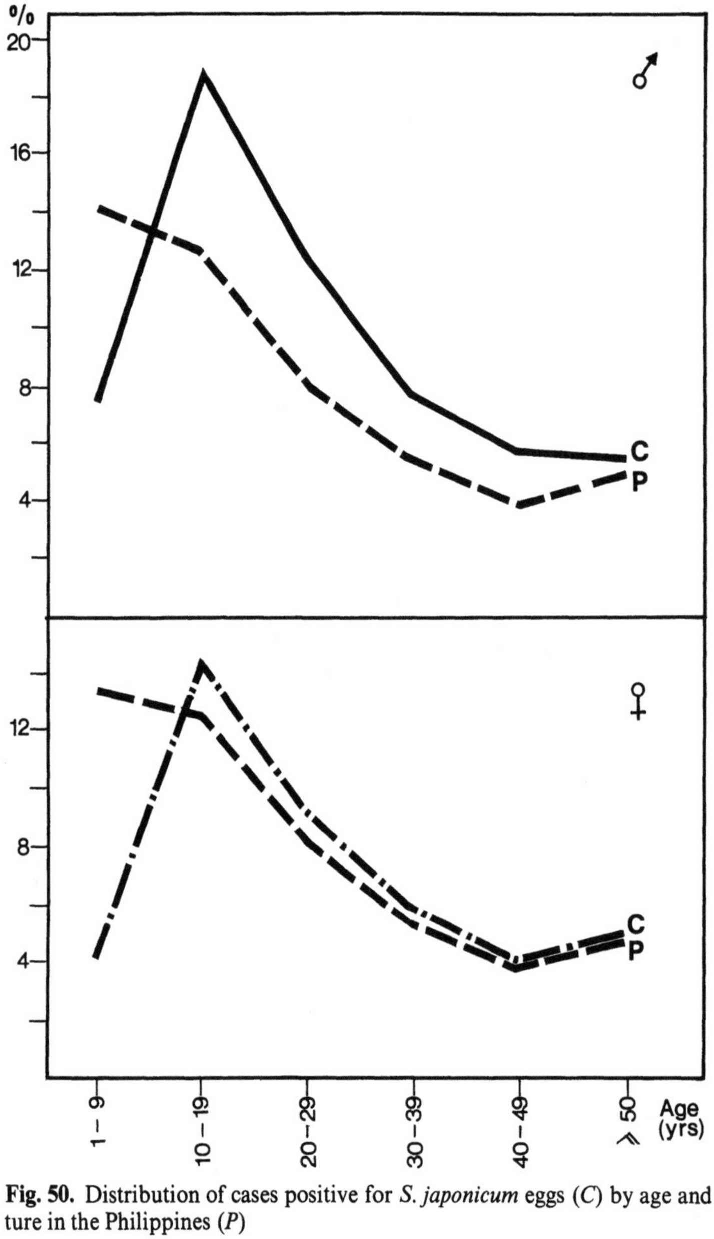

Fig. 50. Distribution of cases positive for *S. japonicum* eggs (*C*) by age and sex and population structure in the Philippines (*P*)

the rats used in the examination had been selected by way of genuinely random samples.

Since the prevalence rates of rats vary from one endemic area to the next, it is not possible to draw conclusions in respect of the situation in other parts of the country on the basis of results obtained in Leyte. Apart from this, established prevalence rates vary according to the examination method. Faeces examination alone registers markedly fewer positive animals than the additional examination of internal organs for egg deposits [cf 141]. However, in the process of transmission only rats with eggs

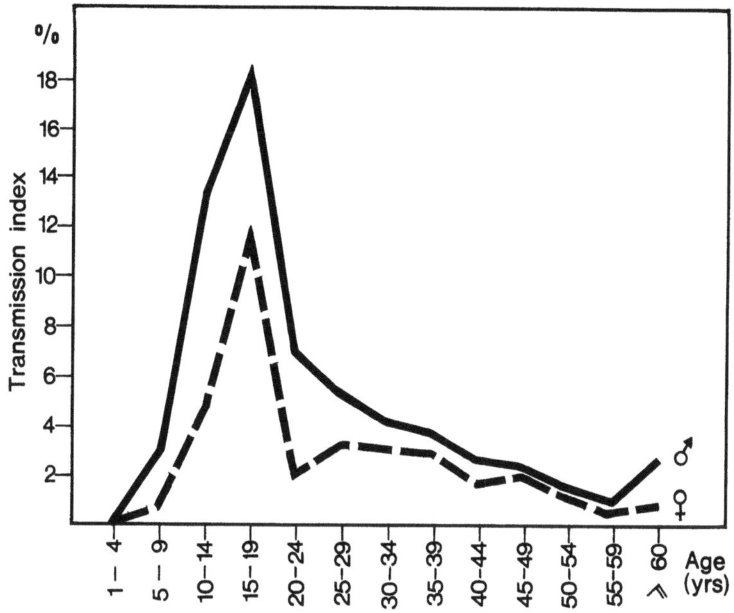

Fig. 51. Relative transmission index of *Schistosoma japonicum* by age and sex in the Philippines (average of data) [Sources: 277, 456, 577, 838]

in their faeces play a role. Moreover, seasonal fluctuations in prevalence rates have been observed [433–435]. They attained their highest values (95%) during the months of lower precipitation frequency (April–July), as opposed to an average of 76% during the remainder of the year. This seasonal rhythm is said to be linked to a complementary situation among the intermediate host snails: a high prevalence of rats during the "relatively dry" months leads to a high degree of contamination of the waters, with the result that intermediate host snails become infected. The release of cercariae from the snails during the "relatively rainy season" takes place at an even greater rate, which in turn leads to the infection of rats, so that these – while considering the prepatent period – will show once again high prevalences during the ensuing "relatively dry" season. This factor has therefore also got to be taken into consideration when evaluating the transmission potential of rats.

Over and above population density and prevalence, the number of schistosome eggs passed out with the faeces also presents an important parameter in classifying the epidemiological significance of an animal species; it is expressed by the transmission index (cf 3.2.1). The literature presently available on this subject shows considerable variations, however: Pesigan et al. [606] and Dumag et al. [281] give a low rate of egg-passing of 21 (13) per animal per day, and a hatchability of 10.6% and 10.7% respectively – i.e. the daily production of an infected rat accordingly amounted to only 2 or 1.36 hatchable eggs respectively. By contrast, this figure was set markedly higher – at a value of 222 – by Hairston [357]. Dumag et al. [281] calculated a large rat population for their area of over 20,000 ha, arriving (like Hairston [357]) at high relative transmission indices of 43.0% and 52.5% respectively, whereas Pesigan, Farooq et al. [606], on account of the smaller number of rats in the area under investigation, quoted a value of merely 1.5%. The contradictory estimates preclude a conclusive assessment of the significance of rats in the persistence of schistosomiasis in the Philippines at the present time.

Relatively high prevalences have also been noted in dogs and pigs (Table 27). Their significance as reservoir hosts is thus undisputed, but their real share in the overall process of transmission cannot be expressed precisely because of the same unresolved contradictions in the case of rats. After rats these two domestic animal species supply the largest numbers of reservoir hosts for *Schistosoma japonicum;* the number of dogs is about 6–10%, that of the pigs about 15–20% of the population living in the endemic area of Leyte. Transmission indices are given as 14.4% or 1.4% for dogs and 1.5% or 1.4% for pigs respectively [281, 606].

Although only a limited number of cattle are kept, at Palo for instance, they showed a high prevalence (38.2%) together with a high rate of egg-passing and high hatchability; in consequence Pesigan et al. [606] calculated a transmission index of 5.7% for them. However, in more recent investigations only one animal out of a total of 61 proved to be infected [281, 297], which suggests that cattle are of subordinate importance in the epidemiology of schistosomiasis.

On the evidence of the comparatively very low prevalence rates, the carabao, too, is only of subordinate significance as a reservoir host (Table 27) although, thanks to the high rate of egg-passing and a hatchability of almost 30%, infected individuals could possibly contribute significantly to the infection of intermediate snail hosts. Other mammals which could be suspected of acting as reservoir hosts have been investigated; three out of 141 goats proved to be infected, whereas all 19 cats proved negative. They are therefore of very little or no significance for the persistence of human schistosomiasis.

In recapitulation it may be stated that a variety of mammal species contribute to the persistence of schistosomiasis. The extent of their contribution to the process of transmission varies from one species to the next, however, and for every species from one area to another. As more recent results are only available for some municipalities on Leyte, it is not possible to make any statements on other endemic parts of the country.

3.5 The Endemic Areas

In the Philippines the distribution of schistosomiasis extends to 22 provinces situated between 6°20′–13°20′ N and 121°15′–126°25′ E. Within these limits it is above all the climatic and orographic factors which decide whether schistosomiasis occurs or not. The amphibian snails which act as intermediate hosts can only persist in climatic zones with high precipitation throughout the year. In addition, these snails are adapted to waters as their habitats which have a very low flow velocity. This is the case only in areas with very low gradients over longer stretches. The distribution of *O. h. quadrasi,* and thus of schistosomiasis, is consequently restricted to lowland plains and plateau surfaces in the climatic zones mentioned above (cf Map Plate 3).

Thanks to the results of wide-ranging mass examinations which had already been published in 1951, due to the work of the schistosomiasis control teams and other bodies, as well as detailed investigations into specific aspects, the distribution of schistosomiasis in the Philippines is very well documented. Appropriate data, including estimates published by the Schistosomiasis Control Council, have been included in Table III; they form the quantitative basis for the discussion of the different endemic areas from north to south.

3.5.1 Sorsogon Province

The only schistosomiasis focus on Luzon is in the extreme south of the Bicol Peninsula in the Sorsogon Province. Suspicions of endemicity having already been raised by the results of a survey conducted by Africa and Garcia [16], Pesigan [593] was finally able to succeed in proving the presence of *Schistosoma japonicum* infections in man in the Irosin area. Mass examinations carried out by E. G. Garcia [309] at a later date served to confirm Pesigan's findings and to extend the area of distribution to the municipality of Juban. In its annual report for 1979 the Schistosomiasis Control Council [705] reckoned with a prevalence of about 9.3% for Irosin and Juban and an affected population of 48,115, i.e. a total of about 4,500 infected persons.

This core area of schistosomiasis distribution in the mainly mountainous Sorsogon Province is a flat valley through which the Irosin River flows (Fig. 46b); it is characterized by high precipitation (c. 3,500 mm), with extreme values from October to January (Fig. 52). The resident population lives mainly on agriculture, in which rice plays the dominant role apart from coconut and abaca. Discharging into the Sorsogon Bay at Juban, the Irosin River along with its numerous tributaries is the habitat of *Oncomelania hupensis quadrasi.* During his investigation of streams, irrigation canals, springs and rice-fields in this area, E. G. Garcia [309] was able to identify 23,580 m out of a total of 62,530 m as being snail-positive. The snail population fluctuates between 10 and 100 animals per m^2, the rate of infection being between 2% and 26%. The favourite habitats were flat sections on stream banks. A special feature in the Irosin-Juban endemic area appeared to be spring outlets providing snail habitats: "often the banks around such a spring, though rather steep, and even sandy, are continuously wet from seeping water. Such places may be very small. Omagom spring ... forms an *Oncomelania* habitat whose maximum dimensions are 10×2 m" [609].

The additional discovery of snail colonies in the municipalities of Casiguran and Bulusan in 1963 and 1964 gave rise to the possibility and the eventual confirmation by B. D. Cabrera et al. [181] of schistosome infections outside the core area of Irosin-Juban; applying the formalin-ether concentration technique (FECT), they were able to prove the presence of eggs in the stool of individual cases (prevalence 0.27%) at Casiguran, Bulusan, Barcelona, Bulan and Matnog. However, this low prevalence rose substantially (to 1.63%) when the circumoval precipitin test (COPT) was introduced as a diagnostic method and proved another two cases at St. Magdalena. From their findings B. D. Cabrera et al. [181] developed a hypothesis of radial distribution of schistosomiasis in Sorsogon Province, and proved it with the aid of the prevalence rates they had found; these showed the highest values at the centre of the endemic area (Irosin), and decreased from this centre with increasing distance from it (Fig. 53).

All in all, the Schistosomiasis Control Council reckoned with an average prevalence of 6.5% in the affected municipalities of Sorsogon Province in 1979. 96,821 (21.7%) of the 447,000 inhabitants of Sorsogon lived in the endemic area and were

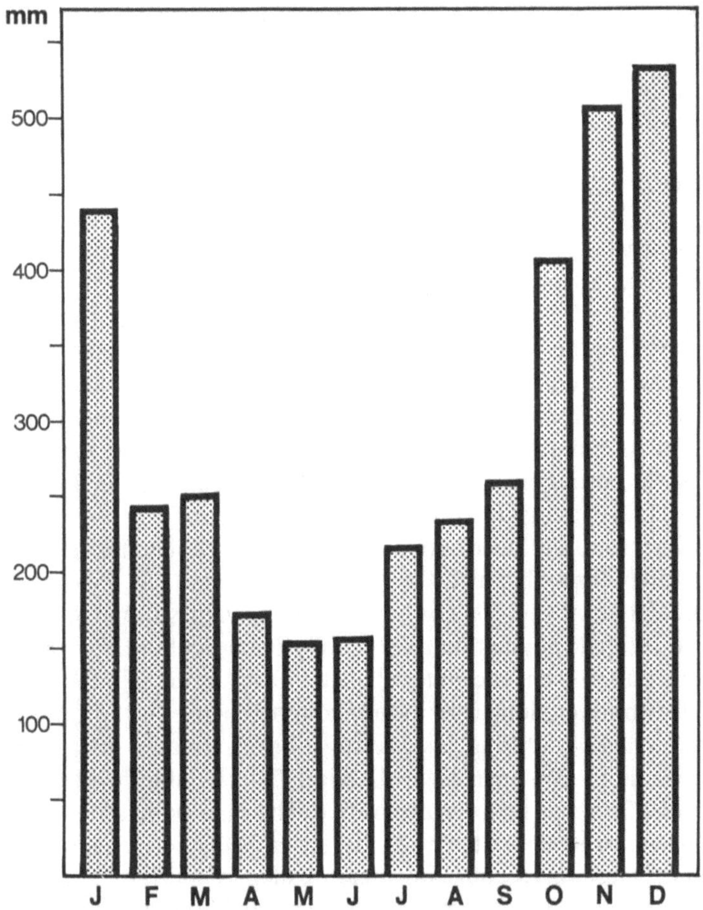

Fig. 52. Average monthly rainfall in Juban, Sorsogon [Source: 827]

thus exposed to the risk of a *Schistosoma* infection. The actual number of cases was estimated to be approximately 6,300. Assuming the population density in the endemic municipalities to be the same as in the province at large, i.e. 208.3 per km^2, the case density will be a mere 13.6 per km^2. However, if one takes into account not the total area but the more realistic part described as being "alienable and disposable" by the 1980 Philippine Statistical Yearbook when calculating the population and case density, values of 258.5 inhabitants and 16.8 cases of schistosomiasis per km^2 emerge for the endemic area in Sorsogon Province. Sorsogon is thus to be found in the lower range of values for the individual provinces (for correlation of population densities and cases cf 3.5.5).

The average prevalences for the municipalities in Sorsogon Province recognized as being endemic listed in Table III and represented in Fig. 53 [705, 706] should not, however, hide the fact that there are considerable differences within the area. The investigations carried out by E. G. Garcia [309] had already demonstrated rates for the individual barrios belonging to Irosin which ranged from 1.4% to 19.4% (average

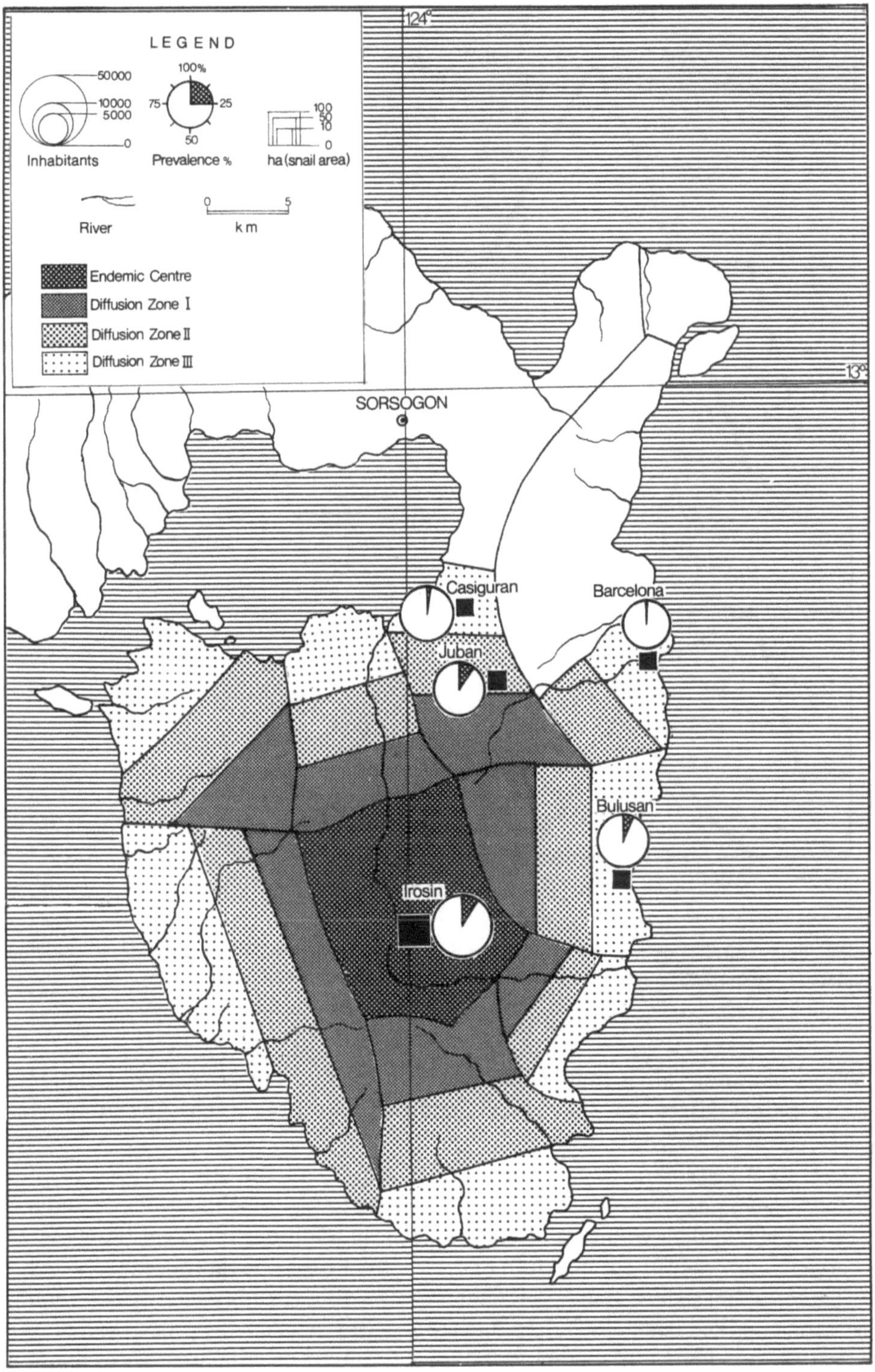

Fig. 53. Schistosomiasis and snail areas in Sorsogon Province [Sources: 181, 705]

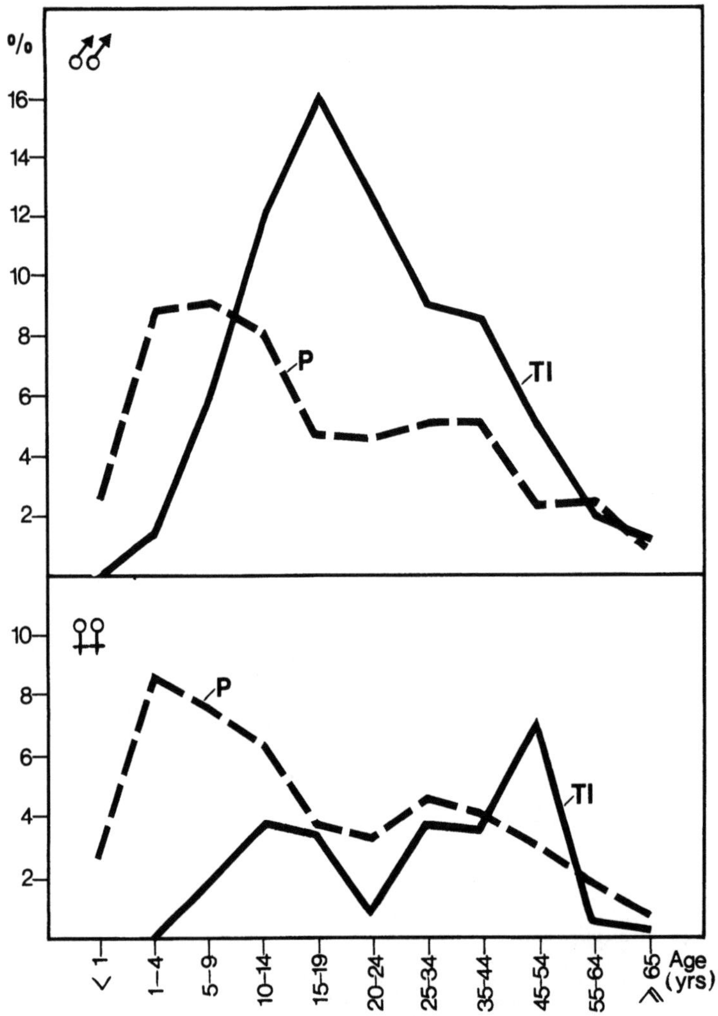

Fig. 54. Relative transmission index (*TI*) of *Schistosoma japonicum* by age and sex and population structure (*P*) in the barangays of Bolos and Gumapia, Iròsin, Sorsogon [Source: 838]

5.66%). That these values could be considerably higher in places emerged in more recent investigations of the Bolos and Gumapia barangays, which showed an overall prevalence of 43.2% [838].

In both of these barangays in which 755 of the 795 inhabitants had been examined for *Schistosoma* eggs in the stool, men and boys proved to be much more frequently infected (54.0%) than women and girls (30.2%). In addition, similar differences were observed in the intensity of infection (♂♂: 120, ♀♀: 81 eggs per gram of faeces).

These findings show that 75% of the relative transmission index fall to the male sex, with a marked peak (16.2%) among the 15–19 year-olds, the proportion of whom in the total population amounted to a mere 4.65% (cf Fig. 54). Apart from this it was discovered that, due to the high rate of egg-passing, 9.5% of all the infected persons (i.e. 31 of 326) accounted for 50% of the entire transmission potential. There was, moreover, a marked aggregation in particular households, in that 38.2% of the persons with a high egg-passing rate belonged to 20.9% of those houses in which there was at least one heavily infected person. There was no correlation between intensity of infection and proximity of houses to snail colonies.

In the case of children the Editorial Committee of a WHO Workshop [838] attributed the difference between the sexes in the prevalence and intensity of infection to the fact that boys, thanks to the greater freedom of movement they are allowed, are exposed for longer periods and more frequently than girls. The higher prevalence among men, on the other hand, was seen in connection with fishing and not with rice cultivation, for the majority of the snail colonies are to be found in streams and canals, which are frequently visited by men for fishing. In the Bolos and Gumapia districts it is therefore not the rice-fields but the rivers and other waters that must be regarded as the major source of infection.

The differences noted for the prevalence and intensity of infection were correlated with clinical parameters as well: when asked, 71% of the persons with a passing rate of more than 400 eggs per gram (EPG) indicated the occurrence of dysentery attacks, whereas only 24.7% of the people with lower EPG values were affected in this way. Hepatomegaly was observed more frequently among infected persons (77.9%) than in those who were not (67.3%), occurred more often among all age groups of male sex than of the female one, and more frequently showed itself to be more marked (liver size of 5 cm and over) in persons with high egg counts. Splenomegaly, too, showed a distinct correlation with schistosomiasis, the rate being 15.5% for infected persons but only 6.0% for non-infected ones. On the other hand there was no such correlation for other parameters such as a history of convulsions, fever, other neurological defects, headache, or dizziness, nor for ascites, peripheral oedema, superficial venous engorgement, presence of occult blood, or elevated blood pressure.

3.5.2 Oriental Mindoro

The most northerly schistosomiasis focus of the Philippines is in the eastern coastal region of Mindoro, the large northern island of the Sunda Shelf (cf 2.2.5.3). First indications that *S. japonicum* might be endemic there, arose from the observation of individual cases which had been diagnosed in other places – more specifically in places that had been shown to be free from schistosomiasis – but affected people who had previously spent some time in the vicinity of Lake Naujan on Mindoro [16, 776, 795]. In 1939 this led to an investigation in Barrio Malayas, then part of the municipality of Naujan; it was carried out by Tubangui and Pasco [794, 795] with the support of the National Research Council of the Philippines. Revealing a prevalence of 19.6%, it thus established the endemic occurrence of *S. japonicum* in Oriental Mindoro. At the same time these authors succeeded in demonstrating that dogs and pigs there were reservoir hosts, as well as proving the cercariae of the blood flukes to occur in 3% of the intermediate snail hosts that had been collected. In the next year the results of the schistosomiasis unit that had in the meantime been set up on Mindoro were already in a position to allow further barrios in the area of Lake Naujan to be classified as endemic [702].

A comprehensive survey of the situation in the Lake Naujan focus was, however, only made after World War II when Pesigan organized mass examinations throughout the country; these established schistosomiasis as being endemic everywhere in the region where rivers and streams flowed into the lake or even into the sea [678]. The prevalence rates in man found at the time were 10.4% in the case of Naujan and 3.1% for Pola. The centre of endemicity, situated between the northern banks of the lake and the San Augustin River (= Pinagsabangan River), as well as the lower section of the Buras River as the northern limit (Fig. 55), had a prevalence of 24% (11.6–46.8%). Even the intermediate hosts proved to be maximally infected there – up to 80%. Besides these there were two more foci on the west bank (prevalence 17.9%) and on the south-eastern bank (13.2%) where snail density (up to 130/m²) and their infection rate (up to 41%) also reached very high values.

Radially disposed around these areas of high endemicity were groups of settlements in which schistosomiasis infections occurred markedly less often (total prevalence: 4%; 0.6–12.8%). This area of lesser endemicity included the coastal strip north of Buloc Bay, the area north of the San Augustin River including the town of Naujan, and the towns and barrios along the National Road between Naujan – Victoria-Socorro-Pola as well (Fig. 55).

Recent investigations [198] indicate that, at least in the West Bank area, the situation has remained unchanged over a period of 30 years. For instance, in the case of Malayas (Municipality of Victoria) prevalence rates of 15.8% were established in 1949 [678] as well as three decades later. In four barrios of this municipality Carney et al. [198] found an overall prevalence of 19.9%, which led them to suspect that schistosomiasis might occur there more frequently than had been assumed by the Schistosomiasis Control Council. However, attention ought to be drawn to the fact that Carney et al. [198] undertook their investigations in an area which already in 1949 had manifested prevalence rates distinctly above the average (cf [678], Fig. 55 and Table III).

The Schistosomiasis Control Council [705, 706] estimates that 7.94% of the 118,149 inhabitants of the Naujan, Victoria, Socorro and Pola municipalities (= 30.4% of the total population of Oriental Mindoro) are de facto infected. Assuming that the population density in the endemic area does not differ from that of the province as a whole, the values indicate a case density of 7.1/km² when considering the total area, and of 13.9/km² if the total number of 9,383 estimated cases is related to the "alienable and disposable area". This puts the schistosomiasis focus in Oriental Mindoro into the category of endemic areas in the Philippines having a very low case density (cf Table 29).

According to estimates of the Schistosomiasis Control Council [705, 706] the total area settled by *Oncomelania hupensis quadrasi* in the Lake Naujan region measures 2,690 ha; approximately half of it lies in the Victoria municipality (Borbocolon). In this exclusively agriculturally-orientated area rice cultivation constitutes the most important economic element for the people. Whether the activities in the rice-fields bring about the contact with the water, which in turn decides whether an infection takes place, has not so far been investigated, nor is it known whether the cause of the markedly higher prevalence rates among males is to be found in these activities (♂♂: 24%, ♀♀: 15%).

3.5.3 The Visayan Islands

Situated between the south-eastern tip of Luzon, the Sibuyan Sea to the north and Mindanao to the south, the Visayan Islands offer conditions suitable for the guaran-

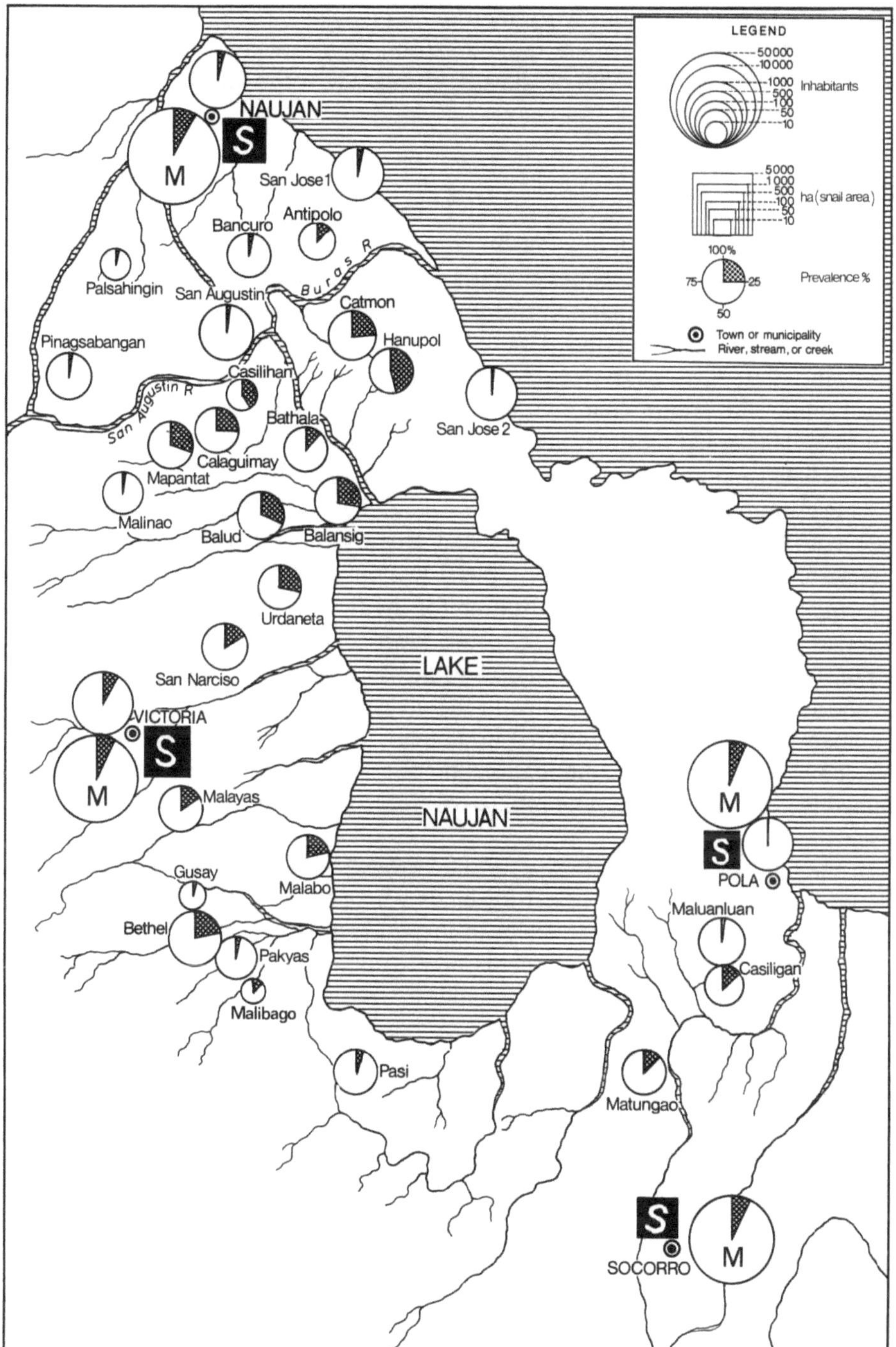

Fig. 55. Schistosomiasis and snail areas in the Lake Naujan area of Oriėntal Mindoro. *M* total population of the municipality, *S* total snail area of the municipality [Sources: 198, 678, 705]

teed persistence of the intermediate snail hosts, and thus of schistosomiasis only in their eastern parts. Infections with *Schistosoma japonicum* are therefore restricted to Samar, Leyte and Bohol, where the Schistosomiasis Control Council [705, 706] assumed 216,774 cases, i.e. 34.5% of all cases estimated for the Philippines.

3.5.3.1 Samar

If the endemic areas of Sorsogon and Lake Naujan, which have been described above, are limited in extent to a few hundred square kilometres Samar may be looked upon as that island where schistosomiasis is endemic in almost all areas below 200 m (cf Map Plate 3). On Samar schistosomiasis occurs over an area of approximately 3,000 km². Some 66.6% of the population of the 1.12 million entered in the 1980 Philippine Statistical Year Book for 1975 live in the endemic areas, with western Samar taking the largest share with 71.4% of the 478,000 inhabitants. As far as frequency is concerned, the average prevalence rates of 12.9–13.1% reveal no differences between the three provinces of the island [705, 706] (cf Table 23).

The examination of prison inmates undertaken by Garrison [337] and Willets [842] at the beginning of the century had already indicated that Samar was to be counted among the endemic regions, but it was only the survey by Africa and Garcia [16] which achieved an exact localisation for the first time; they traced the origins of the cases registered by them back to the towns of Calbayog, Catbalogan and Sta. Margarita (in the West Coast area), as well as to Catarman, Laoang and Palapag on the north coast. Further work aimed at the investigation of schistomiasis distribution in Samar consequently concentrated upon these areas [77, 78, 97, 394, 456–458, 570, 597, 866–870]. Finally, the Annual Report for 1978 of the Schistosomiasis Control Council [704] also included nine municipalities from eastern Samar as being infected by schistosomiasis.

In the light of the present state of knowledge, it appears that schistosomiasis occurs on Samar in all those coastal lowlands where the gradient remains very gentle over long distances. Within the realm of the valley plains their distribution range extends far into the island's interior as in the catchment area of the Catubig River in the north, the plain of the Gandara River and its tributaries in the west, and in the lowlands of the Dolores River in the east (Fig. 56). Schistosomiasis does not occur in the mountainous central part of Samar, nor in the coastal areas where hills and mountains of 200 m and more come close to the shore since these areas are little suited to the requirements of *O. h. quadrasi.* In Samar the restriction of schistosomiasis distribution as a consequence of snail distribution is thus predominantly determined by the orography. Thanks to the high precipitation which falls on all parts of the island throughout the year (Fig. 57), climatic factors permit the occurrence of intermediate hosts everywhere.

As far as the area settled by *O. h. quadrasi* is concerned, the waters investigated by Banzon and Beltran [78] over a distance of 73 km in the Bobon and Catarman municipalities produced positive findings for half of it. In some habitats the density attained extreme values of 500 snails per square metre and over, with a 25% share of cercariae-positive specimens. In 1979 the Schistosomiasis Control Council [705, 706] assumed that the area infested by snails to have attained a total of almost 12,000 ha in Samar's three provinces (Table 23). In the Philippines this extent is surpassed only by the area colonized in the Davao-Agusan Trough (cf 3.5.4.1).

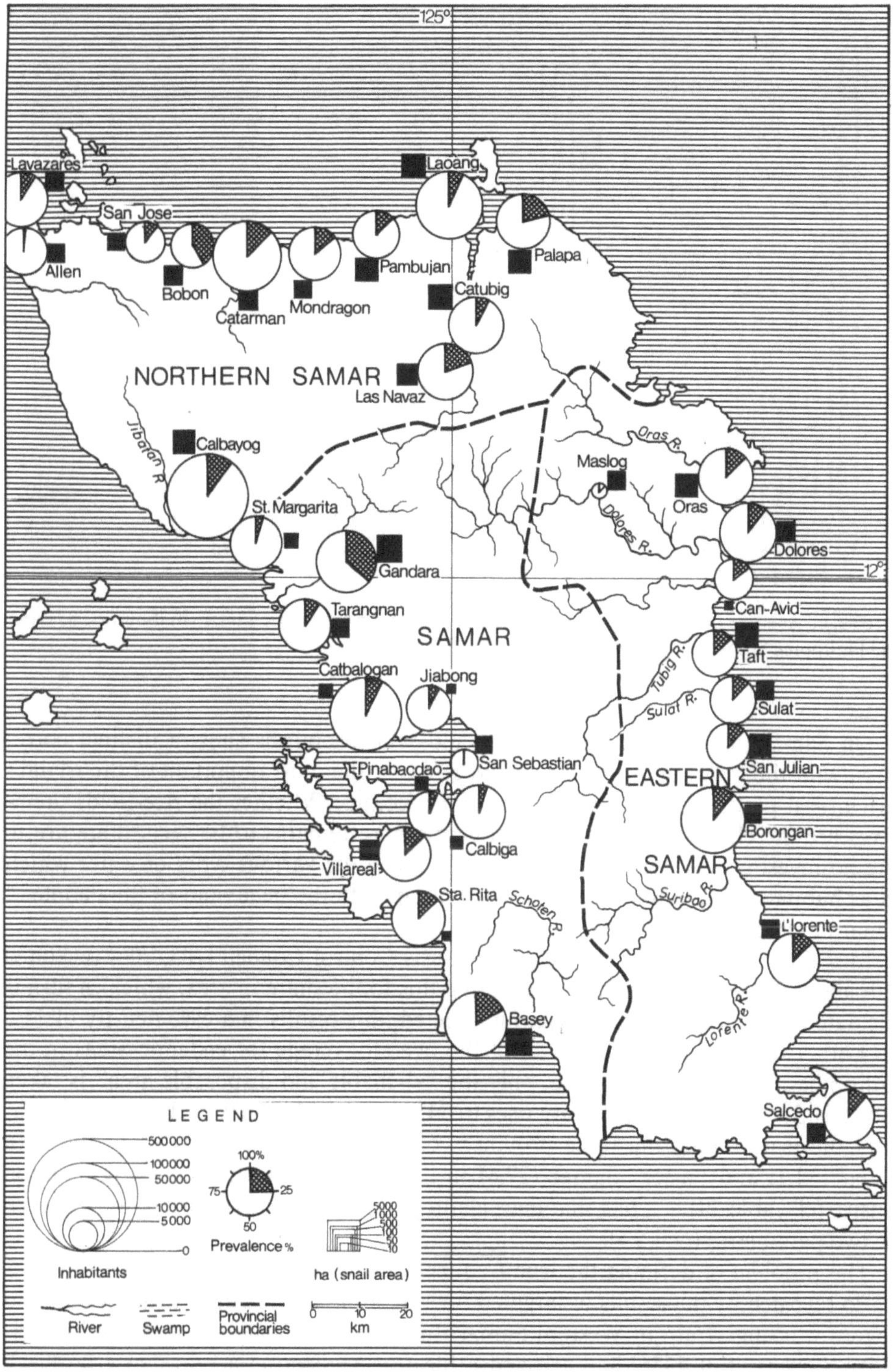

Fig. 56. Schistosomiasis and snail areas in Samar [Source: 705]

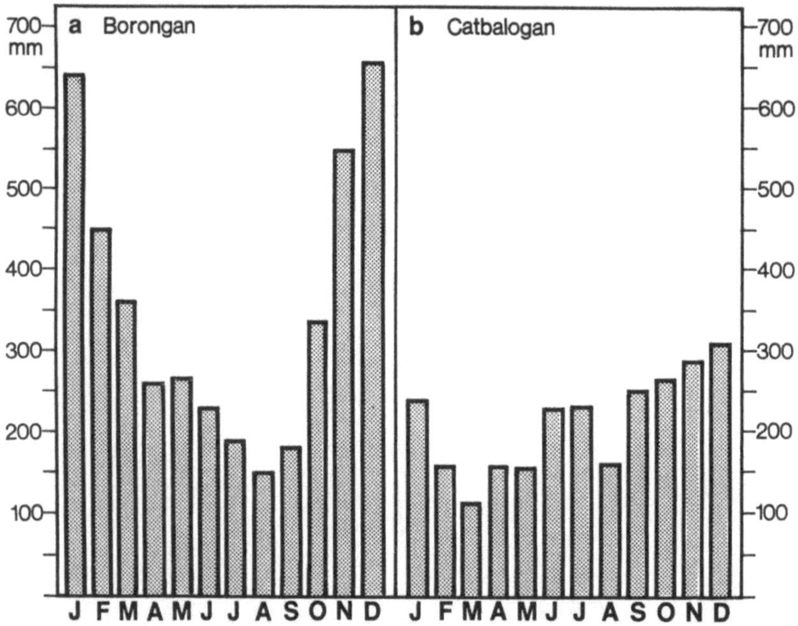

Fig. 57 a, b. Average monthly rainfall in Eastern Samar (**a**) and Western Samar (**b**) [Sources: 831 and 1980 Philippine Statistical Yearbook]

On Samar the distribution of snails and schistosomiasis show a distinct correlation with the growing of rice ("lowland paddi"), the most important of the island's agricultural products apart from abaca and coconuts. Whether the rice-fields themselves are the chief source of infection everywhere remains to be seen; there are to date no investigations which differentiate between the various types of water which act as habitats for the infected snails and link up with the behaviour of the people of Samar in regard to water contact.

In the 1979 report year the Schistosomiasis Control Council [705, 706] reckoned with about 97,000 cases on Samar. Relating this to the share in area, which in turn is calculated according to the population affected, this results in a case density of 11.0 per km². This is a figure below that of the overall average for endemic areas in the Philippines. If, on the other hand, only the "alienable and disposable area" is taken into consideration, the case density rises to 34.0/km². With 39.6/km², Northern Samar takes the lead among the three provinces of the island; it is followed by Western Samar (34.3) and Eastern Samar (27.6) in accordance with the general population density (cf Table 29) – a fact that once again points to a positive correlation between the two parametres (cf 3.5.5).

The problem of defining the true frequency of schistosomiasis in the Philippines is also revealed by the results of targeted investigations carried out by the Lewert and Yogore working group in Western Samar. In Barrio San Antonio of Basey Municipality, Lewert et al. [456] had selected a location registering a prevalence of 40.1% when examined through the application of a modified merthiolate-iodine-formaldehyde concentration technique (MFCT), i.e. a schistosomiasis frequency twice as high as the overall average of 19.0% for the municipality [705, 706]. Distinctly higher prevalence rates, however, resulted when serological tests were applied: in a selected population group including 59.2% of MFCT-positive persons, a rate of 90.2% seropositive cases could be established when the enzyme-linked immunosorbent assay (ELISA) had been used [870]. In the ELISA tests 76.8% of the 106 stool-negative reacted positively. If these findings are taken into account when considering the prevalence of 40.1% in San Antonio's total population – which had been established by MFCT – a de facto prevalence of about 60% emerges. These stool-negative persons detected by serological methods are of great significance for the estimation of schistosomiasis morbidity in the endemic areas, since their inclusion alone reveals the full extent of the problem. For the epidemiology, that is the dynamics of transmission, this group of people plays no role, of course.

To date there has been no detailed investigation as to which waters or which water activities at San Antonio are chiefly responsible for the high frequency of schistosomiasis. The majority of the population is employed in agriculture, and within that employment category rice-growing comes first. Another important contact with water results from the gathering of semi-aquatic plants such as "kang-kong" (*Ipomea* sp.) as a green vegetable – an occupation popular among men, women and children. Notwithstanding this no quantitative survey has been undertaken of activities connected with the risk of infections, nor has the area been investigated in respect of habitats of *O. h. quadrasi.* Random samples taken in 1964 and 1966 had shown snail densities of 80 and 50/m² with infection rates of 12.3 and 8.6 per cent [456].

3.5.3.2 Leyte

In its north-eastern part, Leyte, the second largest of the eastern Visayan Islands, contains what is probably the most important endemic schistosomiasis area in the Philippines. The number of cases in the lowland plain between Carigara in the north and Abuyog on the east coast (cf Map Plate 3 and Fig. 58) is estimated to exceed 88,000. A further 10,000 cases are reckoned to be in the North-West (Villaba and Matag-ob) [705, 706]. This high level of endemicity was also the reason for choosing the North-eastern Plain as a pilot project area for the investigation and control of schistosomiasis in the Philippines after World War II (Leyte del Sur, the southern province of the island is schistosomiasis-free).

Having already been suspected of being endemic at the beginning of this century [337, 842], and confirmed as such when it was proved that snails were infected [781], even before the outbreak of war in the Pacific several places on Leyte were shown by Tubangui and Pasco [795] to have high rates of prevalence. However, the full extent of the problem only emerged in the course of the mass examinations which were begun in 1949 [84, 595, 597], as up to 50% of the population proved to be infected in several municipalities (cf Table III). A total of 35,231 stool samples were examined, of which 14.5% contained *Schistosoma* eggs. The search for snails, which was carried out over a distance of 40 km of waterways, resulted in positive findings for 25.7 km of them, with a maximum density of 200 *O. h. quadrasi* per square metre and an infection rate of up to 54.1% (cf Table 19).

In the light of examinations carried out since then, the Schistosomiasis Control Council [705, 706] estimated that a total of 43.3% of the population of Northern Leyte live in endemic areas and are thus exposed to the risk of a schistosome infection. The average prevalence there is 18.9%; the municipalities of San Miguel (52.2%), Alangalang (41.7%), Santa Fe (47.6%), Palo (33.1%), Pastrana (35.3%), Dagami (36.1%)

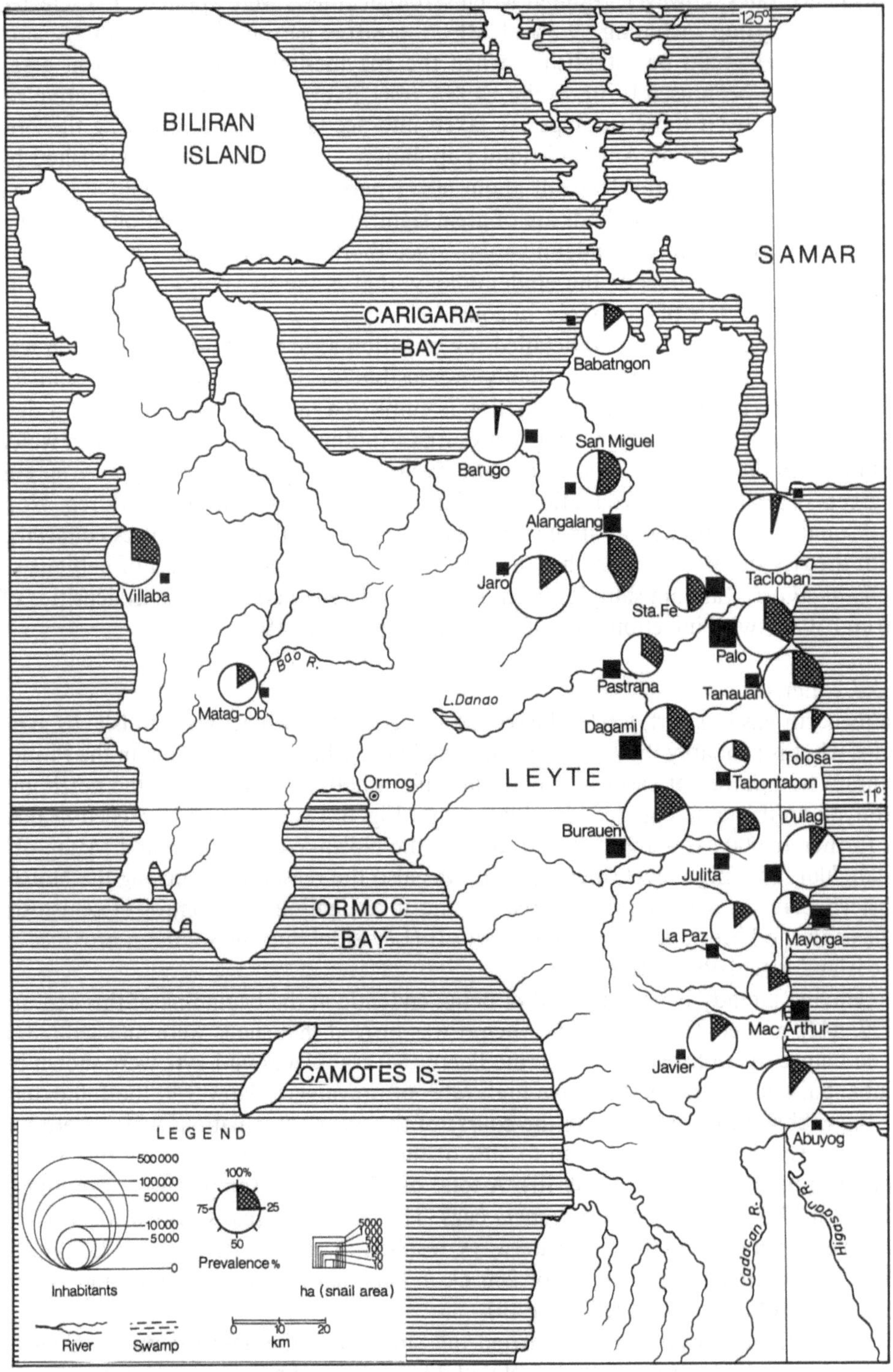

Fig. 58. Schistosomiasis and snail areas in Leyte [Source: 705]

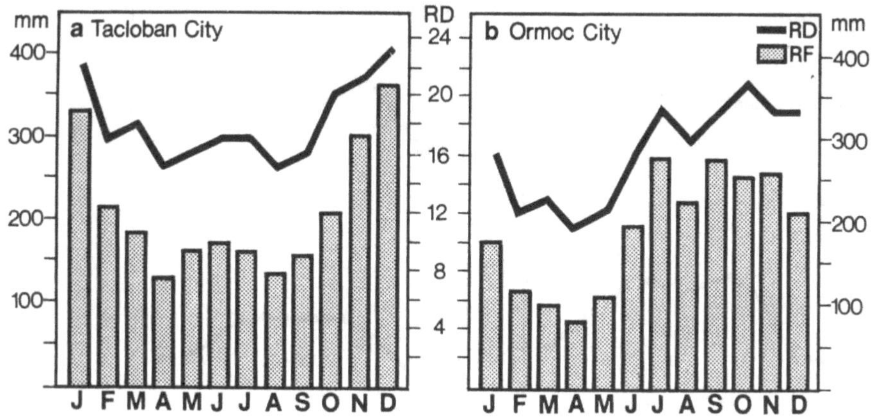

Fig. 59 a, b. Average monthly rainfall (*RF*) and number of rainy days (*RD*) in north-eastern (**a**) and north-western (**b**) Leyte [Source: 606]

and Tanauan (26.7%), which form a semi-circle west and south of Tacloban, the provincial capital, are particularly affected. Municipalities further away from this centre of endemicity show markedly lower average infection frequencies with prevalence rates ranging from 0.8% to 30.6% (Fig. 58).

The total snail area is estimated to be 3,280 ha; of this 1,878 ha (57%) belong to the Municipality of Palo alone, and 2,613 ha (80%) to all the municipalities with particularly high prevalences. The most important snail habitats in this part of Leyte was the Cogon-Anahaway rice-field extending over an area of 12 km² [609].

Extending over a distance of about 90 km, and with a breadth varying from 11 to 20 km, the large lowland plain in the north-east of Leyte covers an area of about 1,200 km², it is drained by numerous perennial rivers, which flow into the Gulf of Leyte in the east, or to the north to Carigara Bay. Flatness of the terrain (cf Fig. 46 a), climate, water network and rice cultivation (lowland palay) as the main form of agricultural land use bring about the conditions for the persistence of the intermediate snail hosts in this lowland plain. Precipitation of 2,500–3,000 mm per year, though varying according to season, amounts to 130 mm even in April, the month with the least rain falling on the average on 15 days (cf Fig. 59 a). The season with the highest precipitation lasts from November to January, with maxima of an average of 364 mm and 23 rain-days in December in Tacloban City. (Extreme values occur in localities near the coast in the extreme south of the plain; in December, for instance, Abuyog has an average precipitation total of 563 mm.) During this season rivers frequently burst their banks and cause flooding. On the other hand, this may have a damaging effect upon the snail colonies of this area, but on the other hand snails transported by the current and deposited elsewhere may settle in new areas. Thanks to the seasonally restricted but regularly recurring floods, all available habitats are likely to have been settled in this way in the past; nowadays, however, this process is regaining its role in areas which control measures had succeeded in clearing, but which are now in danger once again of being re-colonized.

Besides precipitation falling throughout the year, air temperature and atmospheric humidity also favour the persistence of *O. h. quadrasi* in the Leyte plain. Both these climatic parameters are characterized by little fluctuation over the course of the year.

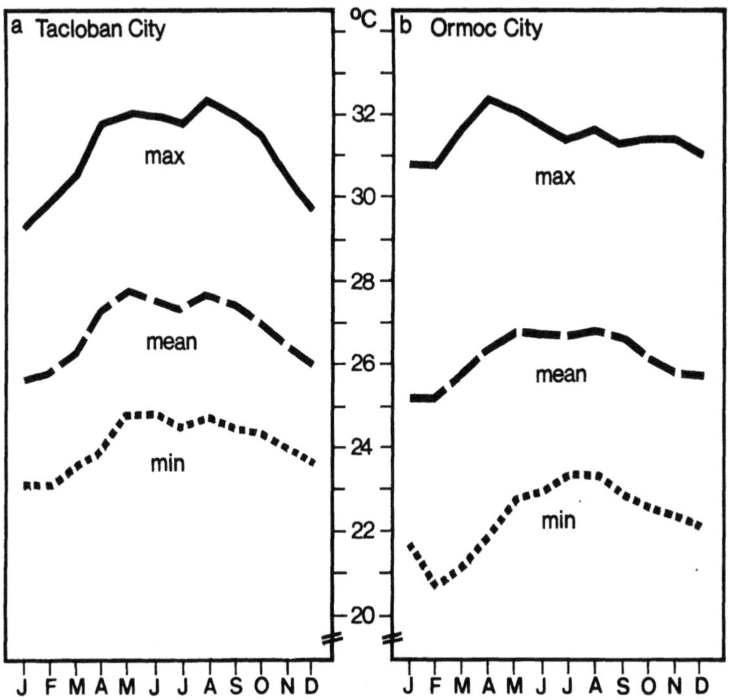

Fig. 60 a, b. Average monthly maximum, minimum and mean temperatures in north-eastern (**a**) and north-western (**b**) Leyte [Source: 606]

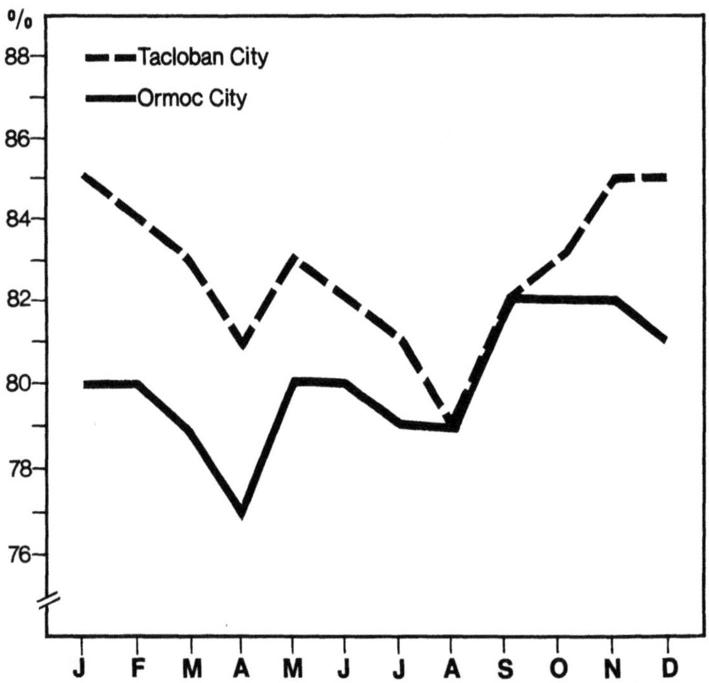

Fig. 61. Monthly relative humidity in north-eastern (Tacloban City) and north-western Leyte (Ormoc City) [Source: 606]

The mean average air temperature remains within the range of 25.6 °C (January) and 27.8 °C (May), with a mean maximum between 29.3 °C and 32.3 °C and a mean minimum between 23.1 °C and 24.8 °C (Fig. 60a). The mean monthly fluctuation of atmospheric humidity is between 79% and 85% (Fig. 61).

The endemic area in north-western Leyte is less extensive; it includes only parts of the municipalities of Matag-ob and Villaba with prevalences of 15.7% and 27.8% in a total population of 40,000 affected people. One assumes the area colonized by snails to be quite small there (15 ha) [705, 706]. Separated from north-eastern Leyte by the Central Cordillera, this endemic area consists of plains in the upper reaches of the Pagsangahan River, which flows into Ormoc Bay, as well probably as the upper Jinagatan River, which runs into the sea on the west coast. Here, too, rice is one of the most important agricultural products, together with maize, but whether rice-fields offer important snail habitats has not so far been established. Santos et al. [693], the discoverers of this focus, found six snail colonies within the region of swamps, creeks and depressions; they did not mention any rice-fields.

The climatic conditions in north-western Leyte differ from those in the north-east of the island, inter alia in so far as the maximum precipitation of 278 mm is reached as early as July (Ormoc City), and remains at approximately this level until the year's end. Two-thirds of the total precipitation (1,510 of the 2,280 mm) fall during this period and only one third between January and June (Fig. 59b). Air temperature (Fig. 60b) and atmospheric humidity (Fig. 61) also differ from those of North East Leyte (albeit only slightly), but even in the endemic area of Villaba/Matag-ob the conditions prevailing throughout the year ensure the persistence of snails.

Though investigations of the behaviour of people on Leyte in regard to their contacts with water (cf 3.2.6) show higher values for the female sex in respect of duration and frequency of contact, men and boys had been more exposed as they remained more frequently and for longer periods in the snail habitats proper. It follows that in the majority of localities on Leyte higher prevalence rates and higher infection intensities were observed among the male sex (cf Table 24). Accordingly 60–75% of the relative transmission index in man is found among boys and men – with the highest proportion among the 15 to 19 year-olds. The higher intensity of infection may, moreover, be the cause of hepatomegaly, which occurs more often among males [277]. However, only in areas with a very high prevalence were an inability to work, weakness, abdominal pain and diarrhoea more frequently observed among infected persons than in non-infected ones [577].

The results of the surveys carried out on Leyte so far (Table III) have provided a solid foundation for the demarcation of the schistosomiasis distribution area (cf Map Plate 3 and Fig. 58). As a result of the establishing of prevalence rates through the application of sero-diagnostic methods, however, the data obtained by direct evidence of schistosome eggs in the stool, which indicate the frequency of this worm infection within the endemic area, are quite obviously in need of revision. Thus the testing for antibodies in school children at Tanauen, carried out with the help of enzyme-linked immunosorbent assay (ELISA) showed a positive reaction in 56.2% of the children examined whereas the application of a modified merthiolate-iodine formalin concentration technique (MFCT) only showed 28.4% of the stools to contain worm eggs [871]. The high specificity (96%) and the sensitivity (98%) of ELISA obliges one to concede that the de facto frequency of schistosome infections had been considerably underestimated so far. This 50 per cent under-estimation, which resulted

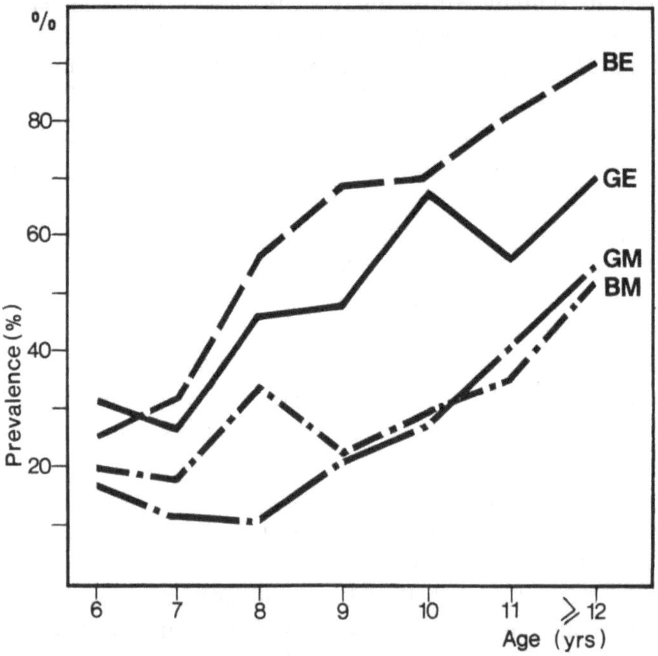

Fig. 62. *Schistosoma japonicum* prevalence rates of boys (*B*) and girls (*G*) in Tanauan, Leyte, as determined by ELISA (*E*) and MFCT (*M*) [Source: 871]

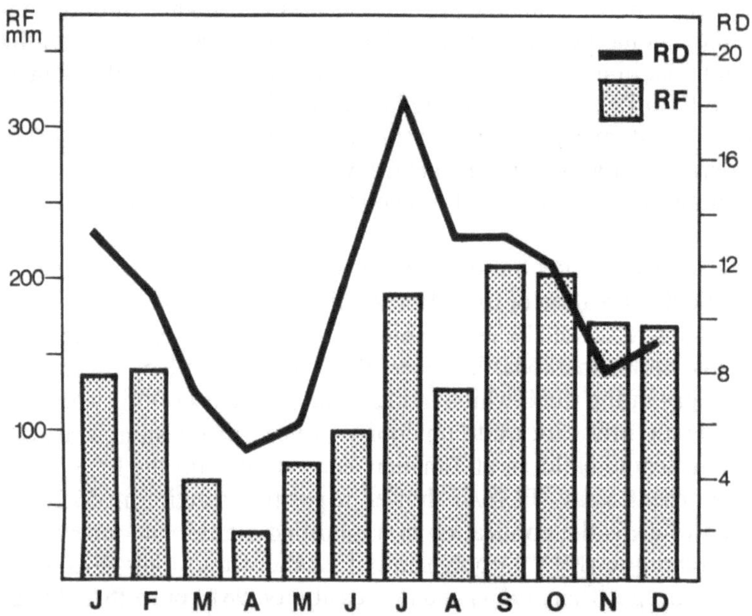

Fig. 63. Average monthly rainfall (*RF*) and number of rainy days (*RD*) in Ubay, north-eastern Bohol [Source: 827]

from the comparison of all the school-children who had been included in the examination did, however, vary according to age and sex (Fig. 62), so that a computer prediction for the entire population cannot be attempted yet on the basis of this result, nor is it known whether areas with varying prevalence rates show the same discrepancy between sero-positive and stool-positive persons or not. If the under-estimating resulting from findings of stool examinations which Yogore et al. [871] have detected, is confirmed, Leyte alone would have a total of almost 200,000 people infected with *S. japonicum*. Further investigation will be needed to clarify whether it is indeed so.

3.5.3.3 Bohol

In the Central Visayas Region (Bohol, Cebu, Negros Oriental, Siquijor) in the west of Leyte, only the eastern part, including North Cebu and Bohol with their precipitation occurring throughout the year (Fig. 63) offer the climatic conditions for colonization by *O. h. quadrasi*. That is indeed the area where the only schistosomiasis focus in the Central Visayas is to be found. It is restricted to the municipalities of Talibon and Trinidad in north-east Bohol. The investigation carried out by Blas and Dazo [101] in this area in 1958, showed a prevalence of 30.5% among the inhabitants of the eight barrios (Fig. 64). Apart from that, six snail colonies were discovered in swampy flats between hills with an average density of 103.6 animals per square metre, 0.9% of which contained *S. japonicum* cercariae. The total extent of the snail area was 12,420 m^2.

The overall distribution of schistosomiasis in the Central Visayas had been largely set out by these results (cf Map Plate 3). The Schistosomiasis Control Council [705, 706] reckoned that in 1978 as well the area colonized by *O. h. quadrasi* on Bohol would not exceed 19 ha. The average prevalence in man was estimated to be 38.45%, the equivalent of 21,198 cases. Being very high for the narrow extent of the Talibon-Trinidad focus, these estimates of infected persons also require correction on the basis of more recent investigations, however. In those barrios which had already been recognized as endemic at an earlier stage (Fig. 64) Carney et al. [197] found evidence of schistosome eggs in the stool of only 5.4% of the inhabitants. These authors also indicated that previous observations made by the schistosomiasis control team established in Trinidad in 1976, had already shown a marked drop in infection frequency. The question therefore arises as to what are the reasons for the restriction of distribution in the barrios situated within the catchment area of the Ipil River, as also to the noticeable decrease in the prevalence rates.

As mentioned above, the potential distribution of the intermediate snail host in the Central Visayas is limited to northern Cebu and to Bohol as a result of the distribution of precipitation in the course of the year (Map Plate 3). However, various reasons make it unlikely that *Oncomelania* persists in north-eastern Cebu. The fact that rice-cultivation, which tends to be closely allied with the colonization of the snail in most of the endemic areas, is insignificant on Cebu, plays an important role in this. Maize predominates there, being the most important agricultural product. In addition, Cebu is one of the most densely settled areas of the Philippines. As early as 1960 the population density was 262/km^2 when taking into consideration the entire surface area, but 888 km^2, when related to the agriculturally utilisable area. Wernstedt and Spencer [831] estimated this to be about 30%. Even then the extreme pressure of population required an intensive utilization of all the land resources on Cebu (besides internal migration), which has by this time increased even further. Intensive culti-

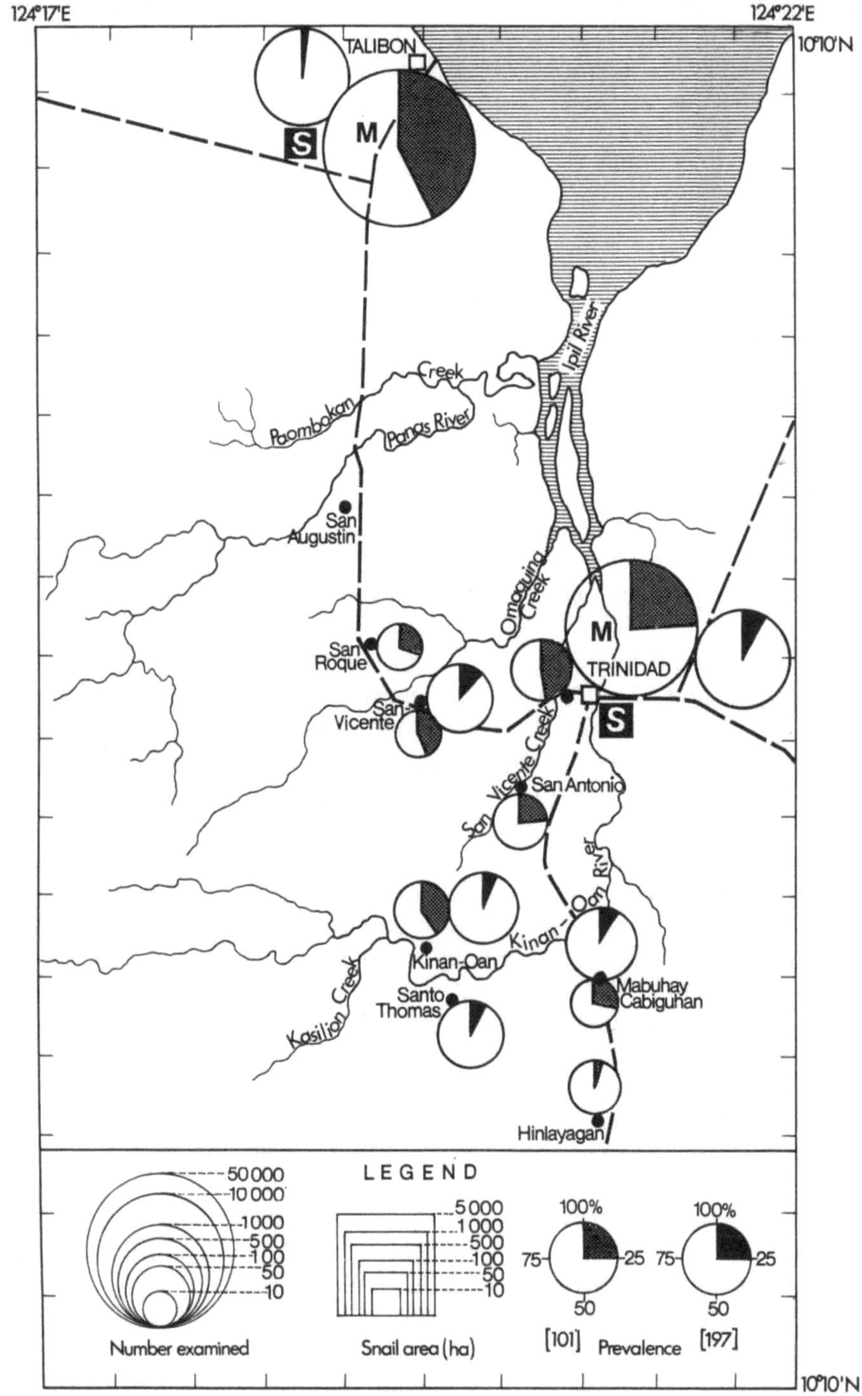

Fig. 64. Schistosomiasis and snail areas in the Ipil River area, Bohol. *M* total population of the municipality, *S* total snail area of the municipality [Sources: 101, 197, 705]

vation of this sort conflicts with colonization by *Oncomelania*, however, for it can only persist in undisturbed environments.

In Bohol the initial situation was different: the cultivation of rice as the most important agricultural product and a low density of population permitted a partial colonization of lower-lying areas by *O. h. quadrasi*, although even these lower altitudes have hilly features which restrict the area of the snails' distribution and thereby of infections by *S. japonicum*. In 1960 the population density at Bohol was 144 persons/km² over the total surface area and 320/km² for the agriculturally-used area. Since then a change has set in, leading to a decrease in the frequency of schistosomiasis. Carney et al. [197] explained the change as follows: "The most reasonable explanation for the spontaneous decline in the prevalence of schistosomiasis in the area is intensive agricultural use of all lowlands available as a result of population pressure. Wet-rice fields are cultivated up to the edges of the numerous hills that characterize this area of Bohol and oncomelanid snails do not thrive in either sloping terrain or constantly cultivated and flooded rice fields. In northern Bohol areas suitable for oncomelanid habitat are the same areas that are now intensively cultivated for wet-rice production."

3.5.4 Mindanao

Thanks to its division into macro-regions (cf 2.2.5.7) Mindanao has several areas in which, divided from one another by mountain ranges, *S. japonicum* is endemic. So too, in Mindanao the endemic areas can be characterized by the same orographic and climatic conditions as previously described for Sorsogon, Mindoro and the Visayas. Once again flat terrain is a typical feature, as this is the precondition sine qua non for the persistence of the intermediate snail hosts. However, on Mindanao, it is not exclusively lowland plains where *Schistosoma* infections are endemic but also areas at higher altitude, such as the Bukidnon-Lanao Plateau. To permit a colonization by snails not only orographic but also climatic preconditions must be met. This is once again borne out in the case of Mindanao, where *O. h. quadrasi* persists only in parts having a high precipitation throughout the year (Fig. 65). In areas where precipitation falls to less than 100 mm a month for periods of over one month, the snails can evidently not persist. For this reason the flats on the north coast of Misamis Oriental and Lanao del Norte (from Gingoog Bay in the east to Iligan Bay in the west), and those on the Zamboanga Peninsula are free from schistosomiasis.

3.5.4.1 The Davao-Agusan Trough

The Davao-Agusan Trough constitutes the most extensive and largely continuous area of endemic schistosomiasis in the Philippines (cf Map Plate 3 and Fig. 66a, b). It stretches from the flats near Bacuag and Gigaquit on the east coast via Lake Mainit southwards to Surigao del Norte, and includes all the flatlands in the catchment area of the north-bound Agusan River, which drains into the Butuan Bay. The lowland plain of Hijo, Saug, Tagum and Libuganon – the rivers flowing into the Davao Gulf – forms its continuation in a southerly direction; adjacent to it is the coastal area stretching from Davao City to Digos and Hagonoy. The linear expansion of this endemic area is more than 300 km; its mean breadth, fluctuating between 10 and 40 km, narrows most in the south where the Olagusan Uplands divide the wide Agusan Val-

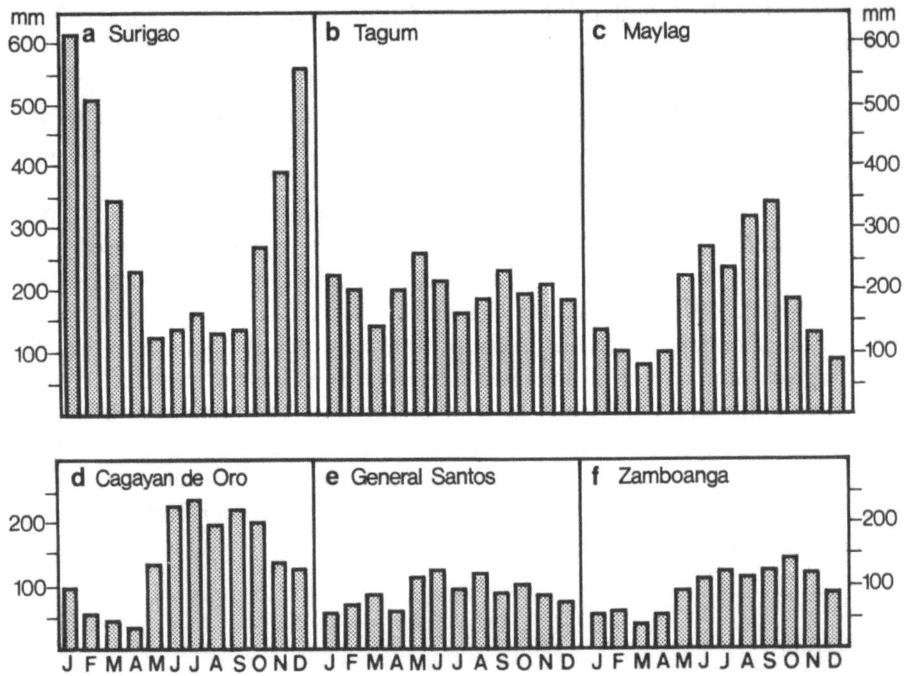

Fig. 65 a–f. Average monthly rainfall in selected areas of Mindanao. **a–c** schistosomiasis endemic areas, **d–f** schistosomiasis non-endemic areas [Sources: 827, 831 and 1980 Philippine Statistical Yearbook]

ley from the Davao Gulf region. Its delimitation is marked in the east by the Pacific Cordillera and in the west by the Central Highland.

Initially established only in the extreme north around Lake Mainit [592, 702, 795], *S. japonicum* endemicity was later found to prevail throughout the Davao-Agusan Trough; mass examinations to this end were conducted by Pesigan, and were later continued by the Schistosomiasis Control and Research Service [540, 595, 597, 610, 704–706]. There are about 1.9 million people in this endemic area of around 6,000 km², more than 260,000 of whom are infected, according to data supplied by the Schistosomiasis Control Council [705]. Because of differences in the population density, which is high in Surigao del Norte, but very low in Agusan del Sur (to cite just a few examples) in regard to the prevalence rates, the affected provinces show considerable variations in the figures for case densities (cf Table 29, as well as 3.5.5).

The Schistosomiasis Control Council [705] estimates the area colonized by snails in the Davao-Agusan Trough to be more than 20,000 ha (Table 23), almost half of which is in the swampy lowlands of the Agusan River near Loreto. (There have been no subsequent publications following the description of the snail habitat of the Manat River swamp in Davao by Pesigan et al. [609] on either its extent, then thought to be over a distance of 20 km with a breadth of at least 2 to 3 km, nor concerning its existence altogether.) The snail biotopes are of different types: besides swamps, streams, rice-fields and canals were found to be colonized by *O. h. quadrasi.* Millar and Putong [540] even described "three canals formed by three continuously flowing artesian wells" for Jabonga on Lake Mainit as being snail-positive.

3.5.4.2 The East Coast

Separated from the Davao-Agusan Trough by the Pacific Cordillera, the flats on the east coast of Mindanao also offer suitable conditions for colonization by *O. h. quadrasi*. *Schistosoma* infections have so far been recorded for two isolated river plains in Surigao del Sur (Map Plate 3 and Fig. 66a). Surrounded by hills and mountains, the wide river basin of the Tago River, which extends westward from the coast for about 20–30 km up to the San Miguel area, contains the northern focus. According to Schistosomiasis Control Council data [705, 706] about 12,500 (= 37.4%) of the over 33,000 inhabitants here are infected. The snail area has been estimated to be approximately 200 ha. There is a second endemic area in the region of Bislig Bay, about 80–90 km south of the Tago River focus. Again it is a matter of a river plain formed by the Bislig river and its tributaries. It must be admitted that the end of the valley formed by the Pacific Cordillera in the west only rises to a little over 200 m. For that reason a possible direct link with the endemic area of the Agusan Valley cannot be excluded. The Schistosomiasis Control Council reckoned with about 13,600 (= 25.6%) infected persons among the over 53,000 inhabitants of the Bislig Municipality. An area of 865 ha was declared as being inhabited by snails.

3.5.4.3 The Cotabato Basin

The Cotabato Basin, an extensive lowland area in the south-west of Mindanao also offers conditions suitable for the endemic occurrence of schistosomiasis. Being a sort of rift formation, which was chiefly infilled by scree material from the Central Cordillera [364], it extends over an area of approximately 6000 km^2. Surrounded by mountain ranges (the Tiruray Highlands in the south west, the Central Cordillera in the east, and the Bukidnon-Lanao Highland in the north), the Cotabato Basin is only open to the Illana Bay in the north west. Even under normal water-level conditions its central section – the Cotabato Valley with the Rio Grande de Mindanao being the most important river – is about half way taken up by extensive swamps, the largest of which are the Libungan Marsh and the Liguasan Swamp, with a combined total area of around 1200 km^2. This central section is flanked by the Carmen Basin (c. 1500 km^2) in the north-east and the river flats of the Alip (120 km^2), the Koronadal (1000 km^2) and the Allah (600 km^2) in the south-east.

The first schistosomiasis focus in the Cotabato Basin to be located was the Pikit-Pagalungan area [610]. Today eight municipalities are considered to be endemic [705, 706] (cf Map Plate 3 and Fig. 66c). The total population to be affected numbers 400,000, of which 50,000 (= 12.5%) are regarded as infected. The case density of 37.3/km^2 is approximately in accordance with the mean value for all the endemic areas of the Philippines (cf Table 29 and 3.5.5). According to data made available by the Schistosomiasis Control Council, *O. h. quadrasi* occurs in an area of only 900 ha.

Since *S. japonicum* in the Cotabato Basin was first shown to occur after World War II, it can now no longer be determined whether these are primary foci of long standing or not, since in the light of internal migration processes in the Philippines it is entirely justifiable to consider the question of a possible introduction from other parts of the country, as the south-west of Mindanao was only sparsely settled at the beginning of this century. The immigration of Filipinos from very densely populated provinces, especially from the Visayas, began in 1913. It attained particularly large dimensions after the Second World War, when the government chose the Cotabato Basin as one of the most important areas for their government-financed resettlement

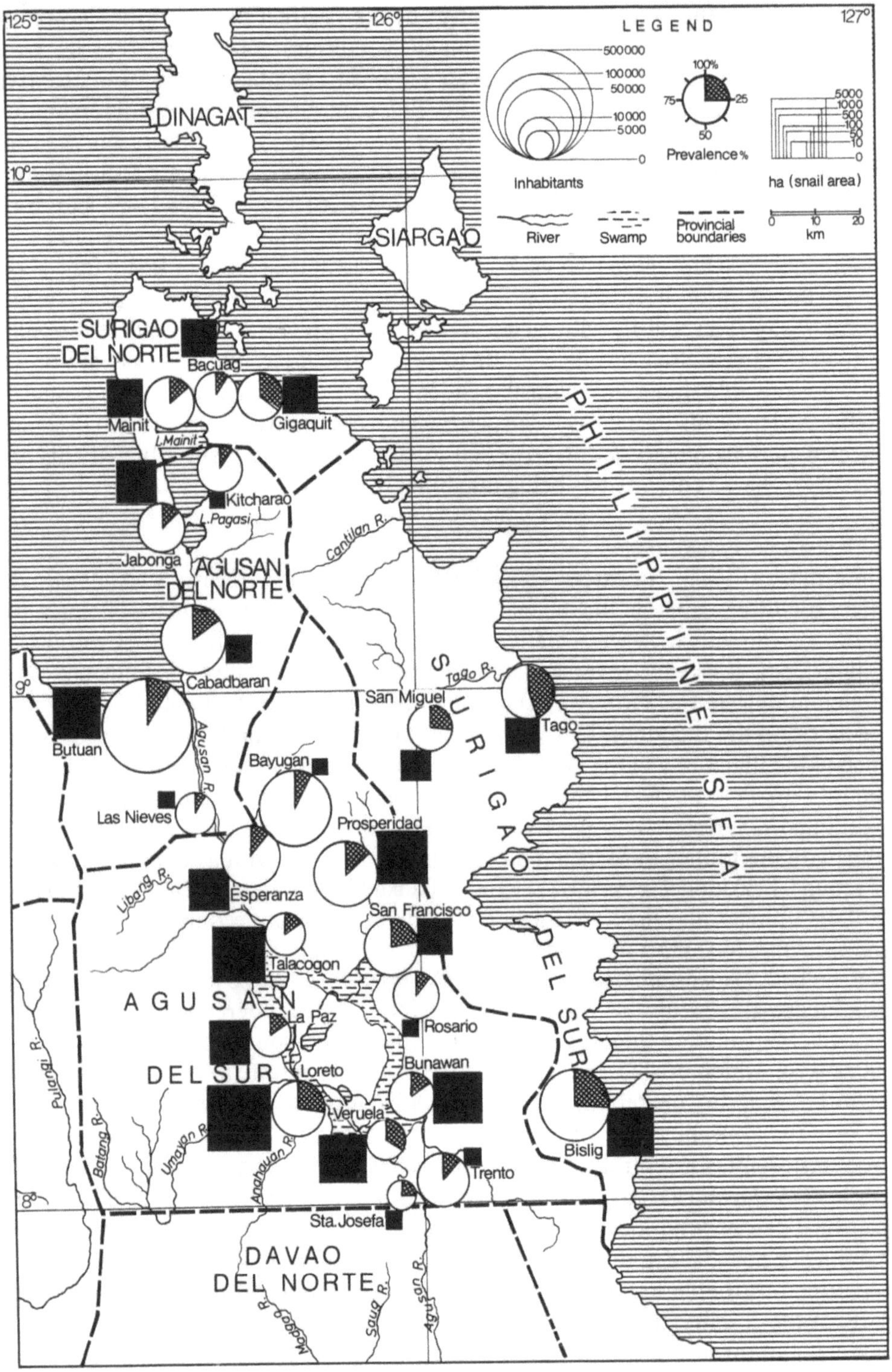

Fig. 66a. Schistosomiasis and snail areas in north-eastern Mindanao [Source: 705]

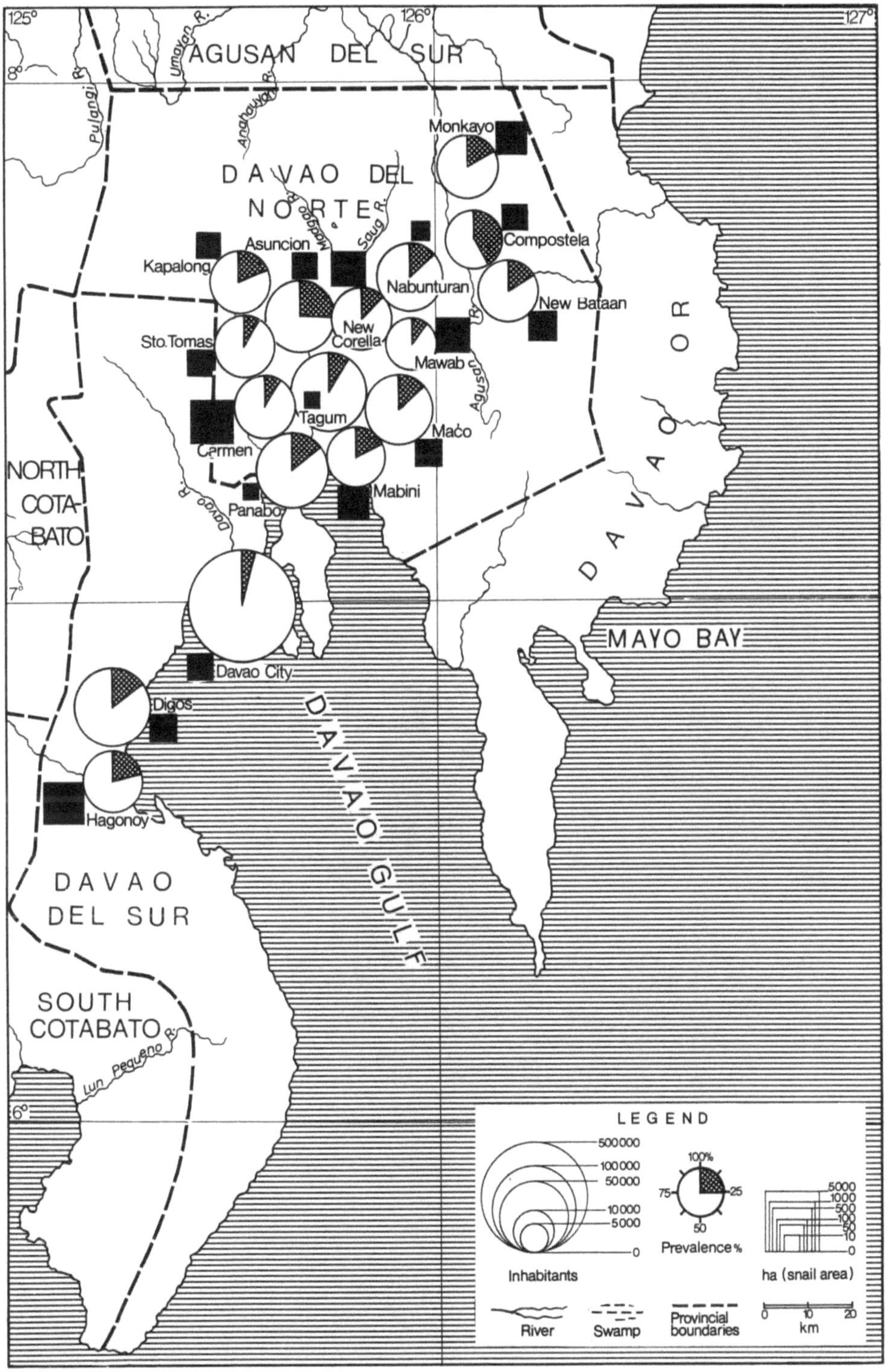

Fig. 66b. Schistosomiasis and snail areas in south-eastern Mindanao [Source: 705]

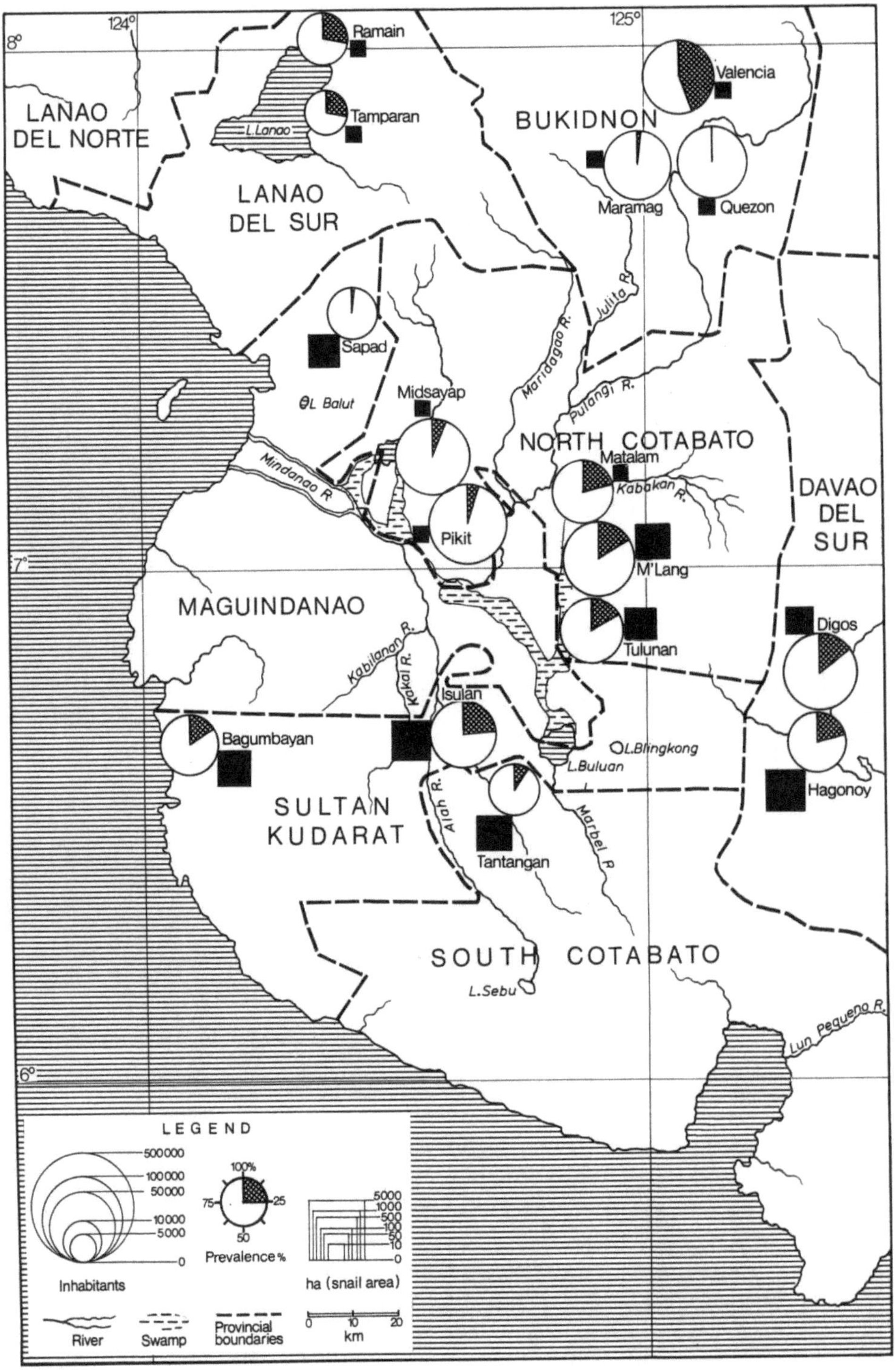

Fig. 66c. Schistosomiasis and snail areas in the Cotabato Basin, Mindanao [Source: 705]

projects. This net immigration amounted to a total of 523,037 migrants and their descendants over the period 1939–1960 alone, according to Wernstedt and Spencer's data [831]. The largest net out-migration [377, 521] next to Cebu, however, came from schistosome-endemic Leyte: i.e. bearing in mind the average prevalence of *S. japonicum* on Leyte as being 8%, an approximate 30,000 infected persons would have been included among the emigrants. Although it is impossible to estimate how many of these infected people would have chosen the Cotabato Basin as their destination, a very small number would have sufficed to bring about the formation of new schistosomiasis foci in all those places in which *O. h. quadrasi* persisted. Thus internal migration may constitute an important factor in the distribution of *S. japonicum.*

3.5.4.4 The Bukidnon Highland

Stretching between the volcanic massifs of the Katanglad and Katlatungan in the centre-north of Mindanao and the Central Highland in a north to south direction, the Bukidnon Highland includes altitudes of 600 to 700 m in its northern and central sections, dropping to below 300 m towards the Carmen Basin in the south. In the Bukidnon Province the distribution of schistosomiasis is restricted to this north-south axis. Besides the deep incisions caused by rivers which are unsuitable for colonization by *O. h. quadrasi,* there are extensive plateaux here which are drained by lesser waters and which – thanks to the slight gradient – thus ensure the persistence of this snail species and the creation of pre-conditions for the occurrence of *S. japonicum* infections. In 1945 the first diagnoses which were in accordance with this were made by Wright et al. [848, 849] for the Valencia area; further municipalities, including Malaybalay, Maramag and Quezon, as well as more recently even Impasugong, had to be added to the list of endemic areas at a later stage [200, 597, 705, 706] (Map Plate 3, Fig. 66d and Table III).

Schistosomiasis thus occurs above all in the area of the right-bank tributaries of the Pulangi River – as the Rio Grande del Mindanao is known in Bukidnon. One of the snail habitats typical of this area has been described by McMullen et al. [525] in the following terms: "In this section of the Pulangui River Valley ... forests remain fairly close to the river. One of these was visited in the area south of Managoc. It is shown on the topographical map as virgin forest but local inhabitants state it as second growth. A populous colony of *Oncomelania* was found in a slow, indistinctly delimited stream which originated from a spring at the base of the mountain". The authors followed up their observations with conclusions which explain the distribution pattern of *O. h. quadrasi:* "After seeing the snails in forest habitats, the pattern of distribution in the areas that have been under cultivation for a long time appears to be more clear. The habitat is apparently along stream origins on alluvial plains or at the bases of foothills in areas that make good rice land. Many of these spots are often difficult to manage as productive land when forest is cleared so they are left as wet, well-shaded "islands" dotting the farming area. These serve as the parent colonies and sources of *S. japonicum."*

The snail area in the endemic municipalities of Bukidnon was estimated to comprise altogether no more than 25 ha, 20 ha of which were situated in the Malaybalay area. However, the highest prevalence of schistosomiasis in man was given as 43.8% for Valencia (cf Table III). Excluding Impasugong, for which no data have so far been made available, the Schistosomiasis Control Council [705, 706] reckoned with more than 45,000 infected persons in a total population of 237,000 in the endemic area of

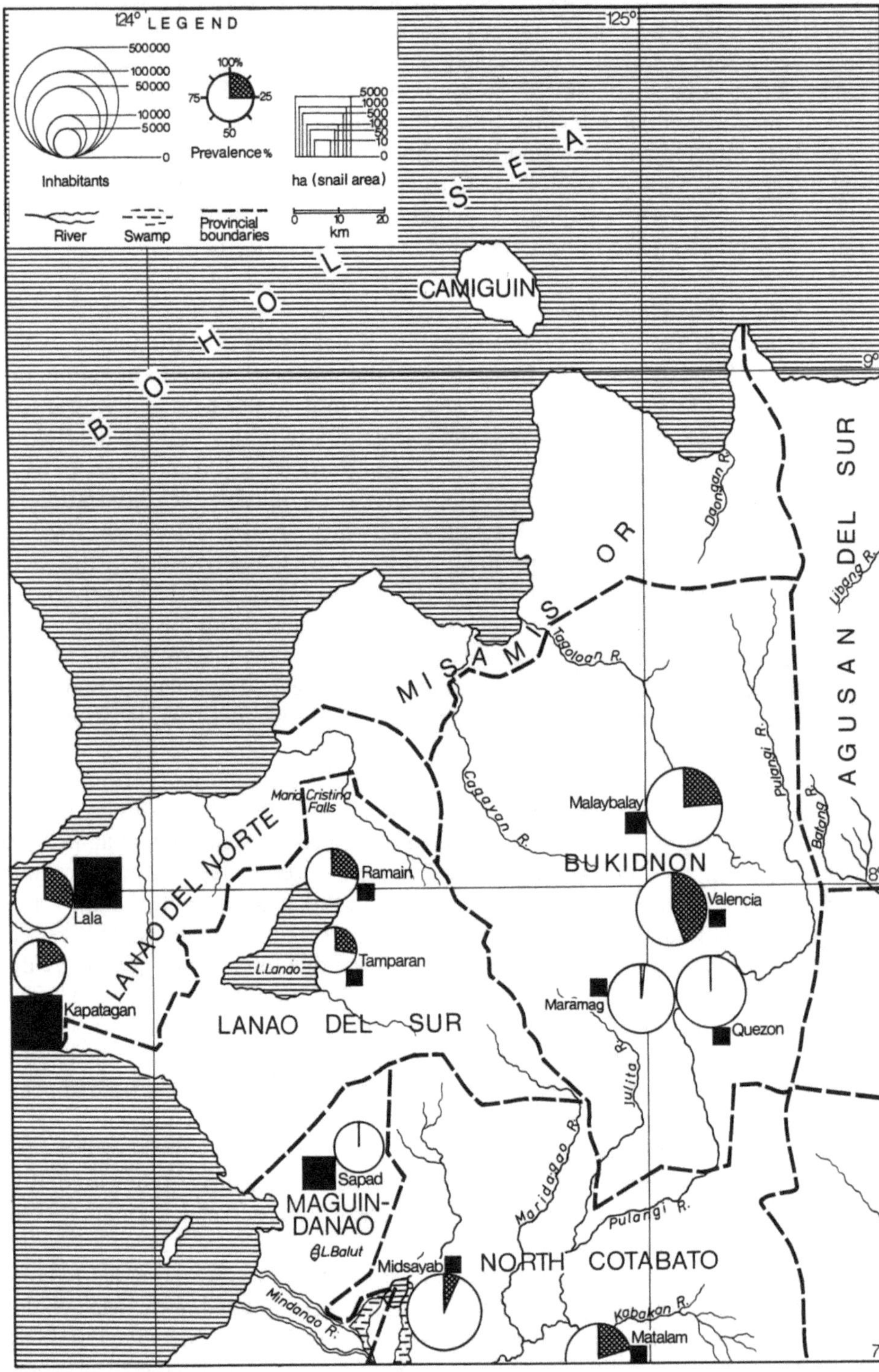

Fig. 66d. Schistosomiasis and snail areas on the Bukidnon-Lanao Plateau, Mindanao [Source: 705]

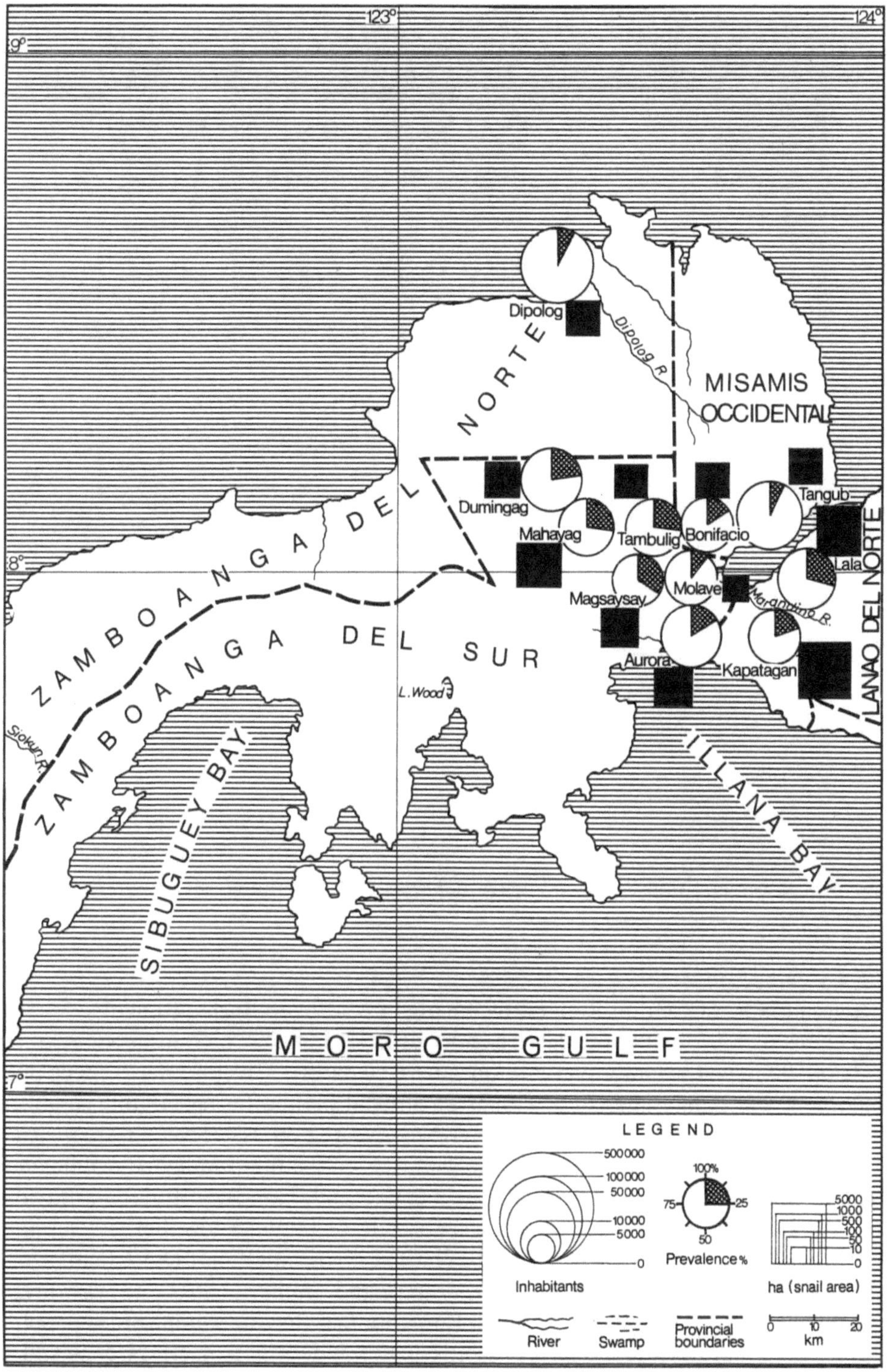

Fig. 66 e. Schistosomiasis and snail areas in the Panguil Bay and Dipolog areas, Mindanao [Source: 705]

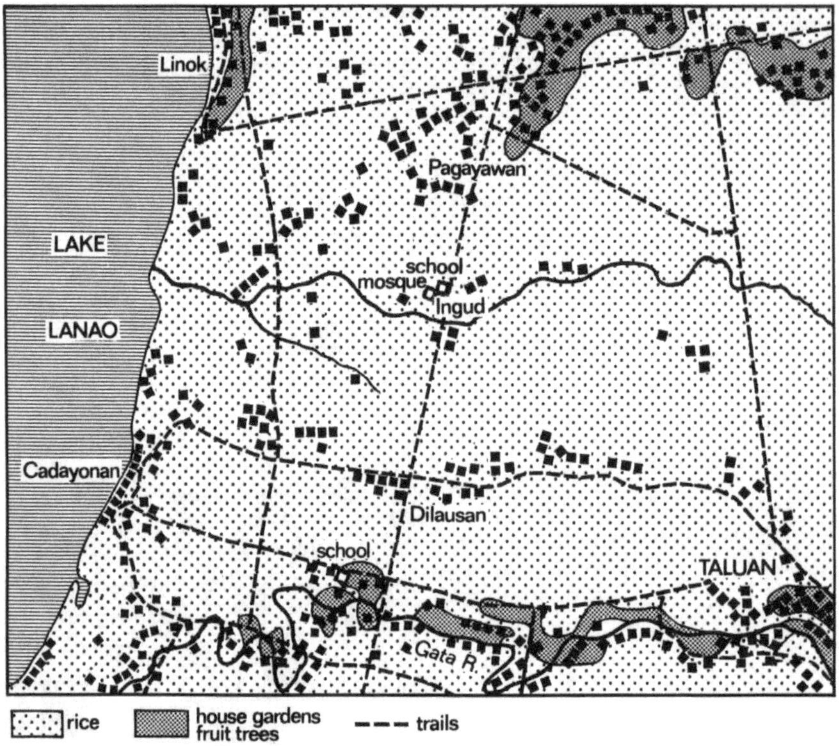

Fig. 67. Land use in Taluan [Source: 364]

the Bukidnon Highland. Case density has attained the mean value of 37.1 per km^2 (Table 29).

3.5.4.5 Lake Lanao

Between the West-Central Ranges (or the Iligan-Midsayap Buried Range), which rise to heights of 1300 m, and the eastern shore of Lake Lanao, which lies here at an altitude of 700 m and extends over 347 km^2, there is another schistosomiasis focus in the Bukidnon and Lanao Highland. This is an alluvial plain offering suitable conditions for *O. h. quadrasi.* Hausherr [364] described this region as a levelled area, rich in water and exceptionally well-suited to wet-rice cultivation; among the indigenous Maranao population it is known as basak, i.e. flat land. Tamparan-Lumba-a-Bayabao and Ramain, the two basaks on the eastern side of the lake together cover a mere 150 km^2 or so, a fact which explains the limited distribution on Lake Lanao. This is all the more the case since the northern, western and southern banks give way directly to hilly terrain, and therefore contain nothing but fast-flowing waters. Since people in the basaks tend to settle along the water-courses and close to their gardens and orchards [364] (cf Fig. 67), it may be assumed that contacts between people and the waters colonized by snails are frequent. They may in consequence form an important focus of infection in addition to the rice-fields.

Although the smaller creeks in the area of the eastern shore of Lake Lanao had to be taken as providing particularly suitable habitats for snails on account of their prolific vegetation and their low velocity of flow, the initial investigations undertaken on a limited scale failed to yield any positive results [849], nor has hepatomegaly been observed among children. Abbot [2] was the first to establish the presence of snails – albeit for the area on the northern banks of the lake; this habitat was later confirmed by Pesigan et al. [610], although after their previous discovery of *O. h. quadrasi* in Tatayan Creek near Tamparan. On the basis of stool examinations, these authors also succeeded in diagnosing infections with *S. japonicum* in 6 out of 78 schoolchildren at Tamparan and Taraka (Table III). Thus the area of the basaks on the eastern shore of Lake Lanao had clearly proved endemic. The 1979 annual report of the Schistosomiasis Control Council [705] finally registered this restricted focus at Tamparan and Ramain (cf Map Plate 3 and Fig. 66d) as sharing an average prevalence of 27.8% among its 31,508 inhabitants. Snail habitats were recorded as extending over 3.5 ha.

3.5.4.6 The Panguil Bay Area

A continuation of Iligan Bay in the south, Panguil Bay cuts deep into north-western Mindanao and, apart from a narrow, 16 km-wide land bridge, separates the Zamboanga Peninsula from the large block of the remaining Mindanao. The mangrove belt surrounding this peninsula gives way to a lowland which, following the valley of the Dipolo River upstream, continues west-north-west into the Zamboanga Peninsula as far as Dumingag (cf Map Plate 3 and Fig. 66e), thus covering an area of c. 750 km^2 in the provinces of Lanao del Norte, Zamboanga del Sur and Misamis Occidental, which meet here. In large part still covered by forest until after World War II, with just a sprinkling of agricultural land-use, the great increase in the numbers of the indigenous population, but above all the influx of immigrants from the other provinces of the Philippines as well, has led to a marked change. It is now largely an agricultural area in which the growing of rice and maize play an important role.

In this lowland around Panguil Bay Wright et al. [849] first showed in the municipality of Lala that infections of *S. japonicum* occur in man as well as in *O. h. quadrasi* collected there. A little later the municipalities of Kapatagan (Lanao del Norte), Aurora (Zamboanga del Sur) as well as Tangub and Bonifacio (Misamis Occidental) were also recognized as being endemic [594]. Mass examinations of 6,147 persons subsequently carried out in this area (including Molave) resulted in the diagnosis of *S. japonicum* infection in 698 (= 11.3%) of cases. The highest average prevalence of 42.7% had then been found at Aurora [588], although some barrios such as Durumawang (Lala) with more than 80% of the inhabitants infected [638] exceed this value considerably. The highest snail density (428/m^2) occurred at Bonifacio on Mitianan Creek. Now the entire plains around Panguil Bay, including the Dipolo Valley, are considered to be endemic. In the year 1979 the Schistosomiasis Control Council [705] reckoned with around 55,000 *S. japonicum* infections in a total population of c. 291,000 inhabitants in this area, i.e. with a prevalence of 18.9%. At 73.3/km^2 the case density thus calculated is almost twice as high as the mean value for all the other endemic areas in the Philippines. The snail area extended over almost 5,500 ha, two thirds of which belonged to the municipalities of Kapatagan (c. 2,400 ha) and Lala (c. 1,200 ha) in Lanao del Norte alone.

The present extent of the schistosomiasis problem in the Panguil Bay Area is closely connected with the population increase after World War II. Hausherr [364] has given a detailed description of this population development with special reference to the Lanao Provinces. The country as a whole registered a population growth of 3.06% per annum during the period 1948–1960, whereas Lanao del Norte had an increase of 8.8% per year – i.e. the population of this province has more than doubled over a period of twelve years. The disproportionate increase is not due to a particularly high birth-rate but to particularly heavy immigration, chiefly from the Visayas. In the census of the year 1960, 42% of the 270,603 inhabitants of Lanao del Norte claimed another Philippine province as their place of origin. (50,919 immigrants came from Cebu alone; on the other hand, the highly endemic islands of Leyte and Samar were given as only 3,876 and 550 persons.)

Forest clearances for agricultural land use resulting from this population pressure took people into the habitats of – and thus into direct contact with – *O. h. quadrasi*, a species which is primarily a forest dweller. McMullen et al. [525] assume that schistosomiasis had been endemic in the Panguil Bay area well before the onset of that wave of immigrants: "In the Panguil Bay and Compostela area some information that seems to be reliable indicates that schistosomiasis was present among aboriginal people and recent settlers got it shortly after arrival." The high prevalence rates noted in this area immediately after World War II also seem to support this hypothesis.

3.5.4.7 The Dipolog Focus

During their search for schistosomiasis incidence in Mindanao, Pesigan, Pangilinan and Sarmiento [610] also carried out some sample surveys in Dapitan and Dipolog in Zamboanga del Norte. They succeeded in proving the occurrence of infections with *S. japonicum* in two sitios of Barrio Nipaan, which is part of Dipolog, and they found *O. h. quadrasi* there as well. Later investigations [597] confirmed Dipolog as an endemic area with a prevalence of 5.6%. The Schistosomiasis Control Council [705] reckoned that in 1979, 4,762 (= 7.7%) of the resident population would be infected. The snail area was estimated to be 150 ha. Apart from this, *S. japonicum* is said to occur at Polanco, south of Dipolog. However, there are so far no data which would permit a more detailed characterization of this evidently unified focus of Dipolog-Polanco.

3.5.5 Summary Consideration of the Endemic Areas of *Schistosoma japonicum* in Regard to Its Inter-relationships With Population Density, Case Density, Internal Migration and the Problem of Introduction

Carney et al. [197] had attributed the marked drop in prevalence rates for Bohol from 30.5% in 1958 to the 5.4% shown by their own checking, to the heavy population pressure; they suggested that it forced the inhabitants to open up all available land resources for agriculture and intensive cultivation (3.5.3.3). As *O. h. quadrasi* can only persist and breed in a largely undisturbed environment, progressive intensification of agriculture would lead to a steadily increasing reduction in snail biotopes, together with the knock-on consequences of a decrease in the prevalence of schistosomiasis in man. The question therefore arises of whether the data of any endemic area indicates such a correlation between the population density on one side and the prevalence and case density on the other. The interpretation of the data published by the Schistosomiasis Control Council [706] does not indicate such a development, however. The only positive relationship to be found was between population density and case density (cf Table 29). It is possible that the situation observed by Carney et al. [197] in the Bohol endemic area presents an exception. In the case of Leyte, which is very densely settled as well, there is so far no indication whatsoever concerning such a reduction in prevalence as a result of population growth.

Nevertheless, it must be assumed that the problems resulting from the considerable increase in population in the Philippines have not been without an effect upon the frequency and distribution of schistosomiasis. In this context the internal migration processes, which tend to move in two directions, ought to be mentioned: the greater Metro Manila area and Mindanao. Though of significance for the spread of various diseases, the strong influx of people into the Manila area does not affect schistosomiasis since, thanks to climatic conditions, *Oncomelania* does not occur in Central Luzon. The situation in Mindanao is quite different. A sparsely settled immigration area with endemic schistosomiasis, it has received several million people from other parts of the country, especially the Visayas, and this has been mostly since World War II. These immigration processes and their possible effects on schistosomiasis distribution are therefore to be described in greater detail by examining the population development in the Eastern Visayas and on Mindanao.

The imbalance of the population growth in the Eastern Visayas and on Mindanao is clearly discernible when viewed in comparison with the overall development in the Philippines. In the period from 1918 to 1975 the total population increased four-fold, from 10.3 to 42.1 million, whereas the number of inhabitants on Mindanao multiplied by 9.4 during the same time period, and those of Leyte (del Norte) and Samar by 2.7. Contrary to expectations based on the overall development, the population of Mindanao had not grown to 3,716,000 but to 8,592,000 inhabitants, whereas Leyte and Samar only reached 2,324,000 instead of the expected 3,492,000 inhabitants. Whilst characterized by a linear rise in the Eastern Visayas the population development on Mindanao shows a distinctly exponential increase (Fig. 68a). Since an approximately equal birth-rate may be assumed for both these areas, the calculations indicate significant regrouping as a result of internal migration: Mindanao shows a surplus of 4,876,000 inhabitants, who must have immigrated, for the period 1918–1975, whereas the Eastern Visayas have a deficit of 1,168,000 persons due to emigration. That this estimate keeps within realistic limits can be confirmed by the data supplied by Wernstedt and Spencer [831] who quoted a net out-migration of approximately 434,000 persons for Leyte and Samar, and a net in-migration of c. 1.8 million for Mindanao during the period 1939 to 1960. The share of the Eastern Visayas in the total internal migration (emigration) was 14%, that of Mindanao 59% (immigration) (Tables 30 and 31).

This extraordinarily great regrouping of the population can, however, not have been without effect on the frequency and distribution of schistosomiasis. If, in accordance with the data of the Schistosomiasis Control Council [705, 706], an average prevalence of 8.4% is assumed for the Eastern Visayas, the emigrants estimated to be 1,168,000 must have had about 98,000 people infected with *S. japonicum* among them, 59% (= 57,800) of whom settled in Mindanao. With the main emigration movement from Leyte and Samar (c. 96% of 1,168,000 persons) only setting in in 1948, an estimated 55,500 people who were infected with blood flukes moved to Mindanao between 1948 and 1975. Those provinces of Mindanao that received the largest quota of immigrants (Table 31) also happen to have climatic conditions which ensure the oc-

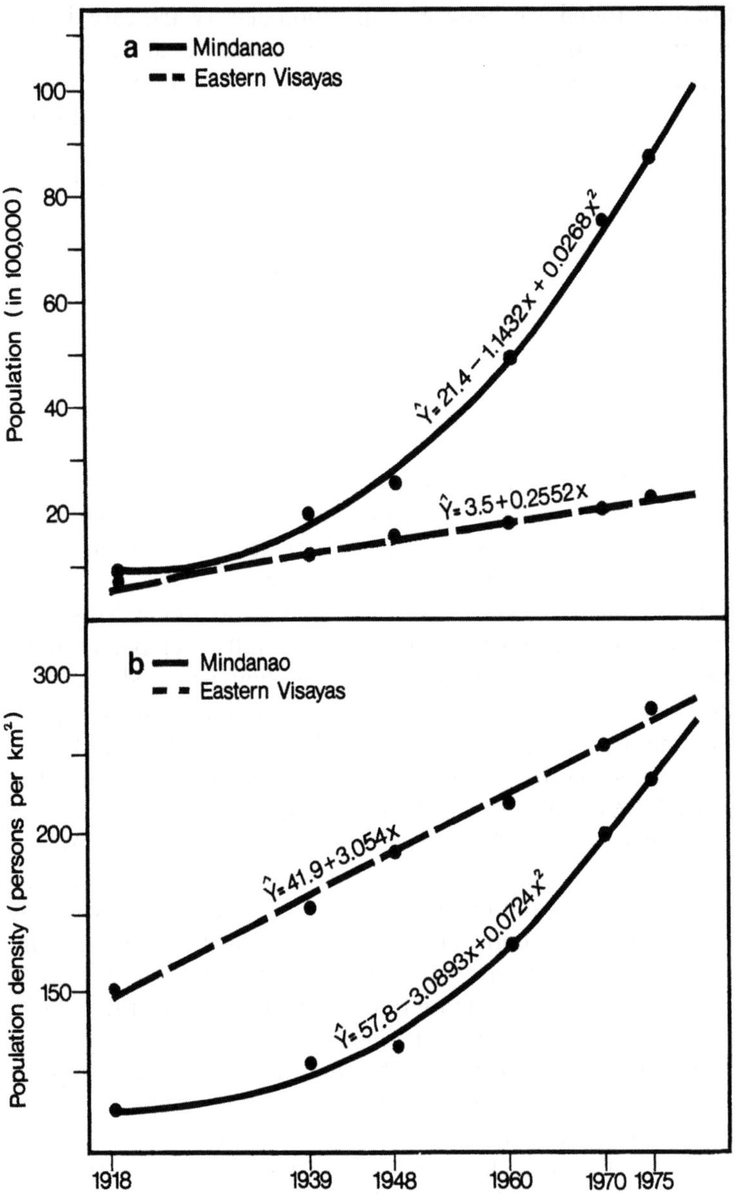

Fig. 68 a, b. Population growth (**a**) and increase of population density (**b**) in Mindanao and the Eastern Visayas, 1918–1975 [Source: 1980 Philippine Statistical Yearbook]

curence of *O. h. quadrasi.* Forest clearances in the search for new land were bound to lead the immigrants into the vicinity of snail habitats, so that the parasite cycle was set in motion in areas that had previously been free from schistosomiasis. It must therefore be assumed that the present pattern of schistosomiasis distribution consists of two at least partially overlapping components: on the one hand the primary foci already present, and on the other the endemic areas created by fresh importation.

As far as a prognosis of future developments is concerned, one factor – that of population density – suggests that the significance of Mindanao as an immigration area for the Eastern Visayas might become less in the future. In Leyte and Samar population number and population density rise in a linear fashion, whereas in Mindanao the increase is exponential (Fig. 68 b).

In relation to the area classified as "alienable and disposable" in the 1980 Philippine Statistical Yearbook, the population density in the Visayas increased from 102.5 to $278.2/km^2$ over the period 1918 to 1975, but in Mindanao this was from 24.6 to $232.1/km^2$. If the trend that prevailed until 1975 were to continue, population densities in both these areas would have become the same by 1982/83 and be followed by a gradual decline in one-sided migration.

However, allowances must be made for the fact that besides these internal migrations other events might have contributed to the further distribution of schistosomiasis in the Philippines. A major force here was the Pacific War which took Filipinos as soldiers or prisoners-of-war from endemic to non-endemic areas and vice-versa. The movements cannot be quantified, however, so that their significance for the schistosomiasis problem eludes assessment.

The question arises in this connection of whether an importation of *S. japonicum* from Japan or China could have taken place in recent times. Okabe [576] estimated the number of schistosomiasis cases in Japan as being 100,000, i.e. a prevalence of 0.1% in relation to the total population. Apart from modern Japanese mass tourism, which does not extend to those areas of the Philippines in which *O. h. quadrasi* occurs, there have been two events which have brought larger numbers of Japanese into actually or potentially endemic areas: the intensification of abaca cultivation in Davao (cf 2.2.3.3) to a lesser degree, and the occupation of the Philippines during the Pacific War to a greater extent.

In 1939 there were almost 18,000 Japanese in the town of Davao alone, 18 of whom would have been *S. japonicum* carriers according to formal calculations. Since Davao was one of the endemic areas (cf 3.5.4.1 and Table III), which implies the occurrence of *O. h. quadrasi* there, the conditions for the establishment of flukes from Japan were in theory provided. This applied to a much greater degree to all the Philippine endemic areas where c. 80,000 Japanese troops had stayed during the war [531]. In the light of our present knowledge it is unlikely, however, that such an importation has taken place, since experimental investigations had shown *O. h. quadrasi* to be refractory to the Japanese sub-species of *S. japonicum* [225, 271], and *O. h. nosophora*, the specific intermediate host of this parasite sub-species, does not occur in the Philippines.

The most significant minority in the Philippines, however, are the Chinese, and *S. japonicum* is indeed endemic in their country of origin. W. H. Wright [847] estimated the number of people in China who are infected with *S. japonicum* to be 32,777,630, the equivalent of an average prevalence of 4.7%. It follows from Kolb's work on Chinese migration [440] that there was already a steady stream of immigrants from southern China during Spanish colonial times; between 1600 and 1850 the minimum number of Chinese living in the Philippines was 5,000–6,000, their maximum 20,000–30,000. For the year 1886 finally, the Spanish colonial administration estimated their number to be 66,000, or 1.1% of the total population. Both before and after the Second World War there were more than 40,000 Chinese in the Philippines. Distinctly higher values were being registered for the time after World War II: c. 121,700 for 1948, and c. 181,600 for 1960. However, the great majority of them had settled in Luzon – as may be noted from the map included by Kolb [440] – and only 10,000–15,000 Chinese lived in the endemic areas of the Philippines in 1960.

The available data do not, however, permit a break-down of numbers according to direct immigrants and those who were born in the Philippines as descendants of immigrants and do not therefore figure as a source for importation, nor is it known whether schistosomiasis was endemic in the areas of origin of these immigrants or not, for the vast majority came from southern Fukien (c. 85%) and southern Kwangtung (c. 15%). Although the seven endemic "hsiens" quoted from these two provinces by Mao [509] are not identical with those from which the immigrants almost exclusively originated [cf 440], schistosomiasis – as shown in the map by Mao and Shao [511] – was evidently occurring widely in the coastal district of southern Fukien and in central Kwangtung.

Thanks to the large although not closely ascertainable number of Chinese who have entered the Philippines during the course of three and a half centuries, it may be assumed that there have been persons with *S. japonicum* infections among the immigrants. Nonetheless, it must be regarded as improbable that such persons contributed to the further distribution of this helminthiasis in their country of adoption, since the Chinese sub-species of the parasite is adapted to *Oncomelania hupensis hupensis* as intermediate host and is under experimental conditions not transferable to *O. h. quadrasi* [225, 271]. The present state of knowledge therefore excludes the importation of schistosomiasis from China to the Philippines in more recent times.

3.6 Synopsis: Geofactors as Preconditions for Schistosomiasis Distribution in the Philippines

It has been shown in the preceding sections (3.2–3.3) that there are many ways in *Schistosoma japonicum* infections in which geofactors impinge on the chain of infections: man–intermediate host–man. Their influence in this is directed solely to the free-living stages of the causal organism and when in the intermediate host, and to the intermediate host itself. Only in conditions where none of the orographic, edaphic, hydrological or climatic factors exceed the limits of tolerance common to all the miracidia, sporocysts, cercariae and intermediate hosts, is the cycle of the parasite able to continue and be completed outside man; only in these conditions can the persistence of *Schistosoma japonicum* infections in one area be achieved.

An important precondition for this persistence is the presence of perennial, slow-moving waters in the vicinity of which the amphibious *O. h. quadrasi* can persist, which guarantees moreover that neither the worm eggs nor the hatched miracidia and cercariae are washed away as a result of increased flow velocity. However, perennial, slow-moving waters only occur in places with high precipitation throughout the year and a minimal terrain gradient. Those areas whose seasonal climate is characterized by alternate rainy and dry seasons and by a hilly or mountainous topography, are therefore not suited to the endemic occurrence of *S. japonicum.* These reasons alone suffice to restrict the distribution of schistosomiasis in the Philippines to the plains in the south of the Bicol Peninsula, in Oriental Mindoro, the Eastern Visayas, Bohol and Mindanao (cf Map Plate 3).

Another limiting factor is presented by the temperature conditions, which affect the hatching of the miracidia, the development of larvae in the intermediate host, the release and life span of the cercariae and the intermediate host. The common optimum for these occurs when air and water temperatures remain in the range of c. 25–30 °C, such as are found in the endemic areas mentioned above. Deep waters and shallow rice-fields, which fall short of or exceed this temperature range are therefore neither suitable for snail habitats nor as infecting waters because the larval stages of the schistosomes can only live in them for a short period, or become inactivated.

The limits of tolerance for the free-living stages of the parasite, and for the intermediate host with regard to such separate components as the quality of water and soil, are not yet sufficiently known. Degree of turbidity, pH value and salt content of the water are important limiting factors. Waters with a turbidity degree in excess of 10 ppm SiO_2 equivalent are thus less suited to providing snail biotopes than those with clear water. In addition *O. h. quadrasi* prefers a slightly alkaline milieu (pH 7.4–7.6), which presents the optimum for miracidia and cercariae. In values even a little below the neutral point hatching rates of miracidia and cercariae release rates are already markedly lower, just as a higher salt content in the water is not tolerated by snails, miracidia or cercariae. In coastal areas which are prone to flooding with sea-water, and in the lower reaches of rivers which are subject to tides, *S. japonicum* infections cannot persist.

In summary the following hierarchy of geofactors emerges:

- The distribution of precipitation falling through the course of a year determines the overall spatial extent to which *O. h. quadrasi,* and thus schistosomiasis, can occur in the Philippines.
- The orography together with the aquatic plant cover determines the flow velocity of the waters which restrict the distribution area within the overall spatial extent to the plains.
- Temperature conditions, as well as the quality of the water and soil (including pH value and salt content), further restrict the potential areas of distribution within the plains to those parts the conditions of which are tolerated by both *O. h. quadrasi* and the free-living stages of the parasites.

3.7 The Control of Schistosomiasis

In face of the fact that there are in the Philippines around 4 million people living in the endemic areas, 630,000 of whom are actually infected with *S. japonicum,* the control of this helminthiasis presents a task of the highest priority. Legislation repeatedly recognised this view of the problems in the past by passing the Republic Acts numbers 299, 2011 and 4359, and the Presidental Decree No. 893 (cf 3.1). Today the Schistosomiasis Control Council is in charge of the co-ordination of all the control measures. Besides this there is the Schistosomiasis Control and Research Service of the Ministry of Health as the body responsible for the planning and implementation of the control programmes, affiliated to which are the research centre at Palo and 19 control teams in endemic provinces. Research work on the control of schistosomiasis so far undertaken in the Philippines, the transfer of the results to practice, the planning and application of control measures have been repeatedly reported upon in general reviews, as well as in original publications which apply themselves to selected aspects only [cf for example 4, 97–100, 103, 196, 268, 293, 312, 325, 331, 332, 360, 375,

459, 521, 567, 569, 598, 600–603, 607, 681–683, 685–695, 703–706, 762, 763, 854–859].

In the following section attention will be drawn to the most important aspects of planning and application which are illustrative of the difficulties encountered in schistosomiasis control in the Philippines. In principle attempts are being made to apply all the measures, simultaneously if at all possible, so as to disrupt the parasite cycle. This spans a broad spectrum:

- the environmental hygiene component aims at proper disposal of faeces,
- the application of ovicides serves to destroy schistosome eggs passed out in the stool,
- the application of moluscicides and of ecological and biological measures pursues the aim of destroying the intermediate snail hosts,
- the production of cercaricidal substances is directed towards the destruction of the infective stages,
- besides the obvious medical task of healing the patient, chemotherapeutic treatment of infected persons is directed at destroying the parasites in order to stop the passing of eggs, and
- publicity is directed at a comprehensive health education of the population and especially at a change in their behaviour in regard to water contact and in their habits of defecation.

Measures of environmental hygiene are an essential component of the Schistosomiasis Control Programme in the Philippines. They are predicated on three things: the existence of toilets, a hygienically correct disposal and treatment of faeces (such as the application of ovicides), and an enlightened population prepared to alter its defecation habits. Already more than thirty years ago Pesigan had deplored the absence of toilets and the habit of indiscriminate defecation in the open. Even nowadays satisfactory latrines are available on average to no more than 33% of the population [706]. Although Pesigan et al. [607], following house-to-house visits in Barrio Malirong in Palo (Leyte), had persuaded 119 out of 142 families to construct latrines within a fortnight, and thus to achieve a marked reduction in the infection rates in the nearby snail colony, these measures were not crowned by significant outcome in the long term. Even in places with a large number of toilets, only some members of the population made regular use of them – e.g. 41.5% in Mayorga, Sta Fe, Tolosa and Javier in Leyte [769]. According to the inhabitants of Sta Rosa, Barugo (Leyte) the reason for the intermittent use is said to be the lack of water in toilets intended for flushing [277]. Accordingly, not only will further enlightenment be needed, but an improvement in technical conditions will also be required if a fundamental change is to be brought about.

Since schistosomiasis japonica is an anthropozoonosis, even the hygienically correct disposal of all human faeces can only lead to a reduction in transmission, but not to its total elimination. With a proportion of 22–60% animal final hosts are involved in maintaining the parasite cycle, and thus in the persistence of this helminthiasis in the Philippine endemic areas [357, 606] (cf 3.2.1). Above all the high prevalence rates among rats – which, depending on the season, may rise to 100% in places [433–435] – contribute to the continuous contamination of the waters. One can predict to only a limited degree the extent to which the risk of infection would be reduced for man should he himself cease of pass eggs. It should, however, be a percentage in excess of his share in the transmission index, since his failure to contaminate snail habitats would reduce the risk of infection in the rat population and thereby lessen their significance as transmission potential.

The decimation and, if possible the elimination of the intermediate host snail presents a further – indeed, the most important – starting point for the interruption of the parasite cycle. It is the only measure which, if it were successful, would lead to the complete disappearance of schistosomiasis in the Philippines, since it would also result in halting the chain of infection running from reservoir host to snail to reservoir host. Originally high hopes for the application of molluscicides gave way to the realization that the control of *O. h. quadrasi* with such measures would only lead to limited success, due to the variety of habitats and its amphibian way of life. In practice it is not possible to take molluscicides to all the places inhabited by snails. Difficulties also arise in the administration of exact doses designed for lethal concentrations even on the fringes of running waters with irregular boundaries. In addition to this, molluscicides are expensive. "With limited funds, it will be best to apply molluscicides as a terminal measure when the snails are already confined to the drainage and irrigation canals and resistant pockets" [763].

Snail control in the Philippines has therefore got to include predominantly those measures which cause irreversible damage to snail habitats. In the Technical Guide for Schistosomiasis Control [763] the following catalogue appears:

- vegetation removal
- drainage
  - stream channelization
  - seepage control
  - diversion and intercepting channels
  - canal linings
  - drainage and irrigation schemes
- combined vegetation removal and drainage
- earth filling
- flooding and ponding
- improved rice culture
- chemical control
- biological control

However, these control measures are connected with considerable costs and are untertaken with great inputs of labour. Thus, even the removal of the vegetation of a river repeated 43 times within four and a half years did not result in complete elimination of *O. h. quadrasi* [360]. It was only with combined vegetation removal and drainage that a 96% reduction of a snail population succeeded [99]. The success of such measures always depends on continuous repetition as well. On Leyte, for instance, a reduction in the number of snails to almost half the previous level was achieved by twice removing the vegetation on the water's edge, but failure to follow this up with further interference in the river bank area led to the population density returning to its original level within eight months [607].

According to investigations carried out so far, measures intended for snail control, and especially their combined and simultaneous application, may well lead to a considerable reduction in the snail population and even possibly to total disappearance. Santos [683], quoting the example of Palo, showed that a decimation of the number of snails can indeed lead to a decrease in prevalence rates in man. Comprising a total area of more than 46,000 ha [705, 706], divided up into 22 provinces and colonized by snails, high costs alone decide the planned extermination of *O. h. quadrasi* in the Philippines which can be realized only as a long-term project and with international aid. The same applies to the chemo-therapeutic treatment with the very expensive schistosomicidal anthelmintics, amongst which praziquantel now represents the drug of choice.

On the Philippine side the setting up of the Schistosomiasis Control Council created the framework conditions for the execution of control programmes. The authorities and bodies (cf Appendix E) represented on the Schistosomiasis Control Council are called upon to undertake the following tasks:

The *Ministry of Health* is the competent authority for the "health" components in the overall programme. It provides the Schistosomiasis Control and Research Service (SCRS) with, amongst other matters, the task of collecting data and information, the provision of technical assistance and the holding of training seminars. The SCRS is, moreover, responsible for questions of co-ordination, for the application of molluscicides, for diagnostic problems, for research projects concerning the immunology of schistosomiasis and the ecology of *O. h. quadrasi*, and for the health education of the population and the therapeutic application of pharmaca as well. The Ministry of Health is, furthermore, set above the rural health units in all the municipalities; these are the small treatment and information units charged with the task of locating, registering and treating cases, and of promoting hygiene and health education measures.

The *Ministry of Public Works* is responsible for carrying out drainage work, flood protection and safeguarding water supplies through the Bureau of Water Resources. In addition it is required to prevent the construction of badly planned dams.

The *National Irrigation Administration's* task is the provision of adequate drainage in irrigation schemes, of material and technical assistance in the construction of irrigation and drainage schemes for the realization of improved methods of cultivation in agriculture, as well as the co-ordination in the planning of new irrigation schemes and the regular maintenance of existing ones.

The *Ministry of Local Government and Community Development* is to co-ordinate plans for self-help projects at barangay level with the SCRS, and to contribute to improvements in the construction of toilets and drainage, and to the control of stray animals.

The *Ministry of Agriculture* provides technical aid to farmers in adopting improved methods of cultivation, advises them on the utilization of reclaimed snail-infested areas, and promotes agricultural pilot projects.

The *Ministry of Education and Culture* is charged with making use of meetings of parent-teacher associations and other school conferences in endemic areas as occasions for spreading information about schistosomiasis, to "include schistosomiasis in science and health subjects", and to "assist in environmental sanitation campaigns and in other community development projects".

The *Ministry of Social Services and Development* has to grant aid to volunteers recruited for schistosomiasis control, to support campaigns for health education, and to obtain financial means from non-governmental sources.

The *National Economic and Development Authority* co-operates in the integration of the control programme into the national development plan and in the supervision and evaluation of activities which are of significance for the campaign.

The *Philippine Medical Association* acts in an advisory capacity in the implementation and completion of research and control programmes, and actively supports the execution of diagnostic and therapeutic measures at a local level.

The *Ministry of the Budget* has the task of granting support in logistic and budgetary questions, as well as assisting the Schistosomiasis Control Council in administrative matters.

In order to make the work of co-ordination within the framework of the Schistosomiasis Control Council more effective, four sub-committees had been set up at an earlier date, each having its areas of responsibility and tasks firmly outlined. These are: drainage/irrigation infrastructure, environmental sanitation infrastructure, health education, and evaluation and health services. In 1980 the technical staff of the Schistosomiasis Control Council, of which the ministries and bodies mentioned above are members, worked out a 5-year integrated national programme for the control of schistosomiasis in the Philippines, which summarized the planned measures and details of the costs for all the endemic provinces (Table 32). The total costs were estimated at 3.6 billion pesos, the proportion for health measures being 169,618,892 pesos and that for measures aimed at improving the infrastructure a further 3,449,700,716 pesos. Besides this national control programme there were (or are, or were intended to be) irrigation and other development projects which form an integral part of the measures directed against schistosomiasis, or into which a control component is integrated. In the majority of the loan-assisted projects, the World Bank, the Asian Development Bank, and the International Rural Development Bank contributed to their financing. The projects in question were as follows:

- Mindoro Integrated Rural Development Project (MIRDP) in Oriental Mindoro
- National Irrigation Systems Improvement Project-I (NISIP-I) in Leyte
- Second Agusan Irrigation Project (AGUSAN II) in Agusan del Norte and Agusan del Sur
- Second Davao del Norte Irrigation Project (DAVAO II) in Davao del Norte
- National Irrigation System Improvement Project-II (NISIP-II) in Davao del Sur, North Cotabato, Zamboanga del Sur and Misamis Occidental
- Samar Integrated Rural Development Project (SIRDP) in Eastern, Northern and Western Samar
- Bohol Integrated Rural Development Project (BOHOL IRDP) in Bohol
- Bukidnon Integrated Rural Development Project (BUKIDNON IRDP) in Bukidnon
- Allah River Irrigation Project (ALLAH RIP) in South Cotabato
- Tago River Irrigation Project (TAGO RIP) in Surigao del Sur
- Manat River Irrigation Project (MANAT RIP) in Davao del Norte
- 5-year Schistosomiasis Control Project in Sorsogon
- National Schistosomiasis Related Drainage and River Control Project of the Bureau of Public Works in all endemic provinces.

Financing the measures required for an effective control of schistosomiasis continues to be a big problem. Since the first grant of 200,000 pesos was made in 1948 (Republic Act No. 299, appendix A), more and more means had to be made available without them ever being sufficient for bringing about a change. In 1954 McMullen et al. [525] had calculated that one million pesos a year were required for a control programme estimated to run for fifteen years. During that time this sum was, however, never reached. In 1964/65, for example, 453,144 pesos a year were spent on schistosomiasis control [602]. Eight years later this amount had grown to 863,000 pesos; in 1976, however, expenditure for this purpose had already reached US $ 1,088,973 (c. 8,170,000 pesos), or about 0.9% of the health budget [399a]. In 1980 the Schistosomiasis Control Council alone received support amounting to 1,277,000 pesos; provision had been made for an annual rate of increase of 6.95% of this amount. Compared with the total costs of the Integrated National Program for the Control of Schistosomiasis, which had been put at 3.6 billion pesos for a running period of 5 years, the sums which had previously been made available were insufficient for an efficient control of schistosomiasis and could only offer limited results. Nonetheless, the pilot project at Palo, Leyte, together with other institutions, was able to develop scientific and technical pre-requisites for the drawing up of a comprehensive control programme. It remains to be seen whether it has been possible to carry out this programme in the way that had been intended, and if so whether it will lead to the expected results.

4 Food-Borne Helminthiases

A substantial number of heteroxenous helminth species that live in man as parasites are programmed to complete their development to the stage of infective larvae in animals or on plants which are used as foodstuffs for man. However, a human infection can only take place if such animals or plants are consumed in a raw or insufficiently cooked (i.e. heated) state. Apart from geofactors affecting the helminth larvae when free-living as well as during the developmental period in the intermediate hosts, it is above all the feeding habits of the people themselves – specific to every region – which determine the geographical distribution of these helminthic infections. In every case in which man functions as the sole final host his defecation habits, too, are of great significance for the persistence of food-borne helminthiases. This is because an appropriate contamination of the intermediate host's milieu alone enables the parasite cycle to continue. In euryxenous helminths with a broad spectrum of final hosts on the other hand, this factor does not play a role: the final animal host then contaminates the intermediate host's milieu.

An analysis of the geomedical situation will, however, also have to take into account whether the worm species concerned is a di- or triheteroxenous helminth – i.e. whether one or two intermediate hosts are involved in its development cycle apart from the final host, and whether there are additional, paratenic hosts. In consequence, the conditions for the persistence of particular food-borne helminthiases may be structures of differing complexity. This also applies to the situation in the Philippines. The following factors play a role in the dynamics of transmission and in the persistence of these worm infections:

- geofactors determining the distribution of the various final host species
- population density, population fluctuations and mortality of the final host species
- prevalence and intensity of infection among the final host species
- reproductive capacity and mortality of the parasite in the final hosts
- contamination of the intermediate host's milieu (defecation habits; among lung flukes of the genus *Paragonimus* there is also the passing-out of eggs in the sputum)
- geofactors which affect those stages of the helminths (eggs, larvae) which, having been passed out by the final host, are now in the open
- geofactors which determine the distribution of the first or sole intermediate hosts
- population density, population fluctuations and mortality of the (1st) intermediate host
- prevalence and intensity of infection among the (1st) intermediate host
- geofactors which influence the development of helminth larvae in the (1st) intermediate host

- reproductive capacity and mortality of parasites in the (1st) intermediate host
- among diheteroxenous helminths: eating habits of man (and other final hosts)

- among triheteroxenous helminths: geofactors which affect helminth larvae after they have left the first intermediate host for the open
- geofactors which determine the distribution of second intermediate hosts
- population density, population fluctuations and mortality of second intermediate hosts
- prevalence and intensity of infection of second intermediate hosts
- geofactors influencing the development of helminth larvae in second intermediate hosts
- mortality of parasites in second intermediate hosts
- eating habits of man (and other final hosts)

- geofactors determining the distribution of paratenic hosts
- population density, population fluctuation and paratenic host mortality
- prevalence and intensity of infection among the paratenic hosts
- geofactors which affect helminth larvae in paratenic hosts
- eating habits of man (and other final hosts)

The list above shows that the large number of factors bearing on particular helminth infections, especially those whose parasites follow a triheteroxenous cycle and have additional paratenic hosts, make it immensely difficult to arrive at an exact definition of the causes of the persistence in all their details (cf the compilation in Table 33). This also applies to the food-borne helminthiases which are common in the Philippines (Table 34).

4.1 Intestinal Capillariasis

Caused by the *Capillaria philippinensis* nematode, intestinal capillariasis[1] occupies a special place among the helminthiases in so far as it tends to be an acute illness which, if untreated, leads to death in about 6–10% of the affected persons [cf 61, 79, 136, 248, 250, 269, 724, 800]. Regarded as a special case among helminthiases, the progress of this disease is caused by an apparently obligatory endo-autoinfection of the final host, and by the fact that the first generation of parasites is succeeded by one other, and thus a few larvae, ingested orally, eventually produce thousands of adult worms. The males of the species are 1.5–3 mm long, the females 2.3–5.3 mm; both live as parasites in the jejunum, in the lumen and tissue of which they have been established following autopsy. Infection leads to atrophy of the Lieberkühnian crypts; the villi are flattened, the Lamina propria infiltrated by plasma cells, lymphocytes and macrophages.

---

[1] In their review article [248] Cross and Bhaibulaya provided a comprehensive account of the present state of knowledge of intestinal capillariasis

Mechanical separation of the epithelial cells by worms and micro-ulcers by worm secretions [746] are probably responsible for the poor absorption of fats and sugars, as well as for the development of a protein-losing enteropathy. Violent diarrhoeas lead to dehydration; the plasma levels for sodium, calcium, carotene and protein are greatly reduced, so that death may occur as a result of the irreversible effect of the electrolyte loss [for clinical aspects, symptomatology, pathology etc. cf 128, 136, 194, 248, 261, 274, 305, 328, 590, 674, 724, 746, 767, 800, 818, 825, 826, 833].

The findings that have so far been made available suggest that fish from fresh and brackish water, which act as intermediate hosts to *C. philippinensis* and are eaten raw or insufficiently cooked, can result in the infection of man. On the basis of our present knowledge, also a zoonotic cycle which could include fish-eating birds as the final hosts has also to be assumed to be in operation.

4.1.1 Short Historical Review

Intestinal capillariasis first came to public awareness when a Roman Catholic missionary from Tagudin (Ilocos Sur) drew the attention of Senator Raul Manglapus from Tagudin to the large number of persons falling ill and dying in Pudoc West in the year 1966; the latter in his turn brought in P. Garcia, the Minister of Health, who arranged for a team of physicians to be sent to the area concerned. There the team ran into a hitherto unknown disease to which the Ilocanos had given the name "the mystery disease of Pudoc", in the belief that its cause must have been of a metaphysical nature.

In the end, however, it turned out to be a nematode of the genus *Capillaria,* which though identified by Chitwood et al. [226] as early as 1963 in an isolated case from Ilocos Norte, had been regarded as an accidental occurrence. Chitwood et al. [227] then described the parasite as a new species with the appellation *Capillaria philippinensis.*

Originally thought to be a helminth, the distribution of which was restricted to the Philippines, this original classification had to be revised at a later date as a result of findings in Thailand and Japan [94, 248, 503, 629, 630, 640, 679, 766]. Moreover, it must be assumed that, although not recognised as such, intestinal capillariasis had in fact occurred in other areas of South-east and East Asia, and may continue to do so wherever fish is eaten raw or in an insufficiently cooked state.

4.1.2 The Biology of *Capillaria philippinensis*

Our knowledge of the development of *C. philippinensis* in the final host rests largely on experimental infections of the jird (*Meriones unguiculatus*) [cf 242, 244], the area of whose distribution does not extend to the South-east Asian region and which is in consequence out of the question as natural final host of the parasite in the Philippines and Thailand. Besides *M. unguiculatus,* it was possible experimentally to infect a number of monkey species (*Macaca cyclopis, M. fascicularis, M. mulatta*), the gerbil (*Mastomys natalensis*), and fish-eating as well as other species of birds (*Amourornis phoenicurus, Ardeola bacchus, Bubulcus ibis, Nycticorax nycticorax, Ixobrychus sinensis, Gallinula chloropus, Rostratula bengalensis, Columbia livia domestica*) [95, 242,

244, 246, 247, 250]. Although more than 150,000 animals from the fauna of the Philippine endemic area have been examined for infection with *C. philippinensis* up to the present time, in none of them could adult worms be proved [248]. Natural infections with adult *C. philippinensis* have so far been found exclusively in man. Nevertheless, observations on the very sporadic and isolated occurrence of capillariasis in man over recent years, as well as the exceedingly rare proven cases of naturally infected intermediate host fishes allow for but one conclusion: namely, that there is a sylvatic cycle in addition to the one involving man.

As for the natural intermediate hosts, further investigations are also needed in order to establish the range of potential animal species. Only a very few capillarid larvae lacking an assured species diagnosis have been found in fish. More than 150,000 representatives of terrestrial, aquatic or marine animal species harboured no *C. philippinensis* larvae, however [243]. Even the evidence of eggs from this helminth in two species of fish (*Eleotris melanosoma* and *Ambassis commersoni*) [133, 156] did not prove their function as natural intermediate hosts. Only with success in breeding a *Capillaria* infection in *Meriones unguiculatus* by feeding it with fish (to be precise, with *Hypselotris bipartita*), which had been caught in the lagoons of the endemic area, had one link in the development chain of the parasites been established. Experiments in infecting a wide range of fish species were successful, namely:

Hypselotris bipartita	*Cyprinus carpio*
Ambassis miops	*Puntius gonionotus*
Elotris melanosoma	*Aplocheilus panchax*
Sicyopterus sp.	*Gambusia holbrooki*
Poicelia reticulata	*Rasbora borapetensis*
	Trichopsis vittatus

[cf 95, 242, 244, 248].

The experimental infections mentioned above reveal the development cycle as running in the following way: having been passed out with the stool, the eggs of *C. philippinensis* embryonize in the open within 5 to 10 days and are then infective for fish living in fresh and brackish waters. Orally ingested by the fish, the larvae hatch from their egg shell when in the intestinal tract and growing – though not moulting – they attain the infective stage within three weeks. Orally taken up (i.e. through eating) as raw or inadequately cooked fish by a suitable final host such as man, the larvae contained in fish (L_1) undergo four moults before they become adult worms in the jejunum. In *M. unguiculatus* this stage of the development is completed after 10 to 11 days; 13 to 14 days after infection the female capillariae begin to deposit larvae, which will have reached adulthood in the second worm generation 22 to 24 days p.i., then beginning with the production of eggs. The passing of eggs commences 24 days after infection at the earliest.

There is a massive increase in parasites in the final host as a result of endo-auto-infection, usually with the natural corollary of acute sickness (although the experimentally infected *Macaca* species did not, for example, develop any symptoms). Having infected a black crown night heron (*Nycticorax nycticorax*) with 26 larvae, Cross and Basaca-Sevilla [246] were able to complete their experiment on the animal's demise by counting 10,182 parasites only 82 days after infection. The rate of increase in parasites in *Meriones unguiculatus* was still higher: an autopsy carried out 46 and 47 days after the start of the experiment resulted in 2,520 and 5,353 *Capillaria* worms of all stages in two animals, each of which had been infected with only two larvae [248].

Findings from investigations so far available compellingly suggest that in nature the cycle of *C. philippinensis* runs its course between fish as intermediate hosts and fish-eating birds as the final hosts. The first clear evidence of this will have been supplied by Carmen Velasquez if her communication to Cross and Basaca-Sevilla [246] on the evidence of a *Capillaria*-♂ in *Ixobrychus sp.* is confirmed. Nevertheless, the question of whether the infection of man must be attributed exclusively to the consumption of infected fish, or whether there might be the additional possibility of a direct transmission by way of the faecal-oral route will be raised briefly in this context, for an infection of experimental final hosts not only succeeds through oral application of larvae isolated from intermediate hosts but also in cases where the infection material (adult worms and larvae) comes from final hosts [246]. Moreover, it is known that infected people not only pass worm eggs but also excrete larvae and adult worms with their stool and, as a result of diarrhoeas as well as their defecation habits, contribute to a considerable contamination of the environment. However, so far no results on the influence of geofactors on the parasite stages which occur in the open have become available. For instance, it is not known how long larvae survive under varying outside conditions after being passed with the stool, nor do experimental findings reveal whether adult worms, larvae or both constitute the infective agents in transmissions from final host to final host.

Although not confirmed by later investigations, the retrospective study carried out by Detels et al. [269] in Pudoc West does provide a further indication of the possible existence of different infection routes. The authors noted that in the case of the period from August 1965 to January 1968, the male sex with a proportion of 6.9:1 was markedly more infected at the start of an epidemic than was the female one, and that less than 8% of the cases occurred among children below the age of nine (while constituting 32.1% of the total population). There was, however, a noticeable shift as the epidemic progressed, with a proportion of 0.97:1 between the sexes, and the group of under-nines contributing about 24% of the total cases (cf Fig. 69). As a result of their retrospective analyses, Detels et al. [269] had the impression, moreover, that the epidemic had spread from initially infected persons to others living nearby (Fig. 70). Starting from the fact that fishing is carried out solely by men and boys, and that boys are introduced to this activity when they are about 10 years old, the distribution by age and sex at the onset of an epidemic are in perfect agreement with the assumption of an illness set in train by infected fish. Cross and Bhaibulaya [248] in fact described how freshly caught fish are eaten raw by the men on the spot; in addition "some people prefer to bite open the abdomen of the bagsan[1] and suck out the intestinal juices". However, this does not explain the further course of the epidemic, with its large proportion of children numbered among those infected, nor the almost equal frequency among the two sexes [269]. Taking this at its face value, another route of infection without the offices of an intermediate host (a faecal-oral one, for instance) cannot be dismissed with absolute certainty considering that several other *Capillaria* species belong to the monoxenous helminth type [63, 89].

---

[1] *Ambassis miops*

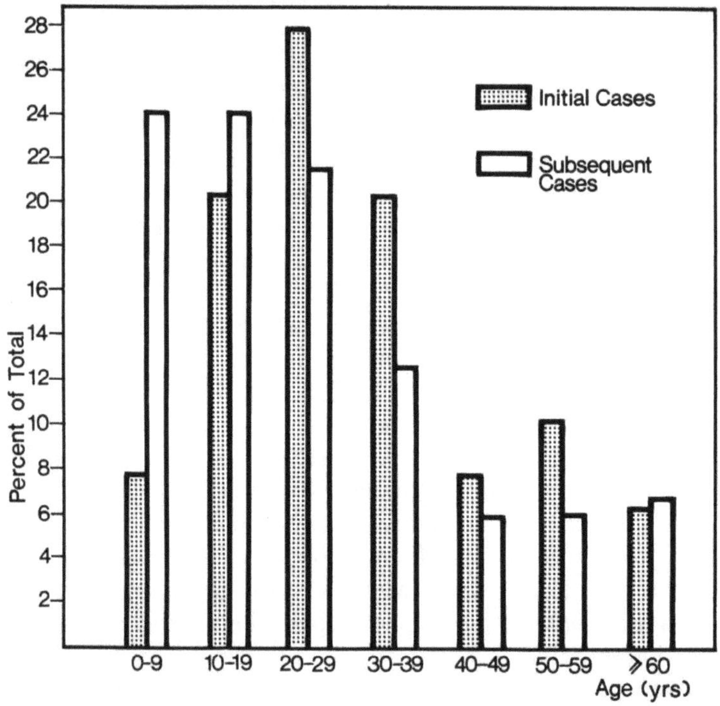

Fig. 69. Age distribution of initial and subsequent cases of capillariasis in Pudoc West, August 1965 to January 1968 [Source: 269]

4.1.3 Geofactors and Capillariasis

There are so far no exact investigations into the significance of geofactors for the distribution of capillariasis. The fact that fish from fresh as well as from brackish water can nonetheless serve as intermediate hosts, permits the conclusion to be drawn that eggs can tolerate certain concentrations of salt. The occurrence of capillariasis in man in the two most widely differing climatic zones of the Philippines (in the north-east of Mindanao, precipitation throughout the year with maxima at the turn of the year, contrasted with a marked distinction between rainy season and dry season in north-west Luzon) also seems to indicate that climatic factors do not have the effect of limiting distribution, at least not in the Philippines. On the other hand, it has been observed that the distribution of cases throughout the year showed a noticeable increase during the rainy season (Fig. 71), which stood out even if the two first epidemic years (1967/68) were not taken into consideration [248]. This, however, does not point to a primary climatic influence. Assuming that the natural final hosts are fish-eating birds of passage or migrants, the impulse to wander – itself a function of hygric or thermal seasons – could lead to seasonally varying population densities in the endemic areas, which would in turn result in a seasonal dependence of capillariasis.

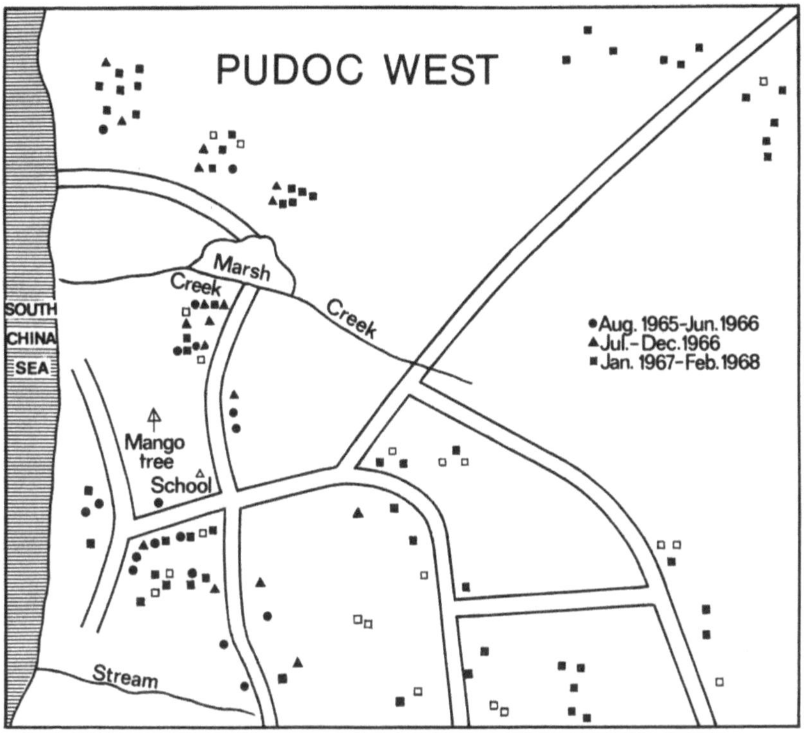

Fig. 70. Sketch map of Pudoc West, showing capillariasis-affected houses from August, 1965 to February, 1968 [Source: 269]

4.1.4 Frequency of Capillariasis

Between February 1967 and December 1981 a total of 1806 parasitologically proven cases of intestinal capillariasis were registered in the Philippines (♂♂: 1269, ♀♀: 537 cases), 108 (= 6%) of whom died. With 1037 cases diagnosed in the period from February to December, the infection had reached epidemic proportions in 1967 [248]. In a retrospective study, Detels et al. [269] found that in Pudoc West alone 31.6% of its 725 inhabitants had fallen ill with this worm infection in the period from August 1965 to January 1968. Later the total figure for all the cases registered in the Philippines fell to a minimum of 5 in 1972, but rose again to an annual average of 39 (21–65) cases. The decline observed immediately after 1967, the year of the epidemic, was at first attributed to the treatment using tiabendazole, which damages *Capillaria* eggs so that they can no longer develop into embryos. The absence of major epidemics in later years was explained by the fact that the affected population was acquainted with the symptoms as well as the treatment with the freely available anthelminthic mebendazole; as a consequence it is customary for people to apply the therapy themselves, which means that many cases are not registered [248].

In contrast to the results of the enquiries made by Detels et al. [269], the data on the distribution of the parasitologically proven cases by age and sex provided by

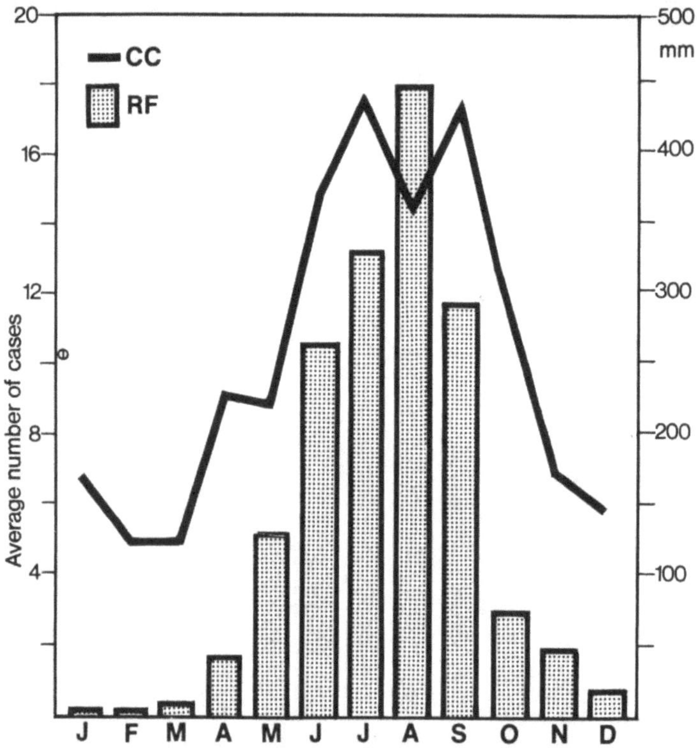

Fig. 71. Monthly averages of cases of intestinal capillariasis (*CC*) and of rainfall (*RF*) in the Ilocano area of Luzon [Sources: 248, 827]

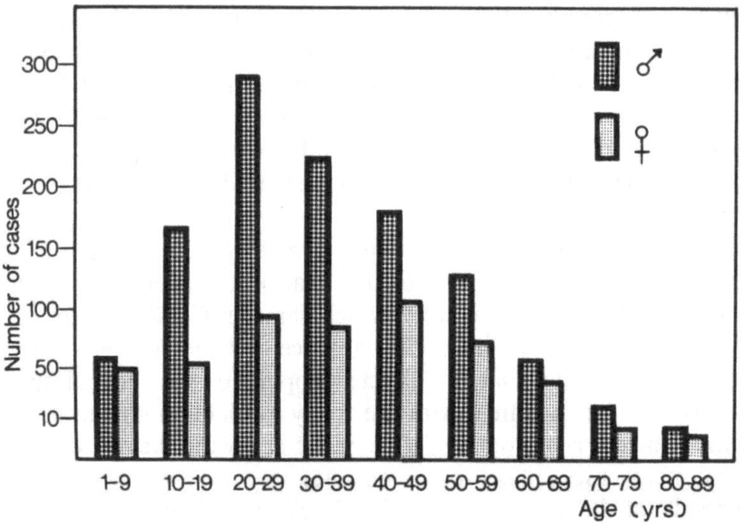

Fig. 72. Age and sex distribution of 1806 cases of intestinal capillariasis [Source: 248]

Cross and Bhaibulaya [248] have remained largely constant during the years since 1967 (Fig. 72). 70% of all cases were attributed to the male sex, which was thus clearly more frequently affected than the female one. The maximum of 18% occurred in the 20–29 year-old age group among men.

4.1.5 The Endemic Areas

In the Philippines intestinal capillariasis in man has occurred autochthonously in 8 different provinces so far. Six of these constitute an almost unified endemic area. Inhabited by the Ilocanos, they are located on the north-east and north coast of Luzon. Additional capillariasis foci have been discovered at Santiago on Lake Pagusi in the province of Agusan del Norte, Mindanao [199] and at St. Bernard, Southern Leyte[1]. Though the factor of eating habits, which Cross and Bhaibulaya [248] have described in detail for the Ilocanos, is evidently the only one the populations of the three endemic areas have in common, it appears to be decisive for the occurrence of the infection. The particular preference of these population groups for the consumption of animals living in fresh or brackish water or the sea in a raw or semi-cooked state is responsible for the presently known pattern of capillariasis distribution in the Philippines. Geofactors do not seem to be of any importance.

4.2 Angiostrongyliasis

Caused by the rat lungworm (*Angiostrongylus cantonensis*), angiostrongyliasis is primarily a zoonosis of rodents of the genera *Rattus, Bandicota* and *Melomys,* with several snail species acting as intermediate hosts [cf 93]. If man becomes infected by eating intermediate host snails which contain larvae, he may develop an infection known as eosinophilic meningo-encephalitis (eosinophilic meningitis) which belongs to the group of the so-called larva migrans. In the natural final host, as well in man, an important stage of the development of the parasite takes place in the central nervous system, prior to moving into the pulmonary arteries and the right half of the heart as the final locations in which the adult worms settle. Being an abnormal host, development in man tends to come to an end some time after the entry of the infective larval stage into the cerebral hemispheres and the olfactory lobes; the pulmonary arteries of man have so far only on one occasion been proved to contain adult worms [859a].

4.2.1 Short Historical Survey

Based on the hypothesis advanced by Alicata [39], Madagascar is held to be the country of origin of murine angiostrongyliasis. The giant African snail (*Achatina fulica*), which is known to harbour up to 30,000 *Angiostrongylus* larvae in one individual [40], is thought to have played an important role as an intermediate host in the zoonotic

[1] Cross, J. H.; Basaca-Sevilla, V.: Biomedical surveys in the Philippines, Manila 1984; quoted from Tropical Diseases Bulletin 82, 433 (1985), abstract no. 1489

cycle there. However, its distribution is said to have also led to the distribution of *Angiostrongylus cantonensis* infections among rodents, with the result that wherever the intermediate hosts and/or paratenic hosts of this parasite were eaten raw by man or in an insufficiently cooked state, eosinophilic meningo-encephalitis was likely to occur. The distribution of *Achatina fulica* is primarily the doing of man: "he has willfully or unintentionally introduced the snail to various areas in shipment of soil, vegetation and war salvage equipment" [40]. This snail had already been known from Madagascar from the time before 1800; around 1900 it was found in Sri Lanka, in 1911 in Malaysia, from whence it reached Formosa in 1932 and finally the Philippines during World War II (1942) [527]. Only after the occurrence of *A. fulica* was *A. cantonensis* shown to occur in rodents and eosinophilic meningo-encephalitis in man in the countries concerned. Whether the first diagnoses of eosinophilic meningo-encephalitis in the Philippines, which were made during the period 1946–1949 [727], were indeed the result of an infection involving larvae of the rat lungworm is, however, just as uncertain as the description of further cases on subsequent occasions [59, 60, 739, 740], as it was not possible to isolate parasites. The infection of rats with *Angiostrongylus* on the other hand was first parasitologically established for the Philippines by the investigations of Nishimura and Yogore [562].

4.2.2 The Final Hosts of *Angiostrongylus cantonensis*

Five different *Rattus* species are so far known to act as natural final hosts of *A. cantonensis* in the Philippines. With 75 (3.55%) out of the 2,113 animals found to be infected, the brown rat (*Rattus norvegicus*) showed the highest prevalence, followed by *Rattus rattus* with 25 cases (1.87%) among the 1,337 that were examined. In addition 1 of the 67 *Rattus exulans,* 2 of the 22 *R. everetti* and 2 of the 8 *R. lutiventris* were found to be infected, but the small number of specimens does not permit a more detailed evaluation of their significance. This also applies to *R. benguetensis, Mus commissarius* and *Suncus luzonensis,* among which no infections have so far been discovered [60, 350, 426, 453, 562, 672, 673, 832].

4.2.3 The Intermediate Hosts of *Angiostrongylus cantonensis*

The first evidence that even in the Philippines *Achatina fulica* serves as an intermediate host was supplied by Arambulo and Yogore [57a]; they succeeded by feeding snails to laboratory rats, from which adult *Angiostrongylus cantonensis* were subsequently isolated. The spectrum of intermediate hosts now known to exist in the Philippines includes the following slug and snail species [60, 672, 673, 832]:

Achatina fulica (prevalence: 4.5–58.0%)
Hemiplecta sagittifera (96%)
Helicostyla macrostoma (12%)
Vaginulus plebeius (5.6%)
Veronicella altae (18.2%)

It was not possible to establish with certainty whether *Cyclophorus sp.* and *Chlorea fibula,* which had been found to be infected with nematode larvae, could also be included among those carrying out the function of an intermediate host, as experiments in transmission failed [832]. All other species of snails that were investigated in the Philippines proved to be negative, namely:

Chlorea dryope	*Pila luzonica*
Macrochlamys sp.	*Lymnaea* spp.
Vivipara sp.	*Melania* sp.
Thiara sp.	*Brotia asperata*
Thiara aspersa	*Kaliella doliolum*

The same applies to the mussel species *Corbicula manilensis*.

4.2.4 Paratenic Hosts of *Angiostrongylus cantonensis*

One of the peculiarities in the development of *A. cantonensis* is that so-called paratenic hosts can be included in the cycle, i.e. infective larvae from the snails are orally ingested by other than the specific final hosts; the larvae then persist in such a paratenic host without developing any further, but do not lose their ability to infect the final host either. Invertebrates (such as predatory land planarians, fresh water prawns and shrimps, land crabs) as well as vertebrates (e.g. frogs) may serve as natural paratenic hosts. In addition, experiments with sea-snakes and freshwater fish showed these to be suitable, whereas pigs and calves played only a short-term role, since the larvae died-off inside them after a very few days [40, 93].

From the wide spectrum of potential paratenic hosts in the Philippines only shrimps and the crab species *Sundathelphusa philippina* have so far been examined and these proved to be negative [60]. Further investigations are therefore required in order to establish the significance of paratenic hosts as a source of infection for specific final hosts as well as for man.

4.2.5 The Importance of Geofactors for the Persistence of Angiostrongyliasis

The influence of geofactors operating during the development cycle of *A. cantonensis* primarily affects the first stages of the larvae passed out by rats along with their faeces. For the survival of the larvae humidity presents an essential factor; in droppings that are drying out they will die within a few hours, whereas stagnant or fresh water allows them to remain alive and infective for one or two weeks [653]. As far as temperature conditions are concerned, the first-stage larvae survive about two weeks at 24–28 °C, but for only one week if temperature fall below 6 °C [62]. Salazar and Cabrera [673] who, on the other hand, kept *Angiostrongylus* larvae under varying conditions through temperatures ranging from 25–29 °C, arrived at only 5 to 6 days.

Under the temperature conditions in the Philippines infection in the snails is only possible if larvae, which have been passed out by rats into water or into a damp milieu along with their faeces, are orally ingested at most within 1 to 2 weeks, or given the opportunity for active percutaneous penetration (both infection pathways are possible). If the faeces are dropped in dry places or in places which allow them to dry out fast as a result of exposure to the sun, the infection of the intermediate hosts will have to take place within a few hours. More favourable conditions for the persistence of angiostrongyliasis are therefore to be expected in those climatic zones of the Philippines which are characterized by high precipitation throughout the year (cf Map Plates). Since the majority of the investigations so far undertaken were concentrated in the central part of Luzon, the possibility of such a correlation between climatic factors and prevalence has not yet become a certainty.

Although results from this are still outstanding, it may be assumed that, taking the climatic conditions of the Philippines into consideration, geofactors play only a subordinate role in the development of parasites in the intermediate host snails. Geofactors can, however, become important when third-stage larvae, which can infect the final host, leave dead snails and come out into the open or are passed out by the snails along with their slime [239, 367, 460]. In freshwater such larvae survive for at least one week [653]. The significance of temperature and the composition of the water in this regard remains to be investigated.

4.2.6 The Distribution of Angiostrongyliasis and of Eosinophilic Meningo-Encephalitis

Apart from the area of Metro Manila, *A. cantonensis* is so far known in 13 different provinces of Luzon as a parasite of rats and/or snails. These are:

Batangas	Pampanga
Bulacan	Pangasinan
Cavite	Quezon
Ilocos Norte	Rizal
Laguna	Sorsogon
Mountain	Tarlac
Nueva Ecija	

Evidence for the Visayas (Negros Oriental, Leyte) and Mindanao (Lanao del Norte) is meagre by contrast. According to E. G. Garcia [314], it is highly probable that *A. cantonensis,* on the other hand, is enzootic throughout the archipelago.

The few cases of eosionophilic meningo-encephalitis in man in the Philippines, which had been mentioned in the introduction, rest solely on clinical diagnosis. Direct evidence of larvae or sub-adults has not so far been proved in any case. Considering the high rates of prevalence in final and intermediate hosts on the one hand, and the eating habits of some ethnic groups on the other (cf 4.1), it must nonetheless be assumed that a substantial number of infections of man must have occurred in the past, although they remained unrecognized. In the same way, further cases must be expected in the future. It is not the eating habits alone, though, which can lead to infection through the consumption of intermediate and paratenic hosts containing the larvae in raw or inadequately cooked food. Infection in man may also occur as a result of a contamination of the drinking water, and of vegetables with infective larvae which have left dead snails or have been passed out by living animals in their slime. Moreover, skin lesions can possibly act as a point of entrance for the parasites if they are brought into contact with foodstuff or drinking water containing larvae – although this has so far only been demonstrated in animal experiments [cf 370].

4.3 Gnathostomiasis

Cases of gnathostomiasis so far diagnosed in man have revealed a clear focus of distribution in the countries of East and South East Asia. The cause of this illness is predominantly if not exclusively an infection by larvae of *Gnathostoma spinigerum* [258]. In the adult state, this triheteroxenous parasitic worm lives in tumour-like knots of the stomach wall of Canidae and Felidae, and completes its development with the aid of cyclopids as the first intermediate host, and numerous species of vertebrates as the second intermediate hosts or paratenic hosts. Man tends to become infected through the consumption of raw or inadequately cooked meat from the second intermediate or paratenic hosts; the main source of infection are fish harbouring advanced third-stage larvae. Although *G. spinigerum* can mature to the adult worm stage in man as well, the disease picture is marked in the main by the migration of larvae and sub-adult worms; depending on localization, a distinction is to be made between cutaneous and visceral gnathostomiasis, in which the infection of the central nervous system, followed by eosinophilic encephalomyelitis may pose a serious health problem.

4.3.1 Short Historical Survey

The first helminth species to be reported from the Philippines – *Gnathostoma spinigerum,* under the name *Filaria radula* – was described by the German parasitologist Anton Schneider [707] in the year 1866 in his monograph on the nematodes from the stomach wall of *Paradoxurus philippinensis.* Even at that time, Schneider drew attention to the striking similarity with *Cheiracanthus robustus* Dies., which together with *F. radula* was later unambiguously identified as *Gnathostoma spinigerum* by Baylis and Lane [88]. Since 1917 this species of nematode has also been known in the Philippines as a parasite of dogs [834]; later on also cats were found to act as final hosts, as were fish as second intermediate hosts [34, 35]. This was followed by the successful experimental infection of cyclopids as first intermediate hosts [648]. The first case of visceral gnathostomiasis in man in the Philippines was reported by A. B. M. Sison and Florentin [725], as well as by Yogore and Juliano [865], although it is not possible to be certain that it was an autochthonous infection. That applied first to the case drawn to public attention by Chanco et al. [217].

4.3.2 The Final Hosts of *Gnathostoma spinigerum*

Though predominantly a parasite of Canidae and Felidae, *G. spinigerum* has also been found in Mustelidae and Viverridae. Apart from the original description in *Paradoxurus philippinensis* [707], evidence in the Philippines is restricted to domestic dogs and cats [35, 46, 47, 643, 648, 778, 785, 828, 834, 881, 882, 884]. Due to the meagre data available so far, prevalence among dogs is put at 8% and among cats at 39%.

4.3.3 The Intermediate and Paratenic Hosts of *Gnathostoma spinigerum*

In the Philippines first intermediate hosts are known merely from experimental infections successfully conducted by Refuerzo and Garcia [648] using *Eucylops ser-*

rulatus and *Cryptocyclops bicolor* (syn. *Microcyclops bicolor*). Up to now no observations on natural infections of these or any other cyclopids have been published.

From the wide spectrum of secondary intermediate and paratenic hosts [cf 258] only four different species of vertebrates have hitherto been found to be naturally infected in the Philippines. The three species of fish appear to occupy the most important position as a source of infection for man; among these, the eleven *Glossogobius giurus* ("bia") which were examined all proved to be infected, whilst prevalences were markedly lower among *Ophicephalus striatus* ("dalag") with 25% (2 out of 8) and *Therapon argenteus* ("ayuñgin") with 13% (3 out of 23) [34]. Clearly, the six specimens of *Hurria* (*Cerberus?*) *rhynchops* are in the class of paratenic hosts, in which Africa et al. [35] proved an infestation with *Gnathostoma* larvae, for this species of colubrid snakes feeds mainly on fish.

4.3.4 The Significance of Geofactors for the Persistence of Gnathostomiasis

Geofactors such as temperature and the chemical composition of the water play a role chiefly in early developmental stages when the larvae develop within the egg shell and in the first intermediate hosts, as well as during the intervening period of the hatched and thus free larvae. However, little is known about their influence. Daengsvang [258] merely noted that at water temperatures of 25–31 °C the first-stage larvae hatch at the very earliest on the 9th day after passing out with the faeces. Development in the cyclopids is evidently encouraged greatly by high temperatures: whilst almost all larvae have reached the third stage on the 7th day, if temperatures are in the range 29–31 °C, this applies to a few specimens only if temperatures are a little lower – 25 to 29 °C. On two grounds a high speed of development, particularly for areas on the banks of stagnant waters, is thereby guaranteed: the highest temperatures prevail there, and *Cyclops* species (*Eucyclops serrulatus, Cryptocylops bicolor*), which are incriminated as first intermediate hosts, live there as inhabitants of the littoral as well. Moreover, the area closest to the banks is most heavily contaminated with faeces from final hosts, which contain eggs – a fact leading to high rates of infection among the first, and subsequently the second indermediate hosts, which are in their turn caught and consumed by fish-eating final hosts in precisely the same area.

As *E. serrulatus* and *C. bicolor* are adapted to the littoral and tolerate only low flow velocities, standing or very slow-moving waters offer the best guarantee for the persistence of infections with *G. spinigerum*. Neither these inter-relationships nor the significance of other geofactors has so far been analysed in any greater detail.

4.3.5 The Distribution of Gnathostomiasis

The small number of investigations so far carried through does not permit statements to be made concerning the distribution of gnathostomiasis in men and animals in the Philippines. Examination of cats and dogs, as well as fish, has been restricted to those originating from Manila and the provinces in the central part of Luzon. As regards infections in man, the few cases (4) do not allow any conclusions to be drawn either. Nevertheless, two patients came from Ilocos Norte [214, 796], where raw fish is considered a delicacy among the indigenous Ilocanos.

4.4 Diphyllobothriiasis and Sparganosis

Among the pseudophyllid cestodes the fish tapeworm (*Diphyllobothrium latum*) is a regionally common parasite in men and fish-eating mammals. Its development cycle is, however, adapted to the lower temperatures of the temperate climatic zones; von Bonsdorff [105] therefore points out that reports of cases from tropical countries require critical checking with regard to diagnosis of the species. Accordingly, this applies to the seven cases diagnosed in man in the Philippines as well [118, 324, 765, 822], as well as to infections in cats and dogs [731, 775][1].

Other species of the family Diphyllobothriidae occur in warm climatic zones as well, however, especially among cats and dogs and other carnivorous mammals. Humans are affected by the larvae of the genus *Spirometra*, causing a disease known as sparganosis which has also been described for the Philippines [317, 330, 404, 875].

4.4.1 The Life Cycle of *Spirometra mansoni*

The most important, probably even the sole species, causing sparganosis in man in the Philippines is *S. mansoni*[2]. The different stages in the life cycle of this triheteroxenous tapeworm resemble that of *Gnathostoma spinigerum* (cf 4.3): eggs are passed out together with the faeces and deposited in water, where the first larvae (Coracidium) hatch in order to continue their development to the second larva (Procercoid) in planktonic cyclopids (1st intermediate host). The second intermediate hosts tend to be frogs which become infected at the tadpole stage. The plerocercoid (= Sparganum) develops in them from the procercoid as the third-stage larva, which is infective to the final host, in which it reaches adulthood. The infection of the final hosts (dogs, cats) can, however, also take place by way of paratenic hosts (snakes, birds), which feed on amphibians. If people consume meat of intermediate or paratenic hosts which contains plerocercoids their intestinal tracts do not become the locus of settlement by adult *S. mansoni;* it is rather the case that man functions as another paratenic host, with the larvae migrating in the tissue and then also encysting subcutaneously.

4.4.2 Final and Intermediate Hosts of *Spirometra mansoni*

In the Philippines cats, dogs and a palm civet (*Paradoxurus philippinensis*) have been identified as final hosts of *Spirometra mansoni* [340, 646, 709, 731, 775, 777, 782, 834, 880, 884; cf also 45–47, 571, 643, 778, 785, 882]. In this cats proved to have higher

[1] E. Y. Garcia and Africa [324] speculated on whether *D. latum* was a tapeworm which had been imported from China in carp (*Cyprinus carpio*), which had been introduced to Laguna de Bay when already infected. Though adopted by Refuerzo [643] and Arambulo [46], this version is nonetheless improbable, since carp do not play the role as intermediate hosts, nor do infections with *D. latum* occur autochthonously in China [cf 105]

[2] Although not certain it is probable that the parasites described as *Diphyllobothrium mansoni, D. erinacei, Spirometra erinacei, Dibothriocephalus* sp. and *Sparganum philippinensis* belong to this species

rates of prevalence than dogs. This is undoubtedly the result of the very different feeding habits of the two animal species.

Ever since Africa [9] undertook experimental infections, *Eucylops serrulatus* and *Cryptocyclops bicolor* have been accepted as first intermediate hosts of *S. mansoni* in the Philippines. Whether other cyclopids fulfil the same function has so far not been investigated, or the information on the species employed in experimental infections is lacking, as in Yutuc [880]. This also applies to frogs, which serve as second intermediate hosts; 2.1% of them proved to be infected naturally [880]. Manuel and Signey [507] are the only persons to cite infected *Rana vittigera* as being the frog species used as starting material for their animal experiments. Plerocercoids were also isolated from a cinnamon bittern (*Ixobrychus cinammoneus*) and two, not further specified, snakes [880] which are to be regarded as paratenic hosts. Manifesting an intensity of infection of 120 and 161 spargana, the snakes clearly demonstrate the function of amphibivorous animals as "carrier" hosts in which an accumulation of plerocercoids occurs.

4.4.3 The Distribution of Infections with *Spirometra* and of Sparganosis

Publications so far available do not allow one to make a final statement on the distribution of *S. mansoni* in the Philippines. However, bearing in mind that the infection of the different host categories, including the sparganosis cases in man, has been established not only for Manila and several provinces in Central Luzon, but also for Albay, Leyte, Negros and Mindanao, it may be assumed that it is an enzootic helminthiasis throughout the archipelago. Whether in the Philippines sparganosis in man is acquired exclusively through consumption of raw or inadequately cooked intermediate or paratenic hosts, remains uncertain, since everyone of the four patients denied such eating habits. O. P. Garcia and Reyes [330] therefore assume the infection source to be cyclopids, which are taken up in drinking water along with the procercoids they contain: the evidence for such a mode of infection is still, however, lacking.

4.5 Taeniases and Cysticercosis

Beef tapeworm (*Taenia saginata*) and pork tapeworm (*T. solium*) are common wherever beef and/or pork are consumed raw or in an inadequately cooked state. Both of these species are found in the Philippines as well, though with regionally differing frequency. Man is the only final host of these diheteroxenous parasites and contributes significantly to the persistence of taeniases, thanks to his habits of defecation: contamination of the milieu with egg-ridden proglottids, which are passed out with the stool, enables intermediate hosts to become orally infected; bladder worms (metacestodes) develop in their organs, and in turn become the infective agent for man.

In the case of *Taenia solium* man can also assume the role of an abnormal intermediate host; the oral ingestion of eggs of this tapeworm species can be followed by the colonization of his organs by bladder worms and thus by a picture of disease known as cysticercosis. However, cysticercosis in man has so far been rarely observed in the Philippines, namely in 6 cases[1] [36, 48, 153, 213, 252, 335].

4.5.1 Short Historical Survey

Cysticercosis among animals (and thus taeniasis in man) had been known to exist in the Philippines as far back as the Spanish colonial times [513; cf 57]. Systematic investigations only began, however, at the time of the U.S.A.'s take-over of the archipelago. According to the information supplied by Garrison [336] it was Strong [742] who first reported *Taenia* infections in man; a statistical investigation of findings based on 1,793 stool examinations and 386 autopsies resulted in two adult taenias, and the remark that *T. saginata* tended to occur among American troops and *T. solium* occasionally among Filipinos. Thus both the *Taenia* species had been shown to be endemic in the Philippines already at the beginning of the century.

Investigations that followed and which have continued up to the present (cf Table 35), finally showed that infections with *T. saginata* clearly dominated in man; *T. solium* was, as it continues to be, the rarer species of adult worm in man as the final host. These findings, however, are in marked contrast to the infection of the intermediate hosts (cf 4.5.3). The extremely rare *T. saginata*-cysticercosis found in cattle and carabao, and the relatively common *T. solium*-cysticercosis of pigs on one hand, together with the greater frequency of *T. saginata* infections and the less frequent occurrence of *T. solium* in man on the other, constitute a still unresolved epidemiological paradox for the Philippines [48, 153].

4.5.2 The Significance of Geofactors for the Occurrence of Taeniases and Cysticercosis

In the development cycle of the two species of *Taenia* it is only the tapeworm eggs which are exposed to the influence of geofactors from the moment they reach the open air together with the stool and until the oral ingestion by their intermediate hosts. The nature of their shell, however, makes these eggs remarkably resistant to the effects of chemical and physical factors. *Taenia* eggs are, for instance, capable of surviving for a fortnight in a concentrated formalin solution (= 40% formaldehyde) without losing their infectivity for the intermediate host. Eggs with pigmented shells as those of the *Taenia* species are also regarded as being very resistant to ultra-violet rays. So too, low temperatures are easily endured; for instance, *T. saginata* eggs survive temperatures of −4 °C for at least 60 days; indeed, a range of low temperatures (2−6 °C) even appears to offer optimal conditions for the survival duration, which was found to be 273−365 days in the case of *T. saginata* eggs. On the other hand, higher temperatures may well have the effect of limiting distribution, since they are less well tolerated, although *T. saginata* eggs survive for at least 10 days even at temperatures of 38 °C (a comprehensive account of the resistance of parasite ova, larvae and cysts against chemical and physical factors was given by Enigk [286]).

---

[1] Due to cross-reactions the results of serological investigations [234] cannot be assessed more precisely

Climatic zones of the Philippines having a marked dry season with very high temperatures may therefore be assumed to allow for shorter life-spans of *Taenia* eggs in the open and consequently for lower rates of prevalence among the intermediate hosts. This could be *one* ground for the higher frequency of *T. solium* cysticercosis of pigs in the Visayas and on Mindanao (4.40‰) by comparison with Luzon (0.17‰). As a rule, even the time that elapses between the dropping of stool containing eggs by man and the oral ingestion of the tapeworm eggs by the intermediate hosts may be shorter for *T. solium* than for *T. saginata* as the pig, an omnivore, is also coprophageous, and thus also a direct remover of human faeces, whilst the infection of herbivorous carabaos and cattle is only an accidental by-product of their food intake. This could be another ground for the higher cysticercosis rate of the pigs in the Philippines. The paradox of the higher rate of *T. saginata* infections in man cannot be explained by the influence of geofactors, however.

4.5.3 The Frequency of Human Taeniases and Cysticercosis of the Intermediate Hosts

Knowledge concerning the frequency of human taeniasis in the Philippines is far from being complete, since the customary single examination of stool carried out in surveys and by routine diagnosis only picks up those infected persons whose faeces happen to contain proglottides, or eggs which have already been released in the intestinal tract. In this respect the number of a mere 1,500 cases which have been published to date over a period of almost 80 years creates a wrong picture of the actual frequency of *Taenia* infections (cf Table 35).

Although distinctions in species are available for only about 20% of all the *Taenia* infections in humans, the examinations do reveal the unmistakable dominance of *T. saginata* infections with *T. solium* proving to be the less frequent parasite. The exact reverse was, however, observed in respect of the infection of the intermediate hosts with metacestodes. In pigs, the frequency of cysticercosis was clearly dominant. As far back as 1903 Calvert [189] reported that 4% of the pigs were "heavily infected" with the metacestodes of *T. solium*. Similarly high prevalences (2.9–4.4%) were later established by Z. de Jesus [415] in pigs for slaughter in Laguna Province. Although infection rates at the abbatoir of Manila over the period 1914–1947 were lower (0.3–1.7%) and even down to 0.02% in 1968 (cf Fig. 73), during the years from 1970 to 1974 the country still had an average infection rate of 0.17% among slaughter pigs (cf Tab. 36). Cysticercosis in pigs occurred with particular frequency at Jaro, on Leyte, where 10 out of 1000 animals had to be rejected because of generalized cysticercosis. Pigs slaughtered at Surigao City showed a similarly high rate and distinctly above-average values were found in the returns from Cebu City, Butuan City and Oroquieta City. Disregarding the findings from Manila, Quezon City and Caloocan City, as the animals taken to slaughter there do not come from the immediate neighbourhood of these cities but from many different parts of the country, the inter-regional comparison reveals a cysticercosis rate of 0.02% for Luzon, and of 0.44% for the Visayas and Mindanao. It is not known whether there is a correlation with corresponding differences in the prevalence of human *T. solium* infections.

Compared with the cysticercosis of pigs, the disease is markedly less prevalent among carabaos and cattle, for since the introduction of the inspection of meat in the

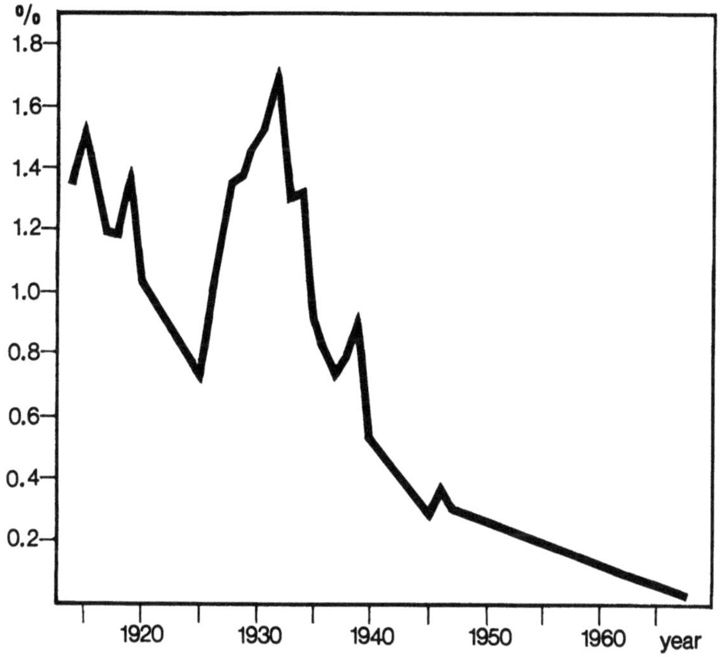

Fig. 73. Prevalence of *Taenia solium* metacestode infections in pigs slaughtered in Manila, 1914 to 1968 [Sources: 52, 644, 714]

Philippines in 1843 metacestodes of *T. saginata* resulting from an autochthonous infection were first diagnosed in cattle in 1949 [644]. Even the specific examination of 1,500 carabaos from all parts of the country did not result in any further positive findings [53]. Only the collation of findings from abbatoirs undertaken by Arambulo et al. [48] recorded further cases: among the animals slaughtered over the period 1970–74 cysticercosis was diagnosed in 8 out of 342,087 carabaos and 13 out of 194,968 cattle.

An attempt to explain this contradiction between the frequent occurrence of human *T. saginata* and the extreme rarity of cysticercosis in cattle and carabao, as well as the reverse conditions in human *T. solium* infections and the bladder worm infections of pigs, was made by Arambulo et al. [48] in their investigations in an endemic area on Leyte [cf 153]. In the local municipality of Jaro two barrios were found to have an average *T. saginata* prevalence of 10.3% [138, 183]. Arambulo et al. [48] arrived at the following conclusions:

- all the tapeworms passed out by 51 patients after they had undergone worm cures were identical with *T. saginata.*
- the experimental infection of calves, caracalves and goats with the eggs of these tapeworms did not lead to a metacestode infection of these animals.
- it was not possible to prove natural infections of cattle and carabaos with metacestodes of *T. saginata* as taking place in the endemic area.
- pigs of the endemic area were infected with bladder worms of *T. solium.* Their transmissability to man was proved through the infection of a volunteer.
- in general neither beef nor the meat of carabaos are often eaten by the population of the endemic area. 92.6% of those asked, however, indicated that they ate raw pork (prepared as "kinilaw" or "sinugba").

- "the cycle of *T. saginata* in the endemic area studied did not follow the known cycle of *T. saginata* type species of man-Bovidae-man. *T. saginata* cysticercosis was never demonstrated in indigenous animals and the sources of human infection could not be ascertained. The question was raised as to whether the local species is the *T. saginata* type species or a *T. saginata*-like species which follows a different cycle. In consideration of our results, we believe we are dealing with a *Taenia saginata*-like tapeworm."

Similar problems concerning *T. saginata* infections in man have been reported from Taiwan [392] and for Sumatra [442]. Further investigations are therefore needed in order to establish beyond any doubt what the intermediate host of the tapeworm species in the Philippines is, and thus to permit an exact description of the epidemiology.

4.5.4 The Distribution of Taeniases

Taenia infections of man have so far been established in many different parts of the Philippines (cf Table 35). However, some areas stand out as having a higher endemicity, on account of the regional variations in eating habits. In northern Luzon they are the provinces inhabited by the Igorots, Ifugaos and Ilocanos [114, 209, 210, 628, 675, 839; cf also 248]). Chamberlain et al. [209, 210] had already noted a prevalence of 12.6% among the Igorots in Baguio, which was far above the average for the country. In Bilibid Prison as well the inmates from the Mountain Province were most frequently affected (15.8%) [842]. Other publications point out specifically that those who were found to be infected with *Taenia* had been in the main Ilocanos. This also applied to 8 of the 16 patients B. D. Cabrera and Yogore [182] had treated for helminths in Manila, for instance. The authors noted in this context that this had to be attributed to the eating habits of the Ilocanos, "who are fond of eating a native dish called 'kilawen', which is raw beef mixed with spices and taken with native wine". The findings of the investigations that are available, however, are altogether insufficient for exactly delimiting the area of higher endemicity in northern Luzon.

A second focus with a distinctly above-average prevalence of 10.3% was discovered by M. G. Cabrera et al. [183] in the municipality of Jaro on Leyte. All the cases there were of *T. saginata* infections; this, and the absence of infections with metacestodes among cattle and carabaos led Arambulo et al. [48] to investigate the epidemiology of taeniasis in greater detail. Yet it has not so far proved possible to draw any definite conclusions from the metacestode infection of the intermediate hosts concerning the frequency of *Taenia* infections in man either in the Jaro area or in the Philippines at large. The higher rates of prevalence among pigs ready for slaughter in the Eastern and Central Visayas and on the north coast of Mindanao (cf Table 36), though indicating a higher endemicity of *T. solium* in these regions, still await conclusive proof.

4.6 Echinostomiasis

The only representative of the trematode family Echinostomatidae in the Philippines is *Echinostoma ilocanum,* which occurs as an intestinal parasite of man. First described by Garrison [338] as *Fascioletta ilocana* – from some Ilocanos in custody at Bilibid Prison – this species of intestinal fluke was correctly classified by Odhner [575] as belonging to the genus *Echinostomum.* By contrast Tubangui's [780] subsequent transferring of it to the genus *Euparyphium* did not meet with approval.

4.6.1 The Life Cycle of *E. ilocanum*

E. ilocanum belongs to those triheteroxenous trematodes which use aquatic snails not only as their first but as their second intermediate hosts as well. Having been passed out with the stool and deposited in water, the eggs develop to the stage of the miracidium at temperatures ranging from 26–31 °C; this takes 6–15 days. 3 to 5 days later the fully developed miracidia hatch out. Planorbid snails of the genus *Gyraulus*[1] are the first intermediate hosts, from which the cercariae swarm out 42 to 50 days after penetration by the miracidia. Their subsequent development into metacercariae can take place inside a number of snail species, among which *Pila luzonica* is the most important source of infection for man [793]. Apart from man, rats (*Rattus norvegicus*) were also shown to be final hosts of *E. ilocanum* [779, 780].

4.6.2 Frequency and Distribution of *E. ilocanum*

Until the year 1983 no more than around 60 cases of *E. ilocanum* infections had been diagnosed in man (cf Table 37). They were quite clearly concentrated in Ilocanos who were either living in or who had originated from the north-western provinces of Luzon, and who had infected themselves by consuming raw or insufficiently cooked *Pila luzonica.* An estimate of the actual frequency of *E. ilocanum* infections among the population of northern Luzon, however, only became possible when the results of investigations which sought to explain the *Capillaria philippinensis* cycle (cf 4.1) had been made available [248]. In 23 barrios of North Luzon, 11% of the 7,417 persons examined proved to be infected with *E. ilocanum.* Among children of up to 9 years of age the prevalence had already reached 7%; the maximum of 17% occurred among the 40 to 49 year age-group.

Among rats at least, *E. ilocanum* infections in the Philippines are not restricted to North Luzon, since Tubangui [779, 780] found 1 in 109 and 5 in 950 *Rattus norvegicus* he had caught in Manila to be infected with this species of intestinal fluke. There are as yet no results from investigations in other parts of the country, nor is it possible to learn from the available publications whether human infections occurred outside the Ilocos Region, since they either contain no information about the origin of the patients [130] or deal with members of the armed forces who were stationed in different parts of the country [863]. In other cases, however, which had been diagnosed at Bukidnon (Mindanao), Victoria (Oriental Mindoro) and North Bohol the diagnosis of the species was uncertain [197, 198, 200]. Since *E. ilocanum* has been shown to occur in other countries of South and South East Asia it may be assumed, on the other

---

[1] It is uncertain whether it really was *G. prashadi* which Faustino [294] had described. This species could be identical to *G. chinensis* (syn. *G. convexiusculus*) [cf 529 a in this connection]

hand, that this species of intestinal fluke as a parasite of animals is widely distributed in the Philippines as well. Nothing, though, is known about the influence of geofactors on their distribution. The relatively frequent occurrence in man in the Ilocos Region is definitely to be attributed solely to the eating habits of the Ilocanos.

4.7 Paragonimiasis

Paragonimiasis is an infection of humans and mammals by trematodes of the genus *Paragonimus,* which tend above all to settle in the lungs of their final hosts. The passing of eggs occurs together with the passing of stool or sputum; the latter is of a rusty-red colour thanks to the addition of erythrozytes. The larvae develop first in snails (Thiaridae = Melanidae), then in crabs and other decapode crustaceans which – if eaten raw or in an inadequately cooked state – in turn present the source of infection for man, amongst other final hosts.

4.7.1 Short Historical Survey

The earliest proof of human *Paragonimus* infections in the Philippines dates back to Musgrave [558], who reported 17 cases, 8 of which were established through autopsies. First indications of the geographical distribution emerged already in the following year (1908) through stool examinations of inmates of the Bilibid Prison, where Garrison [337] diagnosed another 18 cases, 4 of which came from the Bicol Peninsula, 6 from Leyte and 4 from Samar. It was not until much later that Leyte and Sorsogon Province (Bicol Region) were confirmed as being the most important of the endemic centres [860, 862, 864].

More than thirty years passed after the discovery of lung fluke infections in man before – in 1941 – Tubangui [784] recognized a freshwater crab species as the second intermediate host, and thus as the source of infection for the final host. First classified as a member of the genus *Potamon,* Tubangui [785] corrected his original entry in *Parathelphusa (Barythelphusa) mistio,* which was eventually changed to *P. (B.) grapsoides* [861], and finally established as *Sundathelphusa philippina* by B. D. Cabrera [140] on the basis of the monograph by Bott [106]. (It seems doubtful whether the synonymization arranged by Miyazaki et al. [547] for *P. grapsoides* and *Sundathelphusa picta* is correct; the latter does occur in the Philippines, as mentioned in Bott's monograph; however the author does not indicate that *P. grapsoides* is synonymous with *S. picta.*)

At first the search for the first intermediate host concentrated on investigations into *Thiara granifera,* whose function as an intermediate host was known from other endemic countries, but without positive results. Only when *Brotia asperata,* a member of the same family of snails (Thiaridae = Melanidae), was included in the investigations, did Tubangui et al. [791] succeed in proving the presence of *Paragonimus* cercariae. Later on when it was referred to as *B. asperata* form *dactylus* [873], this snail species was finally classified as belonging to the genus *Antemelania,* and split into the species *A. asperata* and *A. dactylus* [406]. *Semisulcospira libertina* can possibly be considered as another intermediate host as well [885].

4.7.2 *Paragonimus* Species in the Philippines

Four different *Paragonimus* species have so far been described for the Philippines, namely:

P. westermani (Kerbert, 1878)
P. siamensis Miyazaki and Wykoff, 1965
P. philippinensis Ito, Yokogawa, Araki and Kobayashi, 1978 and
P. filipinus Miyazaki, 1978.

Until 1972 all infections of lung flukes diagnosed in man or mammals were attributed to infections with *P. westermani.* Only in 1972 when examining naturally infected cats did it emerge that 7% of the lung flukes belonged to another species, that is to *P. siamensis,* which had hitherto only been known to exist in Thailand [173, 174]. When comparing the development stages of *P. westermani* from Japan with those from the Philippines, the smaller size of cercariae, metacercariae, metacercariae cysts and adult worms in materials collected in the Philippines led J. Ito et al. [407] to describe *P. philippinensis* as a new species, possibly even represented by several sub-species in Sorsogon and Leyte. Apart from this it was possible to establish that the species found in the Philippines develops to maturity in rats, carries a larvae migration in these final hosts that differs from those of typical *P. westermani* and can be distinguished by serological tests from *P. westermani* [876–879]. Only further investigations will show whether these results positively exclude the occurrence of *P. westermani* in the Philippines, and confirm the observation that *P. philippinensis* is the dominant species, with *P. siamensis* as the only other species besides. At least Miyazaki [545] and Miyazaki et al. [547] regard *P. philippinensis* as being identical with *P. westermani.* The findings made by Zhong Huilan et al. [885], on the other hand, argue for the presence of another lung fluke species besides *P. westermani* in the material from Sorsogon which they had examined.

As to the validity of *P. filipinus* Miyazaki [544] described on the basis of a single adult worm, the author himself voiced doubts shortly afterwards ("*P. filipinus* has not been decided whether it is a valid species or not") [547]; this was followed by a transfer to a sub-species of *P. westermani* [545].

To sum up: it is suggested that only two different species of *Paragonimus* apparently occur in the Philippines; the species more commonly encountered as a parasite of man and mammals is either a sub-species of *P. westermani* or otherwise an independent species known as *P. philippinensis,* whereas the rarer species, *P. siamensis,* has so far only been found in cats.

4.7.3 Geofactors and the Development of *Paragonimus*

Observations on the significance of geofactors for the development of the larvae of *Paragonimus* in the open and in intermediate hosts are scanty in the Philippines and restricted to those Garrison and Leynes [341] made on eggs and miracidia in 1909. Accordingly, 15 days are sufficient for mobile miracidia to develop in their egg shells in water of 25–28 °C where they may remain alive for a maximum of 160 days. A rise in temperature to 29–31 °C delays the development; at 37–38 °C it is discontinued and leads to degeneration. Miracidia that have been bred at room temperatures also cease to be mobile within an hour of transfer to water at 37 °C. On the other hand, low temperatures are more likely to be tolerated: though there is no development while temperatures remain between 9–15 °C, it sets in immediately once the eggs are transferred to water at room temperature.

Not only high temperature but also direct sunlight and drying-out for just a few minutes have a deleterious effect upon *Paragonimus* eggs. They degenerate quickly under these conditions and in waters with a salinity of c. 3%. A NaCl content of 0.5% in the water on the other hand allows the development to proceed in approximately the same way as in tap water, which allows the miracidia to hatch after 25 days at the earliest. However, hatching fails to occur if the salt content is increased to 0.5–1.0%, although a level of 1% salt content still results in mobile miracidia in 20–33% of the eggs, and only an increase to 1.5% halts all development [341].

The few findings permit only limited conclusions as to the influence of geofactors on the distribution of paragonimiasis in the Philippines. Tolerance of lower temperatures indicates an adaptation to the habitats of the first and second intermediate hosts, which prefer cooler, fast-flowing waters. These are the places where paragonimiasis infections are to be expected. In waters close to the coast which are subject to sea-water flooding, lung fluke infections cannot persist, thanks to the sensitivity of eggs and miracidia to salinity.

4.7.4 The Biotopes of the Intermediate Hosts

The few data available for the characterization of the biotopes of the first intermediate hosts refer to *Brotia asperata* before they were classified as members of the newly created genus *Antemelania* and split into two species. It can therefore not be decided with absolute certainty whether *A. asperata* and *A. dactylus* inhabit identical or largely accordant biotopes. *B. asperata* has, in any case, been described as a frequent and widely distributed inhabitant of mountain streams in the Philippines, preferring clear, fast-flowing waters that are rich in oxygen and have steep banks and rocky, sandy or even muddy beds. As a rule, such mountain streams are only a few feet wide, 1 to 2 feet deep and well-shaded; their water temperature is about 25–26 °C, their pH value of c. 7.4 tending slightly to the alkaline range [861].

The mountain crab (*Sundathelphusa philippina*), the second intermediate host of *Paragonimus* is adapted to clear, cool, fast-flowing mountain streams as a biotope [137, 539]. Thanks to this adaptation of the first and second intermediate hosts, *Paragonimus* infections can therefore also only persist in places where the topography ensures a greater velocity of flow of the waters, i.e. only in mountainous areas and their immediate environs.

4.7.5 The Infection of Intermediate Hosts with *Paragonimus* larvae

Information on the frequency of infections with *Paragonimus* larvae among the intermediate hosts are so far only available for the Camarines Sur and Sorsogon Provinces of the Bicol Peninsula, and for Jaro on Leyte (cf Table 38). The findings do not, however, allow conclusions to be drawn as to the distribution of paragonimiasis in the Philippines. Whilst first intermediate hosts (*Antemelania asperata, A. dactylus*)

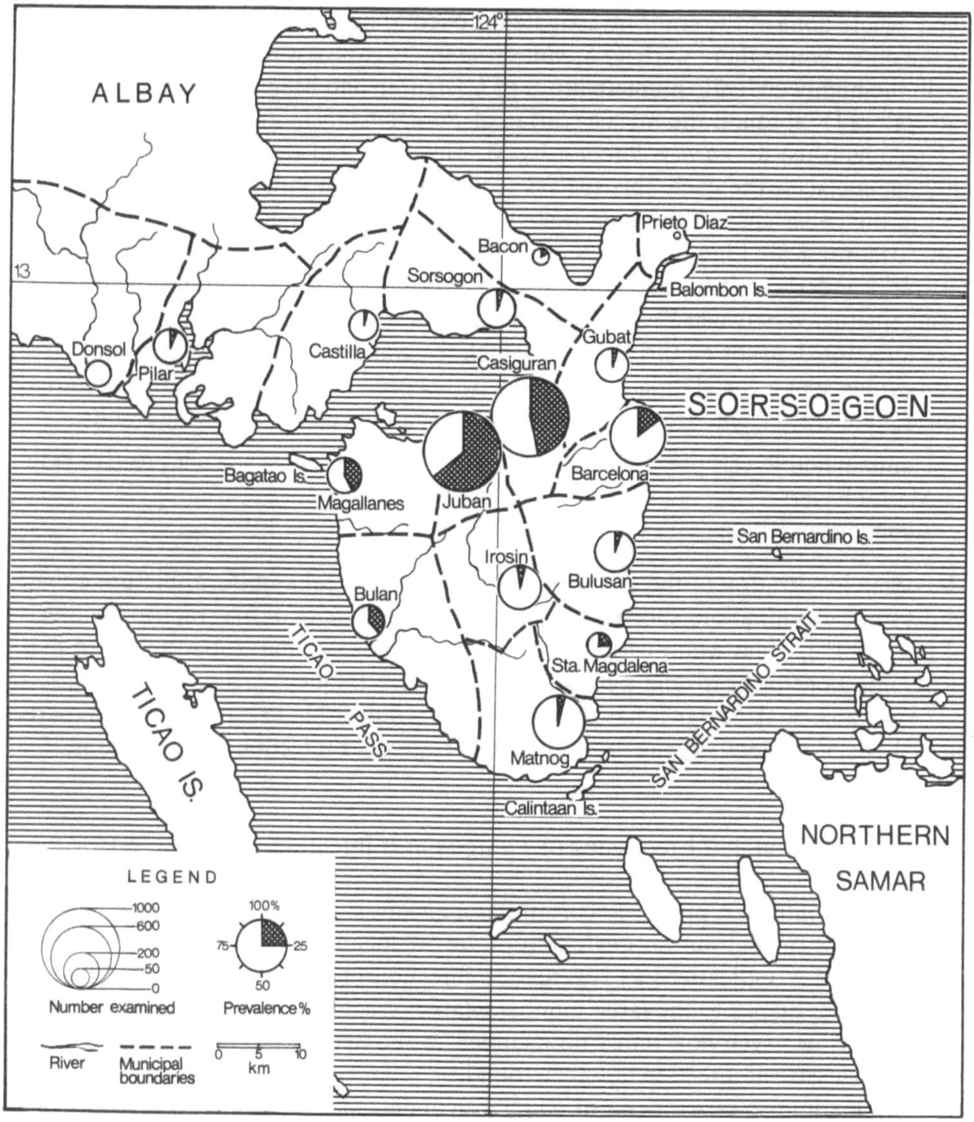

Fig. 74. Prevalence of *Paragonimus* metacercariae in *Sundathelphusa philippina* in the municipalities of Sorsogon Province [Source: 179]

showed only low rates of prevalence, fluctuating between 0.1 and 2.1%, second intermediate hosts (*Sundathelphusa philippina*) showed an infection rate of 100% in some waters.

Having examined crabs from all parts of Sorsogon Province for infection with metacercariae of *Paragonimus*, B. D. Cabrera and Valeza [179] were able not only to note differences in the proportion of waters with positive crabs, but also differences in the prevalence rates of different municipalities (Fig. 74). Juban leads with an average prevalence of 64.3%, and is followed by the adjacent municipalities of Casiguran

(45.8%) and Magallanes (41.5%). The opposite extreme occurs at Donsol on the one hand, where none of the crabs investigated proved to be positive, and at Prierto Diaz, on the other, where there were no *S. philippina* in the twelve water areas selected.

In the highly endemic municipalities of Casiguran and Magallanes, B. D. Cabrera and Valeza [179] found the metacercariae of *Paragonimus* in another species of crab – later identified as *Varuna litterata* by M. Takeda – but as seven crabs contained only one metacercaria each, this species seems to be unimportant as an intermediate host in the persistence of the lung fluke infection. Against that, *S. philippina* can attain extraordinarily high intensities of infection. In ten animals of this species which Miyazaki and Habe [546] had examined at Jaro on Leyte, the number of metacercariae fluctuated between 12 and 4,044 in each. As with the prevalence, so the intensity of infection depends on the waters in which the crabs have originated, though there is no evidence of a correlation between the two parameters. B. D. Cabrera's examination [137] of crabs from three rivers in the Jaro area showed the following results:

Hambabalod River: prevalence	75.4% – 21.2 metacercariae per positive crab
Mainit River:	67.6% – 80.2
Agongongan River:	50.2% – 19.5

The organ localization of the metacercariae does not reveal a uniform picture either: whilst some examinations showed a tendency for the muscles of the crabs to be infected (i.e. 94.4%), other examinations found the gills (i.e. 96.0%) to be the organ prefered; infections of heart and pericardium, too, were frequent (72.2%) in other cases [137, 546, 862, 864]. So far it has therefore not been possible to establish which parts of the body are potentially the most important sources of infection for the final host.

4.7.6 Distribution and Frequency of *Paragonimus* Infection in Final Hosts

Paragonimus infection primarily constitutes a zoonosis. Until now, however, only the investigations carried out on Leyte and in the Sorsogon Province have resulted in – albeit limited – indications as regards the infection of animal final hosts (cf Table 39). Cats and, thanks to the large numbers, rats present the most important reservoir of the parasites; even dogs must be included in the list of reservoir hosts, although only a single case of infection has been found.

Examinations so far carried out on intermediate and final hosts do not lead therefore to a better general view of the overall distribution of paragonimiasis in the Philippines. Proof of its occurrence in man in at least 16 different provinces, even if mostly only in individual cases, argues for the existence of this helminthiasis as a zoonosis over large areas, since single infected persons cannot guarantee its persistence. The occurrence of *Paragonimus* infection in man presupposes the presence of a zoonotic cycle in the Philippines, at least outside the endemic areas of Casiguran and Jaro.

295 cases of the 364 human *Paragonimus* infections (cf Table 39) which have so far come to light permit a geographical allocation. 75.9% of these fall to Leyte alone, where the Jaro Municipality is regarded as the most important endemic area [139, 159].

Situated on the north-eastern foothills of the Central Cordilleras, which run across Leyte from north to south, Jaro offers snail and crab intermediate hosts a large number of suitable biotopes with a rat population, 10% of which is infected with *Paragonimus* [142, 174]. As Jaro is situated further inland, sea fish and other sea foods are available only to a limited extent; instead of these, river crabs and other freshwater animals present an important source of protein. Everybody at Jaro eats such crabs, either roasted or boiled, but at Barrio Pitogo 40% of the inhabitants indicated that they also eat raw crab in the form of a dish called "kinilao".

Yogore [862] describes the preparation of this dish as follows: "As 'kinilao', the crab is eaten in the frankly raw state. To prepare this dish, the sour juice from a tiny local variety of citrus fruit 'calamansi' and the milk from grated coconut 'gata' are merely added to the edible portions of the raw crab. While a 'kinilao' of crabs is also eaten at other occasions in Barrio Pitogo, it is a great favorite with the adults during the local 'cocktail' hour when these people are downing 'tuba' – the local alcoholic drink which is the fermented sap from the coconut tree."

The greater participation of adult men in this "cocktail hour", and the concomitant risk of infection as a result of eating crab "kinilao" finds expression in higher prevalence rates as well. As investigations by B. D. Cabrera and Fevidal [159] had shown, 15.7% of the male population as against 9.4% of the female one in the barrios Pitogo and Buri was infected with lung flukes. Among adult men over the age of 18 years, the prevalence even rose to 19.0% in contrast to 9.1% among boys and youths.

Although responsible for only 6.8% of those infections which can be geographically allocated, Sorsogon Province is held to be another focus of human paragonimiasis. In spite of regionally very high prevalence rates among the crabs, the *Paragonimus* infection appears to occur mainly as a zoonosis. The rare occurrence in man is to be attributed to the fact that the population there does not eat raw crab [864]. Considering the high level of infections among the crab populations, the infection of man with *Paragonimus* in the Sorsogon Province must be regarded as an accidental event. Apart from inadequately cooked crab, crab juice may be an occasional source of infection, since it is used in preparation of a dish called "kinagang". Yogore et al. [864] describe the preparation of "kinagang" as follows:

"The raw crabs are first chopped into small pieces and then ground by mortar and pestle. The ground crab tissue is placed inside a piece of cloth and then squeezed and the crab juice collected. The juice is added to the meat of grated coconut and the mixture is wrapped in some leaves. The leaf-wrapped mixture is then boiled until dry."

It is conceivable that metacercariae get stuck to the hands when the crab tissue is being squeezed, and that they are swallowed together with food that is picked up with the fingers so that the preparation of "kinagang" and not the dish itself could be a source of infection from lung flukes.

Garrison [337] and Powell [628] did diagnose *Paragonimus* infection in ten people from Samar, but due to the lack of specific investigations on the spot it was not found out whether this points to another endemic area of medical relevance. Surveys of infections with *Schistosoma japonicum* and intestinal helminths at least, which were carried out on Samar after World War II, did not produce any further indications of paragonimiasis occurrence either.

It must be emphasized that North Luzon appears to be free from lung fluke infections. Apart from one, unproven, case of an inhabitant of the Batanas which Willets [840] cited in 1913, the northern limit hitherto known is in the area of Subic Bay (Bataan/Zambales); yet even there only one isolated case was diagnosed [298]. Even among the Ilocanos, where *Capillaria philippinensis* and *Echinostoma ilocanum* infections (to name only one or two examples) occur as food-borne helminthiases, *Paragonimus* infections have as yet not been found, even during the investigations conducted by Cross and Bhaibulaya [248] in 23 barrios into intestinal parasites.

In summary it may be stated that human *Paragonimus* infections with significant prevalence rates have so far only been made known for the municipality of Jaro on Leyte. In addition, the municipality of Casiguran in Sorsogon Province may be regarded as another focus, although one with a very low prevalence. Further isolated cases on Luzon with a northerly limit in the area of Subic Bay, on Mindoro, on various islands of the Visayas and on Mindanao indicate, however, a wider distribution of paragonimiasis as a zoonosis. Nonetheless, its appearance need not be expected except in places where relatively cool, fast-flowing mountain streams are available as suitable biotopes for first and second intermediate hosts.

4.8 Other Food-Borne Helminthiases

Besides the food-borne helminthiases mentioned above (4.1–4.7) there is evidence of other worm infections for the Philippines, which follow this mode of infection. They tend to be infection with trematodes, which are mostly brought about by the consumption of fish or water-plants containing metacercariae.

4.8.1 Clonorchiasis and Opisthorchiasis

Parasites of the gall ducts of man and mammals – *Opisthorchis (= Clonorchis) sinensis*, the Chinese liver fluke, and *Opisthorchis felineus*, cat liver fluke, have their foci of distribution in China, Taiwan, Japan and Vietnam, as well as in Europe and the Asiatic part of the Soviet Union. Both of these species have seldom been observed in the Philippines because, according to Arambulo [44], freshwater fish which present the second intermediate host and thus the infection source for man, are as a rule not eaten raw by Filipinos.

Since *O. sinensis* was first diagnosed in a Japanese in the Benguet Sanatorium [113], more than 60 cases of such infections in the Philippines have found their way into the literature [114, 130, 186, 220, 252, 337, 345, 447, 504–506, 558, 714, 842]. In some of the cases it was possible to prove that the patients were Chinese people who had already caught the infection in their homeland. Garrison [337] insisted, however, that in 5 cases out of 11 he had diagnosed a "Japanese liver fluke" the patients were Filipinos, three of whom came from Mindanao and one from Rizal Province. At least 3 of the 8 cases quoted by Willets [842] must have been Filipinos as well. Even if it is assumed that there were indeed autochthous infections, their extreme rarity in man is insufficient to ensure a persistence of this helminthiasis in the Philippines. An animal reservoir would be required, which maintains the *O. sinensis* infection at a positively endemic level, but in the Philippines neither dogs, the most common reservoir hosts in the endemic areas of Asia, nor other fish-eating mammals have ever been found to be infected (the only positive diagnosis being of a dog from China; [cf 883]). Since Velasquez' [814] "List of Philippine fish harbouring larvae of parasitic helminths transmissible to man" does not mention any fish species as being infected with

O. sinensis, evidence of metacercariae in second intermediate hosts is apparently not available either. There is thus no absolute certainty whether there is or is not an autochthous presence of this worm disease in the Philippines.

Similar uncertainty affects the assessment of the situation in respect of *O. felineus*. The only mention in the literature of evidence of this species in man concerns the wife of a Filipino soldier in Camp O'Donnell in Tarlac Province [362]. Two publications establish the domestic cat as a reservoir host in the Philippines: Garrison and Lago [340], who obtained a speciment of *O. felineus* for their helminth collection from a cat, and Wharton [835], who found an *Opisthorchis* species, which he described as *O. wardi*, in more than 20 cats in Manila when conducting post-mortem examinations. However, it subsequently turned out that *O. wardi* is to be regarded as synonymous with *O. felineus* [267, 287, 288]. Neither Wharton [835] himself, who confined his investigations to Cebu, nor any other researchers have since diagnosed *O. felineus* infections in cats in the Philippines.

Another indication of the possible endemicity of *Opisthorchis* infections in the Philippines results from the investigations carried out by Carney et al. [198], in Victoria in Oriental Mindoro, and by Cross and Bhaibulaya [248] in North Luzon. Less than 1% of 1,059, or 1% of 7,417 stool samples contained "opisthorchid eggs" there. Nonetheless, it remains to be stated that further specific investigations are required if the question of the occurrence and distribution of *Opisthorchis* infections in the Philippines is to be answered once and for all.

4.8.2 Heterophyidiasis

Heterophyidiasis is used here to cover worm infections and disease pictures the parasites of which belong to the trematode families of Heterophyidae and Microphallidae. As a rule intestinal parasites of fish-eating mammals (dogs and cats) and birds, these helminths can also infect man, in whom they are said to cause serious illnesses under certain circumstances when their eggs are deposited above all in the heart and in the central nervous system [23, 24, 26–28, 30–33]; cf in this connection the experimental work by Africa and his colleagues [10, 11, 25, 29].

There is no way of knowing with absolute certainty how many representatives of the trematode families cited above did actually occur as parasites of man in the Philippines. Since a number of names of newly described species had to be withdrawn as they proved to be synonymous with species already known, whilst others repeatedly changed their affiliation to the genera, different specific names are given in the concise presentations of animal helminths which can be transmitted to man, as well as in the lists of parasitic worms which occur in the Philippines [13, 18, 19, 33, 44–46, 49, 51, 326, 643, 647, 731, 785, 787, 804–806, 809, 811–815].

This presentation adopts the nomenclature which Velasquez [815] had brought up to date. Accordingly, the following species occurred as parasites of man in the Philippines:

Stellantchasmus amplicaecalis (Witenberg, 1929) Katsuta, 1931,
Procerovum calderoni (Africa and Garcia, 1935) Price, 1940,
Haplorchis taichui (Nishigori, 1924) Chen, 1938
Haplorchis yokogawai (Katsuta, 1932) Chen, 1936 and
Carneophallus brevicaeca (Africa and Garcia, 1935) Velasquez, 1975.

The first intermediate hosts (snails) are known for only two of the intestinal fluke species; they are *Thiara riquetti* for *P. calderoni* and *Melania juncea* for *H. taichui* [812, 813]. Second intermediate hosts or, in the case of *C. brevicaeca*, paratenic hosts are several fish species from fresh water, brackish water and the sea (cf Table 40), which act as a source of infection for the final host.

Investigations of the prevalence rates among intermediate hosts are very few indeed. Having examined 1,711 *Melania juncea*, Velasquez [812] found about 1% to be infected with pleurolophocercous cercariae, which were identified as larvae of *Haplorchis taichui* by experimental transmission. Vazquez-Colet and Africa [806] succeeded in proving the presence of metacercariae in 462 out of 921 (= 50.2%) fishes from three different species. With a prevalence of 71.4% *Hemiramphus georgii* proved most frequently to be infected, the average infection intensity being 226 metacercariae per fish.

In spite of this high prevalence among second intermediate hosts, among dog and cat final hosts infections with heterophyids, which can be transmitted to humans, have up to now been relatively rarely encountered. So, too, for man only about 130 cases have been established in the Philippines, with the diagnoses tending to be based upon evidence of eggs in the stool or tissue [11, 18, 21, 33, 197, 200, 248, 251, 352, 445, 504, 594, 863].

The high prevalence among second intermediate hosts and the broad spectrum of intermediate hosts (Table 40), as well as the small number of heterophyid infections among humans, indicate either that man does not present a very suitable final host for parasitic species occurring in the Philippines, or that Filipinos are not in the habit of eating fish or fish dishes in a raw or inadequately cooked state. (Experimental findings show that metacercariae lose their infectiousness already if they are exposed to a temperature of 55 °C for merely 10 minutes; [cf 316].) The origin of cases diagnosed in man, including people from Luzon (Ilocos Sur, La Union, Manila), Palawan, the Visayas (Leyte, Cebu, Bohol, Capiz) and Mindanao (North-west Mindanao, Bukidnon) suggests that heterophyidiasis occurs throughout the country as a zoonosis.

4.8.3 Fasciolopsiasis

Caused by the giant intestinal fluke (*Fasciolopsis buski*), this trematode infection occurs very frequently in some regions of several countries in South, South-east and East Asia [cf 240]. The source of infection for the final host is aquatic plants. In the process of swarming out from their intermediate host snails the cercariae cling to their surface, encyst and metamorphose into metacercariae. Along with man pigs are important as final hosts.

In the Philippines pigs have never been found to be infected [46, 49, 50]. Even the first-ever mention in the literature is contradictory, with Garrison and Lago [340] stating that cattle (*Bos*) are the host of a *Fasciolopsis* which they had received from the Bureau of Science in Manila. Among humans *F. buski* was diagnosed only once in the Philippines, and this was among several Chinese on Jolo, Sulu [711]. Apart from that only one other finding has become known; this occurred in a Filipino in the U.S.A. [850]. It may therefore be assumed that these few infections were acquired outside the country, and that *Fasciolopsis buski* is therefore not endemic in the Philippines.

4.8.4 Fascioliasis

Fasciola hepatica and *F. gigantica,* two species of large liver flukes, are both agents of fascioliasis and very commonly found in cattle, carabao, sheep and goats in the Philippines [cf 46, 58, 91, 282, 652, 712, 778, 785]. Both use freshwater snails of the genus *Lymnaea (L. philippinensis, L. swinehoei)* as intermediate hosts [417, 418; cf 46]. The metacercariae are found on aquatic plants which lead to infection when orally ingested.

No more than 5 cases of human fascioliasis have so far been detected in the Philippines [114, 209, 210, 445, 558, 561], so that it is a matter of very rare and accidental occurrences, albeit to be expected wherever ruminants are kept. *Ipomea obscura,* a species of aquatic plant which is frequently eaten, is said to be the source of infection [49], but in view of the rare appearance of human fascioliasis this must be taken as mere speculation.

4.8.5 Amphistomiasis

Amphistomiasis is a collective term for the infection of mammals by a number of different species of the trematode family Paramphistomatidae.

The only human parasite so far known is *Gastrodiscoides hominis;* it is endemic mainly in the Indian states of Assam, Bihar and Orissa [443, 537], where pigs are, apart from man, important final hosts. The natural intermediate host, a snail, is not known (though the experiment to infect *Helicorbis coenosus* with miracidia was successful). *G. hominis* infections are brought about by the consumption of aquatic plants carrying metacercariae.

Evidence of *G. hominis* in the Philippines rests on a single case of a member of the Philippine Scouts at Cotabato, Mindanao [798]. Pigs, moreover, have never been found to be infected in the Philippines; the question of this parasite's persistence continues to remain an open one.

4.8.6 Plagiorchiasis

There is very little evidence that representatives of the Plagiorchiidae, another trematode family, which occur very commonly as intestinal helminths of vertebrates are parasites of man as well. The case described by Africa and Garcia [21] has so far remained the only one in the Philippines. In the course of a post-mortem on a Filipino from Ilocos Sur they found five specimens of a trematode, apart from *Echinostoma ilocanum* and *Carneophallus brevicaeca,* which they identified as a new species of the genus *Plagiorchis.* Subsequently named *P. philippinensis* by Sandground [677], it was to remain the only evidence of this worm in the entire world. The event of this infection was attributed by Africa and Garcia [21] to the consumption of insect larvae, something which is very popular in the Ilocano provinces: "It is interesting to note that natives of this locality are specially fond of eating the grub of a particular insect that abounds in marshy places. It is possible that this larva acts as the second intermediate host of this species of *Plagiorchis* here." The development cycle of *P. philippinensis* being still unknown, there is no possibility of this suggestion being confirmed.

4.8.7 Anisakiasis

The so-called herring-worm disease (anisakiasis) of humans is one of the phenomena of the larva migrans visceralis syndrome, and is caused by the larvae of different genera of the nematode family Anisakidae [cf for example 224, 649]. As a result of the invasion of the stomach wall and the wall of the intestinal tract, abcesses develop which are accompanied by symptoms that tend to manifest themselves in the form of an acute abdomen. Only species living as parasites in homoiothermal vertebrates (especially in marine mammals like dolphins and seals) when adult are likely to be as causal agents of human anisakiasis. The source of infection for man is to be found in plankton-eating marine fish (second intermediate host), or carnivorous fish (paratenic hosts), which harbour third- or fourth-stage larvae of Anisakidae.

Among the fish that occur in the waters of the Philippines the following have been found to be infected so far: 12 different kinds of mackerel (Scombridae), tuna fish (Thunnidae), snapper (Lutianidae) and barbed mackerel (Carrangidae) [46, 126, 428, 809, 810, 814]. There has so far been no evidence of anisakiasis in man.

4.8.8 Trichinosis

Trichinosis is a very common zoonosis of carnivorous and omnivorous mammals, which is caused by infection with the nematode *Trichinella spiralis.* Depending on the worm burden, the adult worms (intestinal trichinellae) can be the cause of intestinal symptoms; more serious though is the infection of the striated musculature by larvae (muscle larvae), which may be lethal. In the main man becomes infected by eating raw or insufficiently cooked trichinous pork. In principle, however, the meat of all other mammals functioning as hosts of *T. spiralis* must be taken into consideration.

In view of the fact that pigs and rats present important members of the zoonotic cycle of many endemic areas, research in the Philippines has also been concentrated on the direct evidence of *T. spiralis* in these animals, although there have been no positive findings at all [416, 426, 508, 543, 582, 672, 748, 779]. The results of serological examinations of pigs pointed in the same direction as well. Although Orange [580] had observed a positive result in the Suessenguth-Kline flocculation slide test in 32% of the imported pigs, and in 25.9% of the locally-bred ones, he did not succeed in finding evidence of muscle trichinae in any of them. All the serotests carried out by other authors at a later date were negative as well [54, 55].

As far as man is concerned, Garduño [333] is the only one to have made the clinical diagnosis of "trichinosis" in two cases. Both of the patients had shown the typical symptoms (including high eosinophilia), together with a positive result in the Bachmann-intradermal test, but here, too, there was no direct evidence of muscle trichinae in the biopsy material. There is thus so far neither evidence of a zoonotic cycle of *Trichinella spiralis* nor of the occurrence of this parasite in man in the Philippines.

4.8.9 Rare Human Helminthiases as a Consequence of the Accidental Swallowing of Arthropods

Numerous helminth species, which are parasites in animals, include arthropods as intermediate hosts in their life cycle, which, having been ingested orally, lead to infection of the final host. In some cases it is a clearly defined relationship between a predator and his prey, on others the behaviour of the infected intermediate hosts is responsible for their intake in his food by the final host (as where ruminants eat ants infected with metacercariae of *Dicrocoelium;* [cf 373, 374, 469, 470]). In still other cases ectoparasites function as intermediate hosts which are bitten to pieces by the final host; this may lead to the swallowing of infection stadia (such as dogs crunching fleas which are infected with cysticercoids of *Dipylidium caninum*).

Besides genuine food-borne helminthiases with arthropods as their source of infection (cf 4.7), worm infections in man have been observed time and time again, coming about as a result of the accidental swallowing of arthropods. On occasion, however, it is not possible to distinguish between the two kinds of infection with absolute certainty, since insects (for instance) may be normal components of human food in some regions of the world. It follows that an infection with one and the same parasite may be regarded as a food-borne helminthiasis in one ethnic group but as an accidental swallowing of insect intermediate hosts in another. The very rare occurrence of the human helminthiases described above is a compelling reason for assuming that these infections are indeed accidents.

The dwarf tapeworm (*Hymenolepis nana*) occupies a special place in this context. This parasite is one of the facultatively diheteroxenous helminths the life cycle of which may include an intermediate host (fleas); as a rule, however, human infections do not occur by way of swallowing cysticercoid-infected fleas, but through the faecal-oral route of infection with the oral intake of eggs (direct course of development). Tapeworm infections caused by *H. nana* are therefore primarily to be regarded as geo-helminthiases and will be discussed elsewhere (cf 5.7).

4.8.9.1 Dicrocoeliasis

In the Philippines the lancet fluke (*Dicrocoelium dendriticum*) is known to be the causal agent of the dicrocoeliasis of herbivorous mammals, but there is so far no evidence at all of genuine infections among man as a result of swallowing ants. As a rule, the detection of *Dicrocoelium* eggs in human stools only reveals a pseudo-parasitism in any case, which must be attributed to the consumption of infected livers from ruminants. Another cause of pseudo-parasitism of this kind can be fish infected with trematodes from the family Didymozoidae, for the worm eggs having been eaten together with the infected fish, appear undigested in the stool and, being very similar to those of *Dicrocoelium,* merely simulate an infection. This might well apply to the evidence of "dicrocoelid eggs" in man as well which has been reported from Oriental Mindoro, Northern Bohol, Cebu, Samar and Bukidnon [197, 198, 200, 245, 251].

4.8.9.2 Bertielliasis

The agent of bertielliasis is the tapeworm *Bertiella studeri* (family: Anoplocephalidae); it occurs widely in Africa and Asia as a parasite of monkeys and other primates. Oribatid mites serve as intermediate hosts. In the Philippines *B. studeri* has so far been found only in one girl from Sara in Iloilo Province, in one monkey (probably *Macaca irus philippinensis*) from the Rizal Province and in a dog in Manila [17].

4.8.9.3 Raillietiniasis

Several species of the tapeworm genus *Raillietina* (family: Davaineidae) come into question as causative agents of raillietiniasis, which is very common among animals.

Two species, *R. madagascariensis* and *R. garrisoni,* have been described for the Philippines, although they could possibly be identical. Insects are supposed to be the intermediate hosts and brown rats (*Rattus norvegicus*) the main final hosts. Prevalence rates of 86.0 and 31.7% have been established for them [426, 779], but 9.4% of domestic rats proved to be infected as well [426]. Between 1911 and 1975 a total of 25 human *Raillietina* infections were diagnosed in the Philippines [15, 339, 424, 645, 741]. Apart from the case recorded by Garrison in 1911 [339], all the infections occurred in children.

4.8.9.4 Dipylidiasis

Known as dipylidiasis, the infection of dogs and cats with tapeworms of the genus *Dipylidium* (family: Dilepididae) occurs all over the world. In the case of the Philippines five species [778] have been described, but they are largely identical with *D. caninum* [785]. The infection of the final hosts arises from the swallowing of cysticercoids from chewed fleas (*Ctenocephalides canis, C. felis*) or dog-lice (*Trichodectes canis*). The rates of prevalence established among cats and dogs in the Philippines fluctuated between 3.0 and 81.5%, depending on the method of investigation [56, 571, 646, 736, 774, 775, 834]. In spite of this in part considerable frequency among the main final hosts there is evidence of only two human infections [318, 532, 533], and indications of two more cases [210, 506].

4.8.9.5 Hymenolepiasis diminuta

Cases with *Hymenolepis diminuta* are a very common kind of tapeworm infection in rats, mice and other rodents. Their intermediate hosts tend to be fleas (*Xenopsylla cheopis, Nosophyllus fasciatus*), but other insects, such as flour beetles, do also occur. Investigations carried out in the Philippines showed prevalence rates of 12.9–64.0% among *Rattus norvegicus,* whilst *R. rattus* proved to be somewhat less frequently infested with only 9.9–33.3% [426, 672, 779]. *H. diminuta* infections diagnosed in man are c. 10 cases only [197, 245, 251, 336, 337, 505, 506, 714, 863].

4.8.9.6 Infections Involving Cheilospirura sp.

A nematode which had been removed from the conjunctiva of a man and described by Africa and Garcia [20] as *Cheilospirura* sp., may in fact have been *Acuaria hamulosa,* a parasite in the gizzard of gallinaceous birds, which is also known from the Philippines [785]. The only infection observed in man has been attributed to the accidental swallowing of second intermediate hosts (locusts) or the direct penetration of infective larvae [20].

4.9 Short Synopsis: The Distribution of Food-Borne Helminthiases

There are indications of three distribution centres for food-borne helminthiasis, the most significant of which is the area of the north-west and north coast of Luzon, where the Ilocanos have settled. The endemic helminths there are primarily those which can be transmitted by the consumption of fish and snails, i.e. *Capillaria philippinensis, Echinostoma ilocanum* and also Heterophyidae [cf 248]. Much more closely delimited is the second centre with the Municipality of Jaro, Leyte, where *Taenia* and *Paragonimus* infections predominate among the food-borne helminthiases. Lastly, another focus is to be found in the Bicol Region, where paragonimiasis has been increasingly observed. In the Ilocano Provinces as well as in Jaro the inhabitants have developed a preference for eating certain animals, or dishes prepared from such animals in a raw state. The eating habits of particular population groups and tribes are therefore primarily responsible for the endemic occurrence of the helminthiases referred to above.

Isolated evidence of several food-borne helminthiases in man in a fair number of provinces of the Philippines indicate that these helminthiases are very common as zoonoses, although an increase in human cases is to be expected wherever there is an influx of Ilocanos in the context of internal migration. Next to the Visayas, the Ilocos Region was particularly actively engaged in internal migration (cf Table 30). At the same time it reveals the problem of the introduction of such worm infections, which did not exist in certain areas as a zoonosis or as a parasitosis in man. Infected immigrants may subsequently become the point of departure for the establishment and even the persistence of food-borne helminthiases, which spread to the indigenous population if their eating habits encourage this. However, a pre-condition for this is the presence of intermediate hosts, which are obligatory for the parasitic cycle; essentially this is determined by geofactors. Only specifically-directed investigations in the future will be able to show whether such tendencies can already be observed.

5 Soil-Transmitted Helminthiases

Worm infections caused by soil-transmitted helminths differ from insect-, water- and food-borne helminthiases chiefly by the direct development of most of their agents. Having been passed out with the stool, their eggs or larvae develop on the ground or in its surface layers to the infective stage, either remaining in their shell, being swallowed and causing infection (*Ascaris lumbricoides, Trichuris trichiura,* for example), or entering man percutaneously as a free larva (e.g. hookworms).

As these monoxenous species referred to as geohelminths do not require intermediate hosts during their development cycle, geofactors solely affect the larval stages, the development of which takes place in the open, thereby determining the areas of potential distribution. As a rule thick-shelled helminth eggs like those of *Ascaris* and *Trichuris,* the embryonic and larval development of which occurs within the eggshell, display a high degree of resistance to outside chemical and physical influences – an important cause underlying the cosmopolitan distribution of these species. Thin-shelled eggs and the larvae hatching from them, such as those of the hookworms, show a higher sensitivity towards such outside influences, with the consequence that the distribution of these helminths is limited to the tropics and subtropics. However, the temperature and humidity conditions of the warm climates do also have a favourable effect upon the development of larvae of the ubiquitously distributed geohelminths in the open air and cause them to reach the infective stage in a shorter time than in the temperate latitudes.

Within the potential distribution areas additional factors other than abiotic ones determine the presence and frequency of geohelminthiases in man. A high prevalence of faecal-oral and faecal-percutaneous worm infections is an indicator of a low standard of living and of hygiene. Last, but not least, they explain the great frequency of numerous worm infections in the countries of the Third World.

With their narrow spectrum of hosts, known as stenoxeny, geohelminths present another characteristic which separates them clearly from the water- and food-borne worm species which occur in the Philippines, i.e. man tends to be the only final host, so that animal reservoir hosts need not be taken into consideration for epidemiology and thus for geomedical analysis.

The most frequent causal agents of soil-transmitted helminthiases in the Philippines are large intestinal roundworms, whipworms and hookworms (*Ascaris lumbricoides, Trichuris trichiura, Ancylostoma duodenale, Necator americanus*), followed by pinworms (*Enterobius vermicularis*) and threadworms (*Strongyloides stercoralis*). In addition, less frequent or rare worm infections will also be considered in what follows – i.e. those caused by the facultatively diheteroxenous dwarf tapeworm *Hymenolepis nana* or by the larvae of *Toxocara* and hookworms. Also included is echinococcosis, since man by serving as an intermediate host, receives his infection from dog faeces (faecal-oral infection). On the other hand, cysticercosis has already been treated in connection with the *Taenia solium* infection (cf 4.5).

5.1 Short Historical Survey

There are no reports concerning the occurrence and frequency of soil-transmitted helminthiases in the Philippines for the Spanish colonial period, which continued until 1898. Even the detailed "Topografia medica de las Islas Filipinas" by Codorniu y Nieto [233] from the year 1857 fails to mention such worm infections, although they must have been extremely frequent because of the hygiene situation. Worcester [845] characterized these conditions in a very drastic manner: "I had abundant opportunity to observe health conditions in the Philippines during the Spanish regime and they were shocking in the extreme. There were no provisions for the sanitary disposal of human waste even in Manila. If one had occasion to be out on foot at night, it was wise to keep in the middle of the street and still wiser to carry a raised umbrella."

During the initial period of the American occupation soil-transmitted helminthiases were not much mentioned either. Neither the short report on the "Character of sickness among troops in the Philippines" [223] nor the "Report of a special commission sent to the Philippines by the Johns Hopkins University to investigate the prevalent diseases of the islands" [299 a] contained references to that effect. Thanks to the limited pathogenicity of intestinal helminths, American military physicians evidently at first rated dysenteries and feverish conditions of different origins as being of top priority in their importance for the health of the troops [cf 206, 254–256, 528], although intestinal parasites were known to be very common throughout the Philippines [41, 275, 348].

Though predominantly surgeons, bacteriologists and pathologists, these military doctors soon had to apply themselves to helminthological problems as well since it was precisely the soil-transmitted helminthiases that also occurred among the American troops with increasing frequency. It is understandable that a number of reports and publications singled out these problems; very soon, however, Filipinos in American service (such as the Philippine Scouts and the Philippine Constabulary) and their dependents, as well as the population in the vicinity of recently erected hospitals, including Philippine patients who had been admitted to the hospitals, were included in medical care and thus in investigations into worm infections [67–70, 207, 208, 211, 235, 236, 238, 283, 284, 376, 438, 628, 734, 764, 765, 768, 807, 841].

Other results also contributed to the extension of knowledge of soil-transmitted helminthiases in the Philippines, such as autopsy findings [189, 252, 345], the carrying through of research tours to Baguio and the Batanes Islands [209, 210, 840], and the examination of inmates of the Bilibid Prison [337, 718, 842]. However, particular importance must be attributed to some surveys. Apart from those in Manila and Cebu [108, 343], which were of limited extent, these were chiefly the "Medical Survey of the Town of Taytay" in the Rizal Province [342, 743], the surveys in the Cavite and Rizal provinces and in the Cagavan Valley [655, 839] and the "Sanitary Survey in Mindoro" [559, 560, 822].

The flourishing of helminthological research in the Philippines at the beginning of this century has undoubtedly been encouraged by the setting up of several institutions, such as the "U.S. Army Board for the Study of Tropical Diseases as they exist in the Philippines" in 1900, the Bureau of Government Laboratories in 1901, the Manila Medical Society and the Philippines Islands Medical Association in 1903, as well as the Philippine Medical School in 1907 [522, 599, 723]. The time during and immediately after World War I, however, must be regarded as a period of transition, in so far as research into soil-transmitted helminthiases is concerned, as only a very few publications on this topic appeared [327, 526] (cf also Rozal's bibliography, 1978). Qualified Philippine helminthologists had not been trained until then, and they then took over the research.

The 1920s saw the beginnings of a productive phase, with greater attention being paid to geographical standpoints as well. The external reason for this emphasis on knowledge of the distribution of soil-transmitted helminthiases was the inauguration of the so-called "hookworm campaign" [191, 201, 482, 489, 491, 772]. Although a "Commission for the study of hookworm and other tropical diseases" had been set up as far back as 1909, and been placed under the joint direction of the Bureau of Health, the Bureaux of Science and the Philippine Medical School [115], the problem was taken up again only after the discovery of the effectiveness of carbon tetrachloride against hookworm. The first steps were taken in 1922 with mass examinations and the treatment of infected persons in Cebu. Within two and a half years these measures were extended to 22 different regions of the country. They were not, however, limited to diagnosis and therapy, but included the health education of the population and the building of latrines. Manalang [489] describes the complete programme accompanying it as follows:

"With the treatment campaigns went educational instructions which were carried directly to the home by means of public meetings, moving pictures and microscopic demonstrations of the life cycle and deleterious effects of the parasite, and indirectly thru the school children by means of hookworm pamphlets and their use in the classes of hygiene. The construction of a standard fly proof latrine known as "antipolo system" held an important part in these campaigns. In the majority of the towns ordinances exist which penalize owners of houses without the proper latrines."

From 1921–1928 it was Manalang who above all contributed substantially to the extension of knowledge of ankylostomiasis as well as other soil-transmitted helminthiases. Later on other researchers, such as Tubangui and his collaborators investigated the geographical distribution of these worm infections [788, 789]. At that time already correlation was found to exist between the frequency and intensity of hookworm infection and the Philippines' different climatic zones.

After World War II the investigations into the distribution of schistosomiasis provided extensive information concerning the frequency of soil-transmitted helminthiases in different parts of the country [595, 597]. So, too, the significance of the climate was again emphasized as a cause of regional differences in prevalence [132, 612]. A substantial number of different surveys (cf Table IV), including a comparison between the situation in urban and rural areas, and between different occupational groups, enabled some of the gaps in knowledge to be filled in. American researchers, who were members of the U.S. Naval Medical Research Unit, took part in this [cf 197, 198, 200, 245, 251].

Besides carrying out surveys, other epidemiologically and geomedically important questions – such as the problem of re-infection after previous anthelminthic treatment – were taken into account. The problem was repeatedly followed up through appropriate investigations among selected population groups [143–145, 147, 154, 157, 176, 180, 315, 429, 430]; the results gained in this will be discussed elsewhere (cf 5.2.3).

Specific measures for the control of soil-transmitted helminthiases have hitherto been largely confined to limited therapy programmes [cf 5, for example], or, more recently, to pilot projects within the ambit of Integrated Family Planning, Nutrition and Parasite Control Projects. First begun in several Asian states by the Japanese Organisation for International Co-operation in Family Planning Inc., these pilot projects are concerned with the following matters in the Philippines [436]:

1. the majority of the adult population in the pilot areas should be thoroughly informed and educated about family planning, parasite control and nutrition,
2. all those who are found positively infested by soil-transmitted helminths in the pilot areas should be treated and cured;
3. at least 60% of the married couples in the reproductive age should become acceptors of family planning.

Going far beyond this the so-called Manila Statement of the Seventh Asian Parasite Control/Family Planning Conference on the Integration of Family Planning, Nutrition and Parasite Control (1980) formulated the general aims as follows: "In view of the concern expressed in the preamble, and to implement strategies and mechanisms to deal effectively with the items contained in the Declaration of Alma-Ata, the Conference addressed itself to the following objective:

> The achievement of more effective utilization of appropriate approaches to attain the social and economic well-being of the people, through the experiences gained in the Integrated Family Planning, Nutrition and Parasite Control Project."

Pilot projects, which are intended to run for six years, have so far been started in four different areas of the Philippines: in Caramoan, a remote rural area of the Camarines Sur Province, in Davao City (Davao del Sur Province, Mindanao), in Kawit (Cavite Province) and at San Pedro, a semi-urban resettlement area in the vicinity of Manila [436]. The Caramoan and Davao projects both began in 1976, those at Cavite in the following year. Reports on planning and the outcome have repeatedly appeared in the context of the conferences of the Asian Parasite Control Organization and the Consultative Meetings for Integration [203, 262, 272, 291, 421, 436, 538, 613, 632, 641, 657, 658, 697, 730, 732, 801, 802, 851]. It remains to be seen, however, in how far the experiences gained in Pilot Project Areas can be translated into regional or even country-wide measures.

A survey of the investigations carried out in the Philippines since the beginning of the century which concern infections with soil-transmitted helminths, shows that the results differ markedly from the findings on the occurrence of other helminthiases: whilst schistosomiasis and filariases (to name just a few) have their well defined areas of distribution, there is practically no place without infections by Ascaris, Trichuris and hookworm. For these geo-helminths, regional and local differences manifest themselves merely in the extent of prevalence and the intensity of infection. Depending on the conditions determined by biotic and abiotic factors, considerable differences may exist between neighbouring villages or contiguous parts of towns, whilst entire regions or macro-regions may differ only slightly in frequency. Due to this fact any meaningful mapping of the distribution of soil-transmitted helminthiases cannot be attempted. Nevertheless, such worm infections are of geomedical relevance too, and therefore form part of this monograph.

5.2 Ascariasis

The infection of the intestinal tract with the large intestinal roundworm (*Ascaris lumbricoides*) constitutes the most common human helminthiasis throughout the world. Pathogenicity is generally considered to be low; massive infections, however, may lead to ileus, just as there are isolated cases recurring in inflammatory processes, which are caused by migrating adult worms – for instance, when entering the biliary duct or penetrating the intestinal wall. These complications are amply covered by information in the specialist literature on the Philippines (cf the bibliography of Rozal, 1978).

The mode of infection by *A. lumbricoides* follows the faecal-oral pattern, the basis being a milieu that is contaminated by human faeces. Having been deposited in the open along with the stool, *Ascaris* eggs develop into infective larvae inside the eggshell. If it is ingested orally, as for instance with contaminated foodstuffs, the larvae hatch from their eggshells after reaching the intestinal tract of man, enter the lymph- or blood vessels of the mucosa and return to the intestinal tract by way of the liver, heart, lungs and trachea, where they on average attain adulthood within 2½ months. The daily production of c. 200,000 eggs [cf 502] by the 20–40 cm-long females is one of the reasons for the high degree of contamination of the milieu, and thus also for the high rates of prevalence one finds in many places in the Philippines.

5.2.1 *Ascaris* Development and Geofactors

In the case of a monoxenous helminth like *A. lumbricoides*, the influence of geofactors is limited to the larval development within the eggshell, which takes place in the open; the duration of this development is primarily determined by temperature. The range suitable for *Ascaris* larvae is considered to lie between 15 and 34 °C [554]. At a temperature of 16–18 °C the development into an infective larva takes c. 50 days; it is speeded up by rising temperatures (23 °C: 30 days; 26 °C: 20 days). The optimum is reached with a constant temperature of c. 30 °C: infective larvae already appear after 10 days. A prerequisite for this is, however, access to oxygen, since development does not take place under anaerobic conditions. Temperatures higher than 32 °C have a retarding effect; development ceases completely at 37 °C already; the eggs perish within 6–7 days. A further rise in temperature correlates with a shortening of the survival period, which falls to only 2–3 days at 41 °C.

These results – obtained in laboratory experiments [cf 553, 554] – clearly prove the significance of temperature as a climatic factor for the transmission dynamics of ascariasis; the constancy of largely optimal temperatures of the annual and diurnal course in the tropics permanently ensures a short development period for the larvae, thereby shortening the succession of parasite generations by contrast with conditions in temperate climatic zones, and possibly leading to a more intensive contamination of the soil in the tropics. On the other hand, the optimal temperatures mentioned

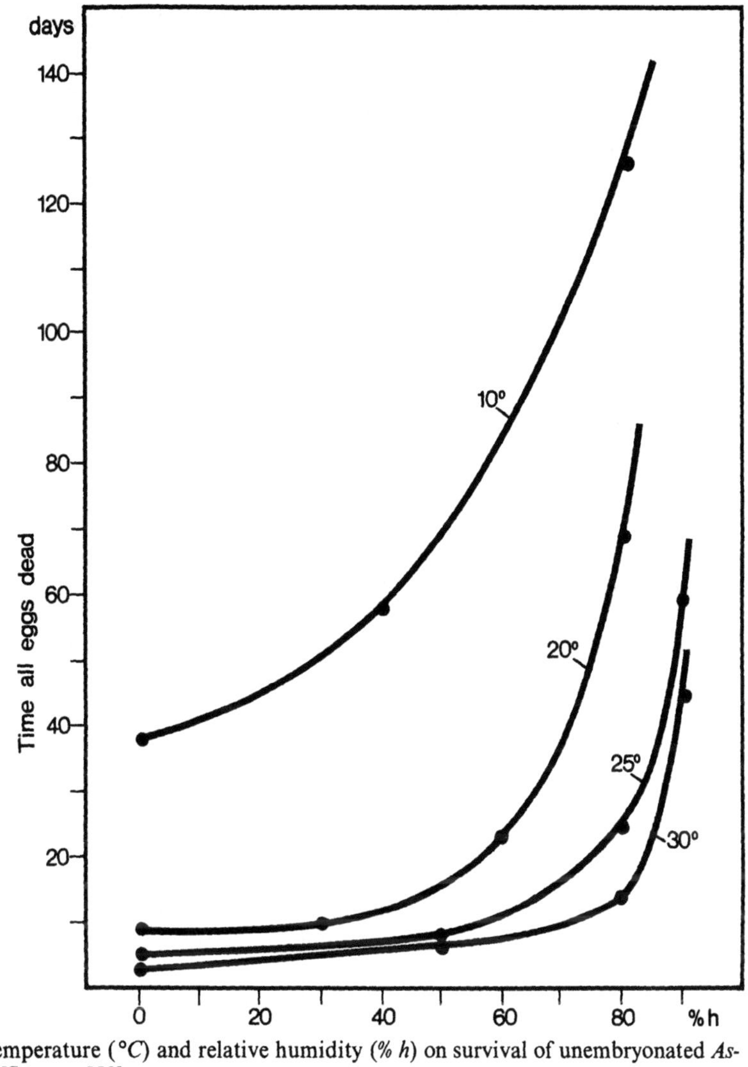

Fig. 75. Influence of temperature (°C) and relative humidity (% *h*) on survival of unembryonated *Ascaris lumbricoides* eggs [Source: 553]

above also exert a negative influence on *Ascaris* eggs, the period of survival of which has a markedly negative correlation with a rise in temperature, though increasingly atmospheric humidity may have a counter-balancing effect (cf Fig. 75). Examinations of non-embryonized eggs have shown that at a temperature of 10 °C and an atmospheric humidity of 80%, all the eggs had perished after 126 days, and that this period was drastically reduced as temperatures increased (20 °C: 69 days; 25 °C: 25 days; 30 °C: 15 days). This correlation applies to humidity values as well. This reciprocal relationship between temperature and atmospheric humidity evidently operates in such a way that low temperatures, together with low atmospheric humidity, enable *Ascaris* eggs to survive approximately as long as in high temperatures and high atmospheric humidity: at a temperature of 10 °C and a relative humidity of 40%

all the larvae had died after 58 days, at 25 °C and 90% relative humidity after 60 days. The attempt at setting up a formula as the basis of the data given by Morishita [553] with temperature (x_1) and atmospheric humidity (x_2) as independent variables, and the period within which all the eggs have perished (y) as a dependent variable, shows that the negative influence of the temperature significantly exceeds the positive effect of the atmospheric humidity ($\hat{Y} = 112.90 - 4.79\,x_1 + 0.60\,x_2$).

The consequence of these laboratory findings for the epidemiology of ascariasis in the Philippines is that in climatic zones with characteristic dry seasons of high temperatures and low humidity, the conditions for the development and survival of *Ascaris* eggs are seasonally distinctly lower than in areas with less high temperatures and precipitation throughout the year. The question of whether there is indeed evidence of this in differences in prevalence will be discussed elsewhere (cf 5.2.3). However, it is not the values for temperature and humidity, which are measured in the shade above ground-level (i.e. the usual climatic data), that are decisive for a comparison of the regionally varying conditions, but the conditions in the immediate vicinity of the ground and in the surface layers of the soil, where the warming may considerably exceed the level conducive to the development of *Ascaris* larvae. Vegetation as a provider of shade may assume a protective function for larvae in this case, since a compact plant cover prevents too intensive a warming up [cf 371].

Direct solar radiation does not, however, have only a negative effect on *Ascaris* eggs as a result of the rise in soil temperature. It also occurs through the proportion of ultra-violet rays. Direct exposure of larvae within their shells causes them to lose their ability to live and develop as soon as it exceeds six hours. 50% of uterine eggs even lose their ability to live after only one hour, and after seven hours or more all of them have perished. In this the sensitivity of *Ascaris* eggs, which do, after all, have a pigmented shell, is even higher than that of the hookworm eggs with their thin and unpigmented shells – to name but one example.

In experiments Velasquez [808] has tested the ability of *Ascaris* eggs to survive under changing external conditions such as occur in the Philippines. She considered the importance of different soil types (red clay, brown clay, clay loam, sandy loam, sand), exposure to sunshine as well as shade, soil on the surface and at different depths, in a moist or muddy state. In addition, the ability to survive in waters of different origin and pH values was included in the investigations, as was their reaction when put on "common rural floorings" (cement, adobe stone, wooden, compact soil of different types). According to specific conditions, survival periods varied from 26 to 567 days. They were generally shorter when exposed to the sun or put into dry conditions, when in mud, brown clay and in water with a low pH value. The abundance of detailed results does not, however, permit precise conclusions to be drawn for a varying frequency of ascariasis in particular areas of the Philippines.

Among the different geofactors that affect *Ascaris* eggs, chemical influences are of only minor importance. Due to the thickness and chemical composition of their shell, these eggs are very resistant to such influences (their infectiveness remains intact even if they are stored in 25% formalin for 7 days; [cf 553]). Geofactors of that kind therefore do not exert any effect which would restrict their distribution. On the other hand, spices added to food cannot prevent ascariasis either. On the grounds of his own investigations, Manalang [499] notes that the impermeability of the eggshell itself towards high concentrations of "common food condiments (salt, vinegar and sugar)" presents one of the causes for the great frequency of ascariasis.

5.2.2 Transmission and Sources of Infection

The chances of acquiring an *Ascaris* infection fan out very widely. However, in general, transmission takes place predominantly in three areas: in faecally-contaminated yards and compounds, at work in fields that have been fertilized with human faeces, and through consumption of contaminated vegetables [295].

As Paguio [584] noted, vegetables offered for sale can be contaminated with protozoan cysts and/or helminth eggs up to 80% and more, depending on the species. He examined 30 samples each of 12 different kinds of vegetables, as well as strawberries, most of which came from public markets, grocery stores, pavement vendors and vegetable gardens in Manila. The average degree of contamination was 31.3%. With 44 positive results (= 11.3%) *Ascaris* eggs proved to be the most frequent single finding.

Ocampo-Paus and Agor [574] achieved similar results when examining 2,311 samples of vegetables (including strawberries), 10.5% of which contained *Ascaris* eggs. On the other hand, it was possible to detect regional differences: vegetables from Baguio City, Benguet Province (13.8%) were significantly more often contaminated than samples from Manila and its surrounding countryside (8.4%). In addition, data collected by Ocampo-Paus and Agor [574] can be used to show the degree of contamination to be noticeably dependent on the hygric seasons, varying from 12.7% for the rainy season as against only 8.4% for the dry season. In this differences in Manila (11.0 as compared with 6.4%) were more marked than at Baguio City (14.7 as against 12.6%).

As far as particular varieties of vegetables are concerned, positive findings were more frequent in those with rough and hairy surfaces offering greater adhesiveness and more protection against solar radiation and desiccation for the eggs, as well as in vegetables that are of low growth habit which, being close to the soil, are more exposed to contamination by watering and manuring. In the investigations of Ocampo-Paus and Agor [574] samples of water-cress from Banguio appeared to be particularly prone to containing *Ascaris* eggs. The reason for this proved to be that streams in which the water-cress was grown, were polluted by sewage from a settlement further upstream.

Although the investigations mentioned above do not provide certainty over whether contamination is caused more by the faeces of pigs or rather by those of man[1], the results do show that vegetables can be an important source of infection. Even the usual washing can only remove a certain percentage of the *Ascaris* eggs adhering [cf 574, 808]. Paguio [584] considers the following causes as being responsible for the contamination of the vegetables:

- use of night-soil as fertilizer for vegetables
- use of polluted ditch or stream water to freshen vegetables
- the vomitus and faeces of filth flies and cockroaches
- dirt from the air as a result of the unprotected exposure of vegetables
- infected food handlers
- infected consumers themselves in the preparation of food

[1] The pig ascarid (*A. suum*) is now regarded as a separate species. Living in man, it may on occasion even reach sexual maturity. The eggs of *A. suum* and *A. lumbricoides* are morphologically indistinguishable

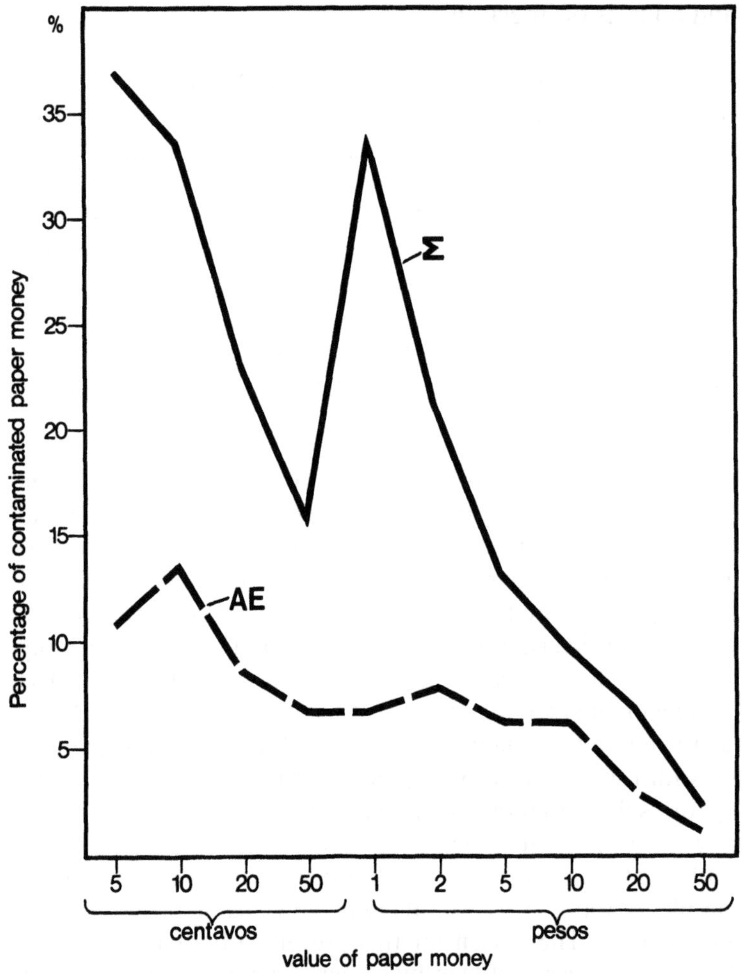

Fig. 76. Parasite contamination rate of paper money of different value. Σ all kinds of parasite cysts, ova and larvae, *AE Ascaris* eggs [Source: 585]

In her laboratory experiments in the Philippines, Velasquez [808] proved that insects can play an important role in the mechanical transmission of *Ascaris* eggs to foodstuffs. *Ascaris* eggs were isolated from the body surface of a number of fly species of the genera *Musca, Calliphora* and *Lucilia,* in cockroaches and ants which had been on contaminated ground, as well as from domestic animals (dogs, pigs, chickens), and from mice and rats. In flies which had been kept in cages containing human and pig faeces for 9–20 hours, even the intestinal tract proved to be egg-positive. After feeding on contaminated bread the faeces of dog, pig and chicken contained *Ascaris* eggs too; dogs and chickens continued to pass such eggs for three days. The significance of vertebrates for dissemination is thus clearly established, since *Ascaris* eggs may survive for several days, even in temperatures which correspond to the body temperatures of these animals [cf 554]. This would mainly apply to pigs, as they tend to eat human faeces.

Velasquez' experimental findings [808] on the possible role of flies in the transmission of worm eggs were confirmed by the results of a small field investigation carried out by E. F. de Leon [454]. The external washing of 62 Muscidae and 25 Calliphoridae, which had been caught in a squatter area of a barrio in Rizal Province, contained a total of 8 *Ascaris* eggs. Of greater significance, however, may be the further distribution through the faeces of these flies, the intestinal content of which was positive for the eggs of *Ascaris* in 8 of the Calliphoridae and in 1 of the Muscidae. Synanthropic insects are consequently to be regarded as an important link in the transmission of ascariasis.

One of the numerous sources of infection is to be found in currency notes contaminated by worm eggs. Examining 1,050 bank notes of different values, Paguio [585] found protozoal cysts, worm eggs and larvae on 19.5% of them, almost half of which were of the infective stage; *Ascaris* eggs headed the list of parasite stages with 7.0%, followed by *Trichuris* (5.0%) and *Enterobius* eggs (3.6%). The degree of contamination showed a distinct negative correlation with the value of the notes: the higher the value the less often there were cysts and eggs on the notes (cf Fig. 76). The reasons for this are the less frequent change in ownership of higher value notes, and the fact that the less well-off population groups, which have particularly high rates of prevalence, have rather less occasion to come into possession of such notes. Another difference arose from the origin of the money. Coming from public markets, stores and restaurants, the overall degree of contamination (31.1%) was markedly higher than in those cases in which it had been in the possession of individuals (16.5%) or banks (10.8%). This, too, is connected with differences in the frequency of ownership changes, though Paguio [585] gave still more reasons for the high frequency of parasite stages in money from markets, shops and restaurants, such as the simultaneous handling of vegetables and other foodstuffs which are themselves highly contaminated, the high humidity of notes making for a highly favourable medium for the development of parasites, in addition to the presence of flies and cockroaches, the significance of which has already been referred to (see above). Thus money notes can also be regarded as an important vehicle for the transmission of ascariasis.

5.2.3 Frequency, Transmission Index, Geographical Distribution and Re-Infection Rate

The high resistance of *Ascaris* eggs to geofactors and a high contamination of the milieu combine to make ascariasis the most common worm infection of man in the Philippines as well. On the basis of numerous findings from investigations in different areas of the Philippines, which have been published since 1971 (cf Table IV), an average prevalence of more than 60% must be assumed for the country as a whole, i.e. in a total population of 50 million at least 30 million people are at present likely to be permanently infected with *Ascaris lumbricoides.* Assuming merely the lowest intensity of infection of one *Ascaris*-♀ per infected person, which produces 200,000 eggs a day, the daily passing of eggs with the stool would be at least 6×10^{11} eggs of this worm species alone in the entire country. In view of the total surface area of the Philippines – 300,000 km² – it would mean a daily contamination of the soil with an average of 20 eggs per square metre.

However, the actual intensity of infection must be estimated as being higher than this. Having examined 100 autopsy cases for the sex ratio of *Ascaris,* E. G. Garcia [310, 311] observed an average infection of 15 worms with a male/female ratio of 1:2.4. Such an intensity of infection would bring in its train an average daily passing of 2.1 million eggs per infected person. Quantitative examinations of *Ascaris* eggs in the stool led to similar results: depending on the groups of persons under investigation, between c. 1500 and 20,000 eggs per gram of faeces were found [612, 788, 863]. Taking a daily total quantity of c. 150 g of stool into account [cf 606], the passing of eggs per day and per infected person would be between 200,000 and 3 million. In addition, the number of ascarids passed out subsequent to anthelminthic treatment can provide important information concerning the intensity of infection. In Barrio Salvacion in Palo (Leyte) for instance, a total of 4,334 *Ascaris* were found in 320 persons who had undergone a single treatment with pyrantel pamoate, i.e. 13.5 per infected person, with a proportion of 10.5 *Ascaris*-♀♀ [154]. Individual passing of eggs thus amounted to more than 2 million eggs, a figure which had to be rated higher in so far as the single treatment had only succeeded in reducing the prevalence from 84.5 to 33.9%; there had evidently been a higher average infection intensity (> 13.5 worms).

The highest prevalence rates and infection intensities were observed among children of up to 14 years of age. Calculated from a number of surveys the mean prevalence already manifests an average value of 70.7% in children up to the age of four, reaching a maximum of 75.9% in the 5 to 9 year-olds (Fig. 77). There is a gradual decrease in prevalence with increasing years: in a generally high level of endemicity

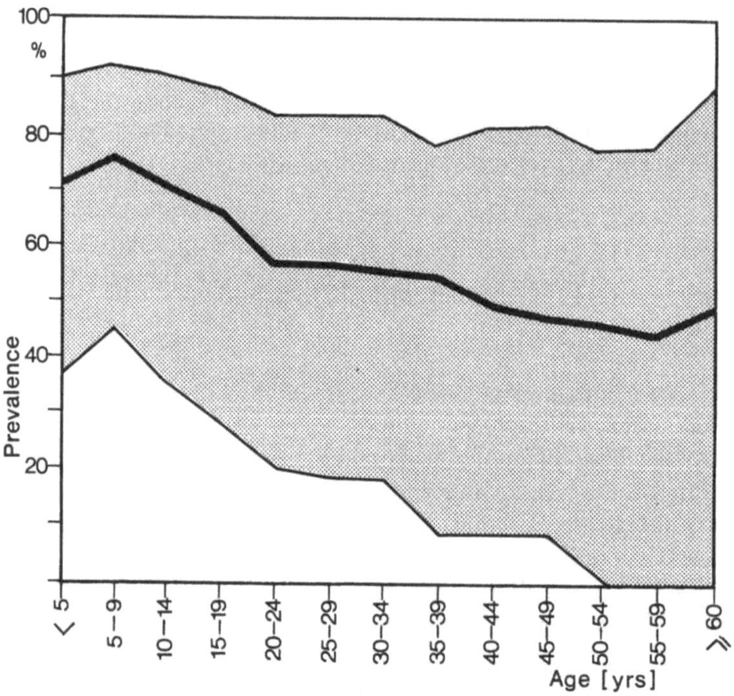

Fig. 77. *Ascaris lumbricoides* prevalence by age: average and extremes in the Philippines [Sources: 86, 154, 166, 197, 198, 200, 205, 248, 251, 568, 606, 612]

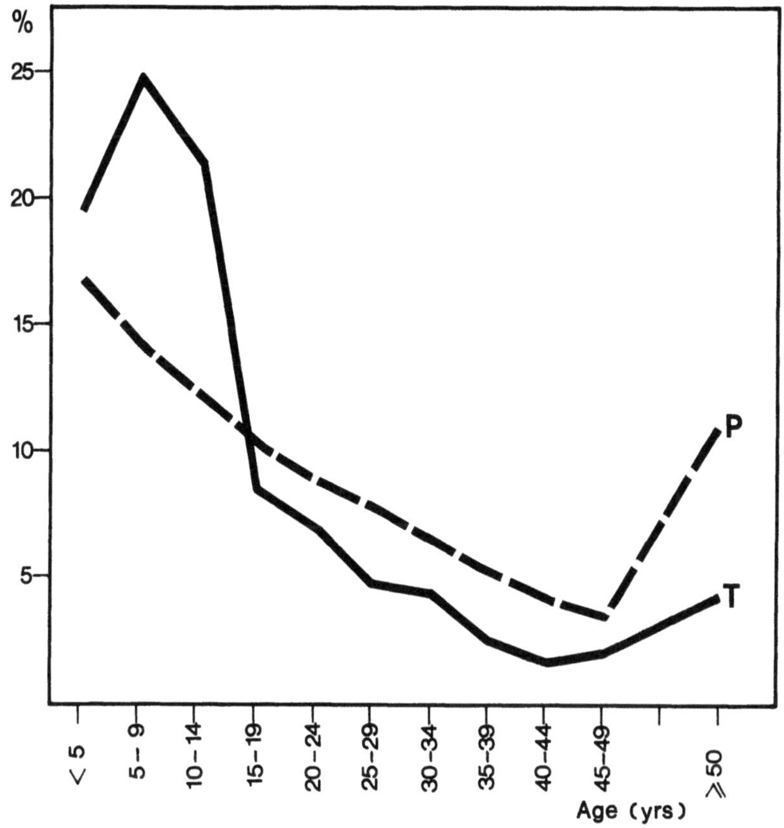

Fig. 78. Relative transmission index of *Ascaris lumbricoides* (*T*) and population structure (*P*) in the Philippines [Sources: 612, 788 and 1980 Philippine Statistical Yearbook]

this correlation is less well marked than in populations with a lower infection frequency. The intensity of infection follows a pattern similar to that of the prevalence in so far as children show a distinctly higher worm burden than adults [154, 612, 788]. Taking into account the proportion of the different age groups in the total population, this results in a situation reflected in a transmission curve – with particularly high indices for children (Fig. 78). Almost two thirds of the relative transmission index falls to the share of those up to the age of 14; for the 5–14 year-olds alone this index amounts to 46.4%, whilst their proportion of the total population is only 26.4%. Thus it is chiefly these age groups that cause the high grade of contamination of the milieu, and maintain the high endemic level of ascariasis. Transmission in yards and compounds which have been faecally contaminated by children, plays an important role in this [cf 295].

On the other hand, parts of the population other than those stratified by age, differ in *Ascaris* prevalence. Social status, related to housing conditions, including sanitary installations, are criteria which exert an influence on the frequency of *Ascaris* infections. Only 1.8% of 166 students, for instance, suffered from infection from large intestinal roundworms [184]. With 73.6% slum dwellers were clearly more frequently infected than people who lived in flats (29.0%) [205]. If facilities for the disposal of excreta were present, prevalence was lower than in places where there were none

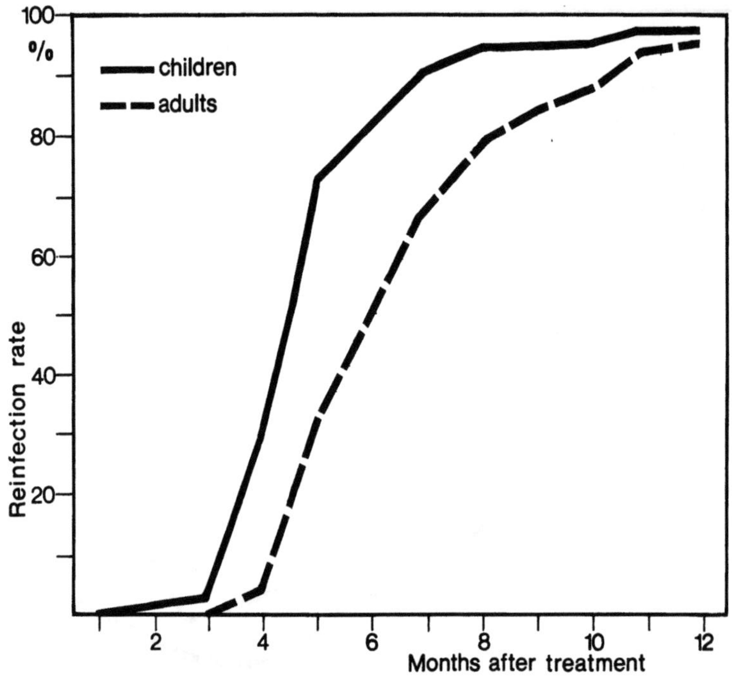

Fig. 79. Re-infection with *Ascaris lumbricoides* after successful treatment in Putingsapa, Juban, Sorsogon [Source: 143]

(43.9% compared to 54.5%) [409]. This is evidently the basis of differences between urban and rural areas to which B. D. Cabrera [132] and Bulatao-Jayme et al. [111] had referred.

The overall picture is one of extraordinary variety and changing frequencies. Within the Philippines the endemicity of ascariasis is clearly more influenced by characteristics Jusatz [432] referred to as "strata" than by "basic conditions". Nevertheless, there are some differences which can be explained directly or indirectly by regionally changing climatic conditions. Thus a compilation undertaken by B. D. Cabrera [132] showed a gradation – however slight – of prevalence rates in the different precipitation zones. In places with high precipitation throughout the year and marked maxima from November to January (cf Map Plates) *Ascaris* prevalence reached its highest value with 90.4%, the lowest prevalence (81.0%) occurring in that climatic zone which is characterized by a dry season from November to March. The results obtained by Bulatao-Jayme et al. [111] in the Visayas point in the same direction: with a prevalence of 70.2%, the Eastern Visayas (Samar, Leyte) belong largely to the range of continuously falling precipitation, with maxima at the turn of the year, whereas the Central and Western Visayas with prevalences of 52.6% and 55% are mainly counted among the two climatic zones with a more or less clearly marked dry season. That the annual course of precipitation is of direct or indirect significance for the survival of *Ascaris* eggs in the open, and for the development of larvae within the egg shell, has already been discussed in detail in paragraph 5.2.1. An unequivocal definition of the connections between the frequency of *Ascaris* infections and the different types of precipitation in the Philippines, which have been hinted at here, requires further specific investigations.

In respect of frequency and distribution of ascariasis, the situation in the Philippines is very stable. Thanks to the high level of contamination of the environment by *Ascaris* eggs, as well as their longevity, re-infections occur within a short time even after successful anthelminthic treatment [143–145, 147, 154, 157, 176, 180, 315, 429, 430]. It can generally be assumed that the original position will have been restored within a year (Fig. 79). Re-infection tends to recur earlier in children than in adults, but after the passing of one year the differences have largely disappeared. Though the prevalence never reached 100% in persons who have received treatment and had previously been *Ascaris*-positive, the overall frequency in the population did not decline because a considerable number of persons who had been pronounced non-infected in the pre-treatment survey does acquire an *Ascaris* infection at a later stage. This applied for instance to 100% of the children within six months according to investigations carried out by B. D. Cabrera [143] in Barrio Putingsapa (Sorsogon Province). The incidence among previously uninfected adults was 48.3% at this time and had risen to 77.4% within the 12 month observation period. Thus frequency had returned to its initial value. In view of such transmission dynamics no change is to be expected in the frequency pattern of the countrywide distribution of *Ascaris* infections in the Philippines.

5.3 Trichuriasis

In the Philippines the helminthiasis caused by the whipworm (*Trichuris trichiura*) occupies second place in the frequency scale of worm infections, with *Ascaris* infections being in the first. Pathogenicity depends on the intensity of the infection; it consists chiefly of damage to the mucosa of the distal section of the small intestine and of the colon by the anterior parts of the worms, which are embedded in the tissue. Massive infections can lead to anaemia. Children may experience a prolapse of the rectum. The transmission of *T. trichiura* follows the mode of faecal-oral infection. Developing in the open inside the egg-shell, the infective larvae hatch after oral ingestion in the intestinal tract and continue their development for about one week in the villi of the lower small intestine and of the colon. They subsequently grow up to become adult worms, the anterior parts of which remain dug into the mucosa. After a total of 6–12 weeks the laying of eggs begins; the average per day and female worm is up to 10,000 eggs, but there may be considerable variations [cf 501].

5.3.1 *Trichuris* Development and Geofactors

The speed of development and the survival period of *Trichuris* eggs and larvae are influenced and directed by the same geofactors as for *Ascaris lumbricoides* (cf 5.2.1). For *Trichuris*, too, temperature is of the greatest significance. Whilst larval development inside the eggshell only leads to the infective stage after 4–6 months at a temperature of 15 °C, higher temperatures shorten the period required for development (26 °C: 3–4 weeks; 30 °C: 17 days; 35 °C: 11 days). A further rise to 37 °C and beyond stops all further development. The optimal temperature for *Trichuris* is thus somewhat higher than for *Ascaris*. *Trichuris* larvae show a strong resistance to desiccation [cf 555]; this factor could be playing a role in the distribution of trichuriasis within the Philippines.

5.3.2 Transmission and Sources of Infection

The mode of transmission and the sources of infection of *Trichuris trichiura* are practically identical to those of *Ascaris lumbricoides* (cf 5.2.2). Of the greatest importance for the transmission of *Trichuris* are – once again – contaminated yards and compounds, fields fertilized with faeces, and the consumption of vegetables grown in such fields. The contamination rate for vegetables, for instance, was 7.9% [584]. Money notes, too, may carry *Trichuris* eggs (5% [cf 585]), and flies may be involved in transmission as well [454]. In respect of the number of positive samples and the degree of contamination, however, *Trichuris* is a clear second behind *Ascaris* in every single case.

5.3.3 Frequency, Transmission Index, Geographical Distribution and Re-Infection Rate

Recent surveys have shown that the average prevalence of *Trichuris* might well exceed 60% in the Philippines (cf Fig. 80), as it did in the case of *Ascaris*. In some way this result is at variance with other compilations; B. D. Cabrera [132], referring to 737,046 stool examinations from the period 1908–1969, as well as Pesigan and Yogore [612], who published the result of a survey by T. Guzman [354] involving 211,217 pupils from 45 different provinces, both arrived at distinctly lower rates of prevalence, that is 33.2% and 32.2%. However, this contradiction might be due in part to the great difference in frequencies which occur from one region to another. In Bangued, Abra, and Malaybalay, Bukidnon, for example, only 10.7% and 11.8% of those examined were found to be infected with *Trichuris* [86, 200], whereas prevalence in the Palo district on Leyte reached almost 95% [167, 568]. In the light of this pattern of regionally varying frequencies the total figure of *Trichuris*-infected Filipinos cannot yet be calculated (cf Table IV as well).

The age-related prevalence curve (Fig. 80) tends to reach its maximum as a rule with the 5–19 year-olds (70.9–72.4%). The subsequent drop, correlated with increasing age, is less characteristic for *Trichuris* than for *Ascaris* (cf Fig. 77). The course of the curve for the transmission index also differs from that of *Ascaris*. Thanks to the low intensity of infection, infants (0–4 years old) play a comparatively minor role in the transmission of *Trichuris* (Fig. 81, cf Fig. 78). The main share falls to the 5–14 year-olds (c. 45%), but among the 15–19 year-olds as well the relative index (c. 14%) is still in excess of their share of the population (c. 8.7%). The contamination of the environment rests primarily on the prevalence, the intensity of infection and the proportion of the population of these two age groups.

Although trichuriasis occurs throughout the Philippines, the influence of geofactors does become evident in the different frequencies of particular climatic regions. The highest prevalence tended to be observed in areas where high rainfalls occur throughout the year but reach their maximum at the turn of the year [111, 132, 612]. In the case of *Trichuris* this frequency pattern can be attributed to the same

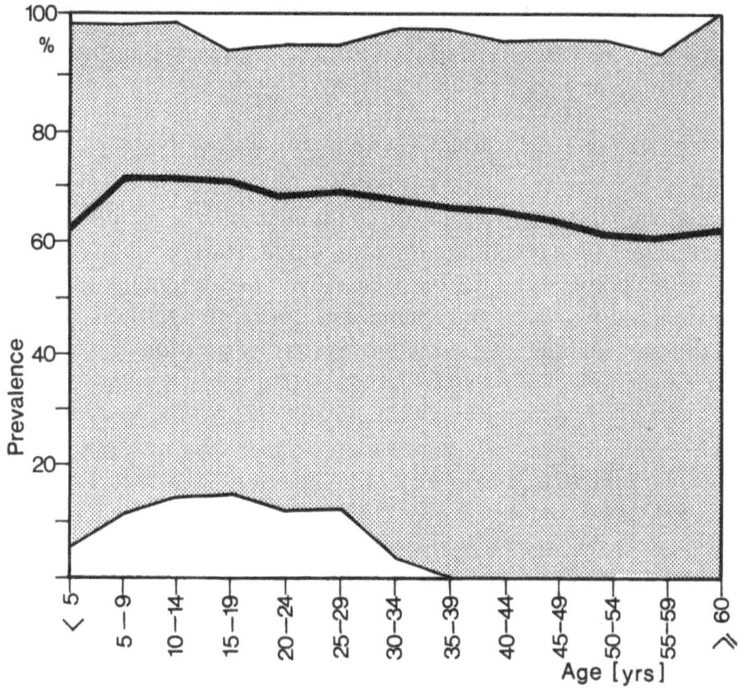

Fig. 80. *Trichuris trichiura* prevalence by age: average and extremes in the Philippines [Sources: 86, 166, 197, 198, 200, 205, 245, 248, 251, 568, 606, 612]

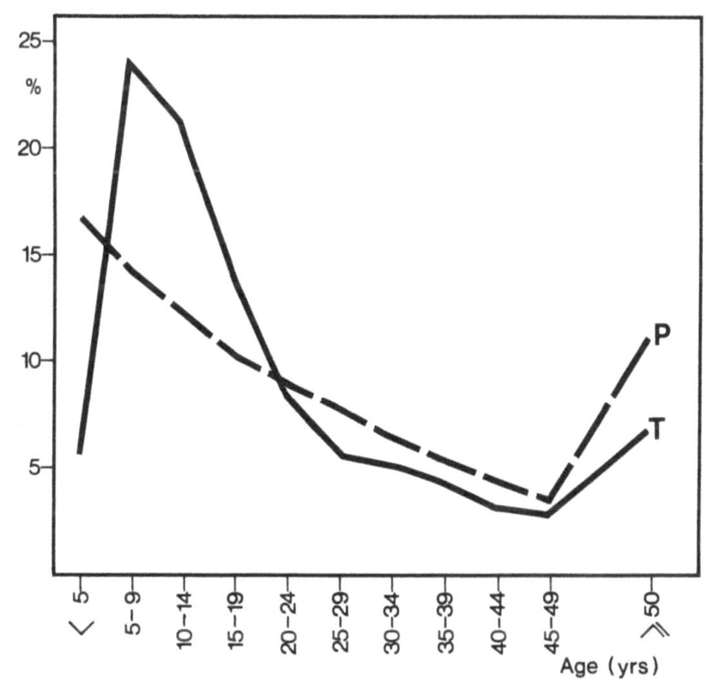

Fig. 81. Relative transmission index of *Trichuris trichiura* (*T*) and population structure (*P*) in the Philippines [Sources: 612, 788 and 1980 Philippine Statistical Yearbook]

causes as in the case of *Ascaris* (cf 5.2.3). Frequency and distribution of *Trichuris* infections may be regarded as relatively stable in the Philippines, since anthelminthic treatment is followed by re-infection within a short time [143–145, 147, 157, 176, 180]. According to investigations by B. D. Cabrera [143] it succeeds even faster than in the case of *Ascaris,* as more than 50% of those who had been treated had been re-infected with *Trichuris* within three months; after 7 months the 90% mark had even been reached. The remaining 8% who had failed to become re-infected after an entire year were compensated for by persons who had originally not suffered from such an infection, but had acquired trichuriasis during the period of observation. Prevalence had thus approximately returned to its starting position.

5.4 Ancylostomiasis

The term ancylostomiasis covers a number of infections with different species of hookworms (Family: Ancylostomatidae). As parasites of the small intestine they feed on blood. The result is that in cases of high infection intensity or of insufficient supply of iron in the food, an infection with hookworms may lead to an iron deficiency anaemia. The general opinion now is that the faecal-percutaneous mode of transmission predominates, i.e. that infection is brought about by the penetration of the skin by infective larvae. On the other hand, there is the possibility that man becomes infected through the oral ingestion of larvae; there are even some indications of galactogenous (transmammary) transmission [572, 573]. Furthermore, mammals must be assumed to act as paratenic hosts, so that hookworm infections could be acquired through the consumption of raw of insufficiently cooked meat of such animals [701].

The eggs produced by the female hookworms are passed out together with the stool. In the open the development within the eggshell initially produces a first-stage larva which, having hatched and moulted twice, reaches the infective stage. After penetrating the skin these larvae are carried with the blood and reach the intestinal tract by way of the heart, lungs and trachea, where they settle mainly in the jejunum, reaching adulthood within 4–7 weeks after having moulted twice again.

5.4.1 Hookworm Species in the Philippines

Ancylostoma duodenale and *Necator americanus,* the two human pathogenic hookworm species, have been known to occur in the Philippines since the beginning of this century [235, 236, 238]. As far back as the 1920s therapeutic measures (treatment with tetrachlorethylene) revealed that *N. americanus* predominated, as more than 80% of the expelled worms were of that species [450, 491, 492]. Geographical differences also emerged in this connection. On average the proportion of *A. duodenale* was highest in North Luzon, but lowest on Mindanao, the Bicol Peninsula and the Eastern Visayas [450, 489].

At a later date the examination of larval cultures largely confirmed this distribution pattern [278, 279]. In Ilocos Norte, Pampanga and Bulacan 86.1% of the positive cultures contained *Ancylostoma* larvae, as compared with 63.6% with *Necator* larvae (A:N = 1:0.7), whereas in Camarines Norte infections were exclusively attributable to *Necator*. In Negros Oriental and Zamboanga del Norte (including Ozamis City) finally, the proportion of *Ancylostoma* to *Necator* was 1:1.1 and 1:2.1, with a large number of mixed infections. On the other hand, different parts of a relatively limited area may show significant deviations. Investigating conditions in the town of Bay, Laguna, Jueco [423] detected a significantly higher proportion of *Ancylostoma* infections (76.2%) in a barrio on the coast of the Laguna de Bay than in a village in a mountainous area further inland (28.2%). *Necator* admittedly also dominated in the coastal area as a result of the large number of mixed or double infections (A:N = 1:1.1).

Besides *Necator americanus* and *Ancylostoma duodenale*, another two hookworm species have repeatedly been cited in the literature as parasites of man in the Philippines: *A. ceylanicum* and *A. braziliense* [485, 489, 492, 816]. Both of them are known in the Philippines as parasites of dogs and cats [260, 486, 710, 775, 884]. As their species validity has been in dispute for a long time, there is no absolute certainty in the matter of the data furnished by Manalang [483, 486, 492] as to which species the human infection observed by him should be attributed; on the basis of present opinion only *A. ceylanicum* is thought to be able to settle in the intestinal tract of man [541], whereas *A. braziliensis* plays a role exclusively as a cause of larva migrans cutanea (cf 5.6). Manalang [489, 492] reported in his time that up to 1% of the hookworms expelled after anthelminthic treatment were *A. ceylanicum*. Later this species was found only once – in a man from Ilocos Norte [816]. In contrast to *A. duodenale* and *N. americanus*, it is of subordinate importance.

5.4.2 Hookworm Development and Geofactors

The potential distribution area of human ancylostomiasis is solely determined by the effect of geofactors on the development of hookworm larvae in the open. Among these factors temperature and humidity are the most important ones. The interval of temperatures between 11–13 °C and 40 °C is generally regarded as the range within which development of larvae can take place [cf 441], with *A. duodenale* tolerating extreme temperatures better than *N. americanus*. In the lower temperature range of 15 °C about 90% of the first-stage larvae of *A. duodenale* hatch within 130 hours, whereas only 60% of the *Necator* larvae had managed to do so at that point. At a temperature of 20 °C, 90% of all the larvae of both species take just 50–60 hours to hatch. A further rise in temperature significantly accelerates the development of larvae: at 25 °C the process of hatching of all hookworm larvae is completed within 30 hours, at 30 °C within 20 hours. Whilst constant temperatures of 35 °C produce a further slight acceleration in the development of both species, the limits of tolerance for *N. americanus* have already been exceeded at 40 °C: all the *Necator* eggs died within 6 hours without hatching, whereas 65% of the *Ancylostoma* larvae survive and hatch. Extreme temperatures of 45 °C have a lethal effect on eggs of both species; all the *Necator* eggs perish within 15 minutes, the less sensitive *Ancylostoma* eggs, however, only do within 90 minutes [419].

Temperatures not only affect development inside the eggshell and the moment of hatching, but also all subsequent steps in the development up to the infective stage, which at 28 °C – to give just one example – is attained by *A. duodenale* after 70–72 hours, and after 80–82 hours by *N. americanus*, always assuming that there is sufficient humidity. Hookworm eggs perish as soon as soil moisture falls below 9%, though larvae show somewhat greater resistance [441]. Exposure to direct sunlight also has a damaging effect on account of the ultra-violet rays contained in it. On ground exposed to the sun, almost all infective larvae perish within 5 days, whilst lasting 33 days in the shade [420]. *N. americanus* proves to be more sensitive to desiccation and to direct sunlight [519, 520].

Oxygen content and pH value of the soil are also of significance for the development of hookworm larvae. Because of the oxygen requirements of the larvae, well-aired soils are considered to be the best; a light sandy loam is regarded as the ideal [295]. Lack of oxygen has a lethal effect on *A. duodenale* larvae within 20 days, and within 12 days on those of *N. americanus*. pH values of 4–7 are not detrimental to the process of hatching, but exceeding values cause increasing inhibition [852]. Filariform larvae, on the other hand, survive values as extreme as pH values of 2.8 and 9.4 for as much as 33–36 days; although this period of survival is clearly more than doubled under conditions of lower acidity and alkalinity (pH 5.4–8.4) [729].

Some of the geofactors mentioned above not only limit and direct the development of hookworm in the open but also determine the coming about of percutaneous infections, for the chances of reaching human skin are much greater for infective larvae if they do not remain on the ground but do so at a certain level slightly above it. A corresponding and probably positively phototactically directed vertical migration of larvae presupposes a low vegetation and a humid milieu. This explains the obvious correlation between the seasonal climates of the Philippines and the coming about of human infections.

In summary it can be said that the inter-relationships of temperature, humidity, exposure to sunlight, vegetation as the provider of shade, and soil conditions determine the frequency pattern of *A. duodenale* and *N. americanus*. The same thing applies to the Philippines. Thus *N. americanus* is less well represented in those climatic regions of the country which have a dry period with a great deal of sunshine and higher temperatures than *A. duodenale*, but dominates in the permanently humid areas. These latter areas are characterized, moreover, by a higher total prevalence of hookworm infections. Although the result of the basic conditions, this basic pattern can, nonetheless, be subject to considerable modifications at the local level. Residential and settlement conditions, occupation and other human factors are important reasons for this.

5.4.3 Transmission and Sources of Infection

The transmission of hookworms occurs primarily in places where eggs, which have been passed out together with the stool, find suitable conditions for development in the open; if the degree of contamination is high, the risk of infection for man is at its greatest here. This applies to the agricultural sector to a particularly high degree. But according to Schad et al. [701] it is "not necessarily these particular occupations that predispose to infection but rather the rural life style with inadequate facilities for the safe disposal of human excrement". However, in individual cases the risk of infection must be assessed as varying widely. If field crops predominate which shade the ground well and keep it moist, conditions for the development of hookworm larvae are optimal, but such fields turn into important places for transmission only when larger numbers of people contribute to the contamination of the soil, or human faeces are used for manuring, and where there is sufficient contact with the contaminated soil. The type of crop, as well as the number and length of stay of people working in the fields are important factors influencing the process of transmission in the agricultural sector. In these circumstances the risk of infection may even be smaller in the fields than in the immediate environs of the houses, especially when the proportion of people not permanently engaged in agriculture is very large; they can contaminate the ground in the vicinity of houses more than a few people working in the fields.

Larvae-containing soils are thus of by far the greatest significance as a source of infection. There has been too little research into the role of drinking water and vegetables in the Philippines to permit conclusions to be drawn from the meagre results that are available. Although Paguio [584] found hookworm eggs on 12 out of 390 samples of vegetables, and other unspecified larvae in 5 of them, these data are not sufficient for estimating the risk of infection with hookworm emanating from them. The same applies to contaminated currency notes [585] and to synanthropic flies [454] which might be involved in transmission, since there was no evidence or identification of infective hookworm larvae in any of these cases.

5.4.4 Frequency, Transmission Index, Geographical Distribution and Re-Infection Rate

In the Philippines infections with hookworm rank third in the frequency of various soil-transmitted helminthiases after ascariasis and trichuriasis (cf prevalence rates in Table IV). Due to differences between particular parts of the population and between different regions, it is not possible to estimate the average prevalence for the entire country. There has been repeated evidence of the fact that ancylostomiasis occurs more often in the rural than in the urban areas, thanks to the more favourable conditions for the development of larvae in the open (cases in point are 18% by contrast with 7%, or 12% as compared with 0.7% [cf 111, 167]). On the whole, prevalence rates in villages are still likely to reach approximately 50%; in some villages even over 70% of the population were infected with hookworm [176].

The age-specific prevalence curve (Fig. 82), however, deviates significantly from the conditions associated with ascariasis and and trichuriasis in so far as children show a markedly lower percentage of hookworm infections than adults. The prevalence rate rises with increasing age to a maximum among the 15–19 year-olds, and remains more or less at the same level among the older age groups. In the case of the intensity of infections matters are very much the same. As far back as 1926, Manalang [492] had already counted positively fewer expelled hookworms after anthelminthic treatment of children. Data on the passing of eggs provided by Tubangui et al. [789] showed very similar results; when comparing particular age groups they observed a drastic rise of c. 350 hookworm eggs per ml of stool from children aged 5 or less to values of c. 2,800/ml among the 5–14 year-olds, and 3,700/ml among the 15–49 year-olds.

An assessment of the transmission process also needs to take into account the differences between the sexes. Almost all the surveys had shown prevalence rates that were noticeably higher for the male sex; there are an average of 0.8 cases of hookworm infection among females for every single case among males. Linked with it are

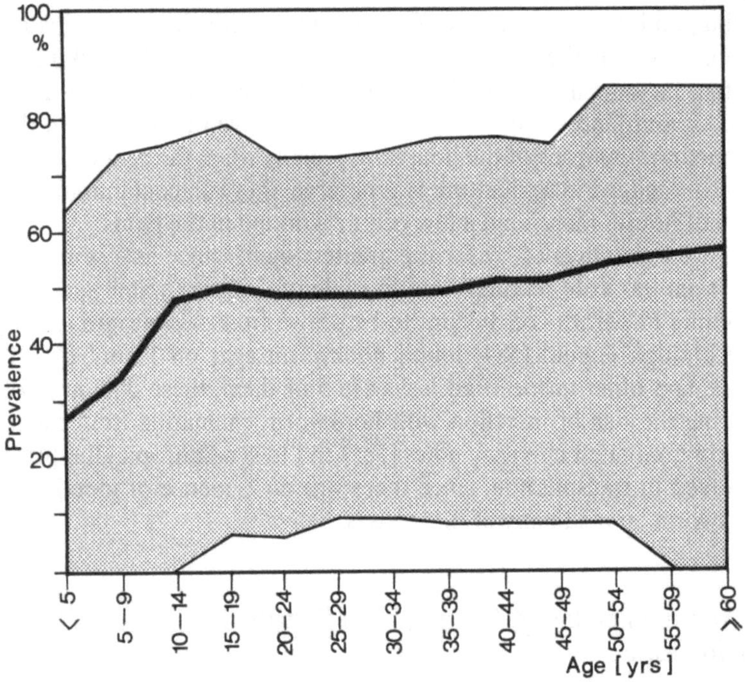

Fig. 82. Hookworm prevalence by age: average and extremes in the Philippines [Sources: 86, 166, 197, 198, 200, 245, 248, 251, 568, 606, 612]

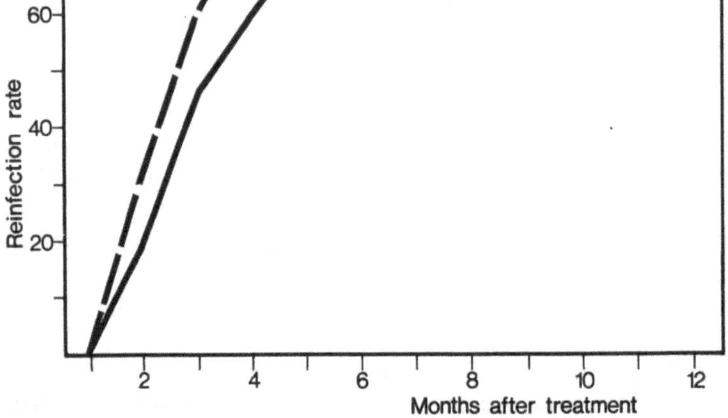

Fig. 83. Re-infection with hookworms after successful treatment in Putingsapa, Juban, Sorsogon [Source: 143]

the differences in the intensity of infection, which are 2–3 times higher in men and boys than in girls and women. The calculation of the relative transmission index from the data published by Tubangui et al. [789] thus shows the very high proportion of 75% for the male population, with almost 52% of the total index falling to the share of the 15–49 year-old men alone. Quite differently from ascariasis (Fig. 78), infants (<5 years) play practically no role in the persistence of ancylostomiasis, in a proportion of around 15% of the population their relative transmission index is a mere 0.8%. On the basis of the age- and sex-specific differences in prevalence and infection intensity mentioned above, a positive correlation between hookworm infection and the extent of agricultural activity of various population groups must be assumed.

As far as the geographical distribution of ancylostomiasis in the Philippines is concerned, the macro-regional view indicates a frequency pattern that is clearly related to climatic regions. In areas with dry periods extending over several months, the prevalence is less than in places where rains falls throughout the year [111, 132, 273, 278, 279, 612, 789]. Decisive for this are the environmental conditions as they affect the development of hookworms in a more or less favourable way (cf 5.4.2). Differences in prevalence within particular regions are due either to different residential or settlement conditions, and their consequences (such as in the urban or in the rural areas), or to human interference in the complex web of effects occasioned by the geofactors (in agriculture or forest clearance, for instance). As long as there is no change forthcoming in these points, the relatively stable situation in respect of the frequency of ancylostomiasis will be maintained. Studies of re-infection undertaken so far indicate this, just as in the cases of ascariasis and trichuriasis. B. D. Cabrera [143], for instance, found that in Barrio Putingsapa, Juban, Sorsogon Province, 50% of the people who had had successful anthelminthic treatment had been re-infected within 3 months; after 6 months all the adult men were once again passing hookworm eggs in their stool. In children, this process of re-infection is slower and passed the 90% line only after 8 months, whilst adults reach that point within 5 months (Fig. 83). In fact the process of re-infection, including that with *Ascaris* and *Trichuris,* is much faster than was shown here, since a re-infection is diagnosed on the basis of eggs that have been passed out along with the stool, and the definition of the time of infection needs to take into consideration the period of pre-patency, i.e. the time the worms require for their development to adulthood. In the case of hookworm this time is about 4 weeks. This means that re-infection proper takes place approximately one month earlier than is indicated by the evidence of eggs in the stool. In the investigations conducted by B. D. Cabrera [143] the total rate of re-infection with hookworm had risen to more than 94% within one year. The remaining 6%, who had not been re-infected, however, did not cause a drop in prevalence by comparison with the initial situation, because they were replaced by a group of people who had originally not been infected and showed a 100% infection with hookworm at the close of the investigation period. In the rural areas at least, frequency and geographical distribution of ancylostomiasis in the Philippines are still thus characterized by great stability.

5.5 Strongyloidiasis

The causal agents of strongyloidiasis are threadworms of the genus *Strongyloides,* the parasitic generation of which consists solely of females living in the intestinal mucosa.

In the Philippines, *S. stercoralis* is the only human parasite. The females, which breed parthogenetically, deposit eggs from which larvae hatch while still in the host; these may lead to endo- or exo-autoinfection after further development in the intestinal lumen or in the peri-anal region, but tend to be passed out with the stool. Development takes place on the ground in the open, and is said to adopt different courses, depending on environmental factors. When soil moisture and concentration of nutrients are low, infective filariform larvae emerge directly from so-called rhabditiform larvae. However, if concentration of nutrients and soil moisture are high, the rhabditiform larvae give rise to a generation of free-living male and female worms, which in turn produce larvae themselves [cf 295]. Other observations, however, argue in favour of this mechanism being determined by the set of chromosomes of the eggs deposited by the *Strongyloides*-♀♀ [529]. According to this the ♀♀ are able to produce three different types of eggs: triploid eggs, from which infective filariform larvae proceed, as well as diploid and haploid eggs, which develop into free-living females and males.

Living in the open, infective larvae enter man percutaneously; they may cause skin conditions of the "creeping eruption" type (cf 5.7.4). Along with the blood, they reach the intestinal tract by way of the heart, lungs and trachea; in the mucosa they grow up to become females. Their passage through the lungs may cause bronchitis and broncho-pneumonia, their infection of the intestines abdominal symptoms (such as diarrhoeas). Massive infections like those described as "hyper-infection" in immuno-suppressed cases can make strongylodiasis assume forms which threaten life.

5.5.1 *Strongyloides* Development and Geofactors

Even if geofactors do not play the role that has been ascribed to them as determinants of a direct or indirect development of larvae which have been passed out with the stool (see above), they are of importance for the process of development in the open. As far as temperature is concerned, *S. stercoralis* tolerates a minimum of 15 °C and a maximum of 40 °C. At constant temperatures of 20 °C the development into filariform larva is completed within two days; the subsequent development into a male worm takes 2–3 days, whereas females take just over 3 days to reach sexual maturity [749]. Even under optimal external conditions the life span of larvae is supposed to be less than 3 weeks. On the whole, however, there has been too little research on the influence of geofactors on the different stages of *Strongyloides* in the open to allow the results to be used for conclusions on the geomedical situation in the Philippines.

5.5.2 Frequency and Geographical Distribution

In the Philippines strongyloidiasis is one of the rare helminth infections (cf Table IV). In his comparative study of six selected areas, B. D. Cabrera [146] only found an average prevalence of approximately 1.2%. The greater frequency he observed among the male sex (♂♂ 1.7%, ♀♀ 0.8%) cannot be explained as being a result of increased agricultural activity as in the case of ancylostomiasis (5.4.4), since men were less often infected (0.7%) than boys (≤6 year-olds: 0.9%; 7–14 year-olds: 2.3%). Results from other authors [248], which diverge from these, do not permit conclusive arguments to be made in respect of more exposed groups of the population either as the prevalences are very low indeed (1–2%).

According to the survey by B. D. Cabrera [146] prevalence rates fluctuate between 0 and 2.3%, depending on the area selected. A correlation of frequency with the different climatic conditions cannot, however, yet be derived from this. It is likely that local circumstances determine the process of transmission more than the regionally effective conditions.

5.6 Enterobiasis

Owing to the special features of its life cycle, the pinworm (*Enterobius vermicularis*) cannot be counted unreservedly as being among the causal agents of soil-transmitted helminthiases. Inhabiting the colon, *Enterobius*-♀♀ leave the anus when their batch of around 10,000 eggs has matured, depositing them in the peri-anal region within minutes of leaving. Under the influence of body temperature and additional oxygen from the air, infective larvae may develop in the eggshells within the very short time of 6 hours. Itching caused – mostly at night – by migrating females, makes the affected persons (mainly children) scratch themselves so that the eggs stick to the fingers or below the fingernails. If taken to the mouth, contaminated fingers can thus present an important source of infection; the so-called anus-finger-mouth route is of particular significance in the case of *Enterobius vermicularis* since it leads to auto-infection. As transmission can also occur through eggs that have reached the ground, this nematode infection will be discussed along with the true soil-transmitted helminthiases.

5.6.1 *Enterobius* Development and Geofactors

Thanks to the special features of the development of *Enterobius* the influence of geofactors plays only a minor role for the distribution of this worm infection; it affects only the smaller part of the eggs, that is those which make their way from the peri-anal region to the bed-linen and eventually to the floor and ground. Once there the temperature determines the speed of larva development within the eggshell. The temperature range within which the development can take place extends from 15 to 40 °C; development fails to take place at temperatures below 15 °C, whilst temperatures above 40 °C cause the eggs to perish. At 20 °C the development of larvae is completed within 20 hours, and temperatures of more than 30 °C cause rapid acceleration (33–34 °C: 6–7 hours; 37 °C: 3–4 hours; 40 °C: 1 hour). Infectivity already begins, however, to recede two days after the completion of the development of the larvae. The maximum period of survival of *Enterobius* eggs is, on the other hand, supposed to be in excess of 40 days [38].

5.6.2 Transmission and Sources of Infection

As was mentioned above, transmission largely takes place by way of the anus-finger-mouth route. Clinging to the fingers, the eggs are not only the cause of auto-infections but are also in the very place to be directly passed on to contact persons, or indirectly, by way of touching objects, to others.

In the course of examing the finger-tips and the dirt lodged under their fingernails in the case of 226 *Enterobius*-positive children, B. D. Cabrera et al. [160] found *En-*

terobius eggs in 12.4% of the samples. Girls (15.9%) were much more frequently affected than boys (8.8%). The authors attributed this to the fact that girls wash more frequently in the contaminated area (e.g. after miction as well), and that their clothes make it easier for them to scratch their anus. Differences in the contamination rate of fingers also appeared to depend on the period of attendance at school: the correlation was a negative one (Grade I: 18.9%; II: 13.2%; III: 8.6%). Chanco's investigations [215] show that the fingers of children may carry a still higher percentage of *Enterobius* eggs: in 18.8% of 109 children the examination result was positive; in the light of an *Enterobius* prevalence of 42.4% it implied that 44.5% of the infected persons had eggs on their fingers.

These findings emphasize the importance of living and residential conditions for the transmission of enterobiasis. The closer the contact between the people living in small quarters, the greater the risk of contact with *Enterobius* eggs. This not only applies to the domestic realm, but also to the school and workplace. Set against that, other sources of infection, such as contaminated currency notes and vegetables [584, 585] are far less important.

5.6.3 Frequency, Transmission Index and Re-Infection Rate

Thanks to its particular transmission routes, *E. vermicularis* occurs all over the world. The results of stool examination do not permit any statements to be made on its frequency patterns in the different regions of the Philippines, since they cover only a fraction of the infected persons. The highly specified method of evidence by means of scotch tape swabs has so far been applied only in Central Luzon (Manila, Rizal, Laguna) and in Panay (Iloilo). In the majority of the cases investigations concerned children only; they resulted in average prevalences of 5.9–75.2% [6, 162, 215, 219, 253, 347, 583, 587, 635, 696, 715, 716, 737]. From a total of 8,016 young children and school pupils, 46.7% proved to be infected.

Children are generally regarded as the age group that is most severely affected, but adults may also show high prevalence rates, as Chanco [212] found when examining 200 cadavers (≥16 years of age), the ceca of which contained *Enterobius* in 41% of the cases. Paguio's results [583] for the slum area of Tondo, Manila, even showed 84.4% of the adults (>20 years of age) to be carriers of *Enterobius* in contrast to 52.3% among children aged 5 or under, and 57.8% among 6–20 year-olds. As a rule though, children have a higher percentage of infection than adults. Prevalence also tends to decrease over the years as children attend school.

In respect of the sexes, girls are usually considered to be infected more often, although the following collation shows this not to be the case:

Boys	Girls	Source
60.1%	63.8%	583
7.9	11.5	6
45.6	53.9	162
62.9	56.9	715
17.8	7.9	347
29.6	31.3	253
34.6	20.4	716
50.9	60.3	696
38.7	38.3	average

However, incomes – and therefore living conditions – do exert a considerable influence on the frequency of enterobiasis. Children who attended expensive schools were infected significantly less often (29.3%) than were others in less expensive ones (55.6%) [cf 162]. Such differences between families with different incomes had also been observed by Ignacio [400]: in middle income groups prevalence was 49.2%, but 60.0% in those with lower incomes – i.e. the residential conditions imposed on families as a result of their economic situation are of great importance for differences in the frequency pattern of *Enterobius* infections.

Presumably the different circumstances which favour or restrict transmission also exert their influence on the process of re-infection, though comparative analyses for this are still lacking. B. D. Cabrera et al. [161] at least were able to show that children who had undergone successful treatment for worms very soon became re-infected if other members of the family did not receive treatment at the same time. Only 5 months later their prevalence had again reached 48.8%, and had risen to 91.1% after 10 months. It may therefore be assumed that *Enterobius* – like other soil-transmitted helminthiases – will continue at a high endemic level.

5.7 Other Rare Soil-Transmitted Helminthiases

Besides the usually very frequent nematode infections some relatively or very rarely soil-transmitted helminthiases were diagnosed in the Philippines, which will be included here, although they are of little geomedical interest. Their causal agents have some peculiarities which distinguish them from geohelminths.

5.7.1 Hymenolepiasis nana

The infection with the dwarf tapeworm *Hymenolepis nana* represents an anthropozoonosis which occurs not only in man but also in rats. The causal agent belongs to the facultatively diheteroxenous helminths whose life cycle may incorporate fleas as intermediate hosts. Generally, however, one tends to take the view that humans mostly acquire *H. nana* infections through the direct oral ingestion of tapeworm eggs.

In many of the countries in the tropics and subtropics *H. nana* attains high rates of prevalence among children [cf 427]. This does not apply to the Philippines: the highest prevalence ever observed there was 1.7% among children of Santa Isabel, Ilagan in Isabela Province [655]. An examination of 211,217 schoolchildren carried out in 45 provinces at a later date resulted in 91 (= 0.04%) positive findings only [354]. The total number of human *H. nana* infections in the Philippines has so far amounted to a mere 170 cases [cf 5, 118, 300, 327, 336, 337, 343, 361, 362, 447, 487, 504, 505, 594, 612, 628, 655, 839, 842, 850]. The reasons for this rare occurrence in man are not known. Infections of *Rattus norvegicus* (1.7–4.9%) and *R. rattus* (0.8%) admittedly remain at a low level, too [426, 672, 779].

5.7.2 Echinococcosis (Hydatidosis)

Echinococcosis (Hydatidosis) is the term for the infection by larvae of the tapeworm genus *Echinococcus.* The adult worms are intestinal parasites of canids and felids; herbivorous and omnivorous mammals, including man, serve as intermediate hosts, in whose organs, especially the liver, the larvae develop. Echinococcosis is primarily a zoonosis. The infection of man (as of the other intermediate hosts) takes place as a result of the oral intake of tapeworm eggs from a milieu that has been contaminated with the faeces of final hosts. *Echinococcus granulosus,* the causal agent of cystic echinococcosis, is the most common species; it is also the only one known to occur in the Philippines.

After the first indications of the occurrence of cystic echinococcosis in man in the Philippines [336], only 5 more cases have so far come to light and the evidence of these is to be regarded as autochthonous in 3 of the cases [43, 449, 455]. Even the infection of intermediate mammalian hosts appears to be an exceedingly rare event, since only one infected cow and one infected water buffalo are mentioned in the literature [709]. The same applies to the infection of the domestic dog as a final host: adult *E. granulosus* have so far been found in only a single case [340].

This great rarity of echinococcosis in the Philippines may be taken as being:

1. that thanks to their size metacestodes of *Echinococcus* cannot be overlooked in the meat inspection of the organs of cattle, carabaos, sheep and other animals brought to slaughter in the abattoirs,
2. though the adult worms in the intestinal tract of dogs are very small, measuring only 5 mm and their eggs in terms of differential diagnosis are not discernible from those of other Taeniidae in faecal examinations, human infections should nevertheless occur frequently, even where prevalences are low, for the simple reason that Filipinos are fond of eating dog meat. Since the *Echinococcus* eggs cling to the fur of these animals, the risk of infection when slaughtering and skinning them is indeed a high one.

As to the occurrence of echinococcosis in the Philippines, the situation is characterized by its extreme scarcity; since it is unlikely that a helminthiasis can persist under such conditions, it must either be assumed that the cases diagnosed so far were not the result of autochthonous infections, or that there is a hitherto unknown sylvatic cycle which leads to human infections in some very rare cases.

5.7.3 Larva migrans visceralis (Toxocariasis)

The term larva migrans visceralis is generally applied to a syndrome brought about by the migration of parasitic larvae in the internal organs and tissues of an abnormal host. These are primarily nematode larvae (cf 4.2 and 4.3 in addition), among which the larvae of the dog ascarid (*Toxocara canis*) are the most significant worldwide. As an adult worm *T. canis* is an intestinal parasite of canids and felids with universal distribution. On being passed out with the faeces of these animals, the development of its eggs follows the same pattern as that of *Ascaris lumbricoides,* as does the infection mode. However, one important difference is that the tracheal migration undertaken by the larvae after hatching in the intestinal tract, eventually taking them back to the intestine, only occurs in young animals. In older animals the somatic migration route is taken; the larvae remain in the tissue and only become activated when their hosts are pregnant; they then make a transplacental migration to the foetuses, in whose intestines they mature into adult worms after the birth has occurred. A further way is by galactogenous (transmammary) transmission from the infected mother to the young animal.

In man, as in other abnormal or paratenic hosts whose oral infections occur primarily in the environment contaminated by the puppies, the hatched larvae penetrate the intestinal wall and are then able to reach a large number of organs and tissues with the blood, where they settle. The symptoms thus span a wide spectrum; however, a persistent eosinophilia is almost always characteristic. The diagnostic difficulties – only the finding of *Toxocara* larvae in the tissue is considered to be evidence – are the reason for what have so far been only a small number of definite cases in man becoming known. So, too, in the Philippines evidence of larvae has only proved possible in a single case [738]. A further four patients suffering from eosinophilia had a positive skin test with *Toxocara* antigen [59], and a fifth showed typical symptoms [466], although it is not certain whether they really were infected with *Toxocara canis* larvae.

In the light of the prevalence rates among Philippine dogs [cf 56, 646, 736, 774], which fluctuated between 4% and 13.6% in adult animals, and did not deviate markedly from values known for Europe [cf 104], a high degree of environmental contamination by *Toxocara* eggs, along with a greater frequency of larva migrans visceralis must be assumed.

5.7.4 Larva migrans cutanea Caused by *Ancylostoma braziliense*

Brought about by infective larvae of *Ancylostoma braziliense*, the skin lesions known as "creeping eruptions" belong to the group of phenomena associated with larva migrans cutanea. They are the result of reactions by the mistakenly chosen host, "man", and a prolonged migratory phase of larvae which have entered percutaneously, but which are then unable to complete their development. Although known in the Philippines as adult parasites of cats and dogs, with a high prevalence [56, 221, 260, 710, 775, 884], corresponding clinical cases of larva migrans cutanea have never so far been diagnosed in man [46, 50]. In spite of the contamination of the environment with cat and dog faeces, and the fact that Africa [8] was able to produce "creeping eruption" with larvae of *A. braziliense* in an experiment with two Filipino volunteers, the problem over why infections in man have up to the present failed to come to notice remains unresolved.

5.7.5 Syngamiasis

Syngamiasis is the term for the infection with different species of the nematode genus *Syngamus*, though those occurring in animals are now counted among the genus *Mammomonogamus*. Their sole representative proved to occur in man as well is *M. largyngeus*, a monoxenous helminth that tends to inhabit the trachea. So far only one case has become known from the Philippines [735], although *M. laryngeus* is a very common parasite in cattle and carabaos [785]. A human infection is therefore to be regarded as an extremely rare and accidental event, resulting from the oral ingestion of eggs from a milieu contaminated by animal faeces.

5.7.6 *Syphacia obvelata* Infection

Syphacia obvelata is a pinworm affecting mice and rats which has only once been found in man – in an American child at Zamboanga, Mindanao [654]. This isolated case is without geomedical relevance.

5.8 Short Synopsis: Distribution of Soil-Transmitted Helminthiases

In soil-transmitted helminthiases, the most important causal agents of which belong to the monoxenous nematodes whose life cycle therefore includes no intermediate hosts, geofactors affect solely those stages of the larvae whose development takes place in the open. Due to the tolerance limits of these stages, the occurrence of soil-transmitted helminthiases is possible in all parts of the Philippines. Unlike other worm infections, their distribution is therefore not restricted to certain regions. Nevertheless, the influence of geofactors is evident in the soil-transmitted helminthiases as well: the varying conditions are reflected in a pattern of graduated frequencies. The macro-regional view recognizes these frequency patterns as being determined by climatic conditions. Because of the sensitivity of the larval stages of large intestinal roundworms, whipworms and hookworms to high temperatures, dehydration and exposure to direct solar radiation their development meets – at least temporarily – with less favourable conditions in climatic regions with seasonally no or very little precipitation than in those which enjoy high precipitation throughout the year.

Within the individual climatic regions the frequency of soil-transmitted helminthiases may well be subject to considerable modifications through human factors. Important criteria in this regard are the differing conditions of urban and rural areas. Although – to name but one example – high residential density of the town favours the faecal-oral transmission of *Ascaris* and *Trichuris* in principle, infections with these helminths tend to be less common there, since better hygiene conditions, except in the slums or on the outskirts of the cities, prevent a substantial contamination of the immediate environs with worm eggs, thus reducing the frequency of contact with a contaminated environment for the city dweller. An increase in the risk of infection, however, may arise in towns through the delivery of farm products from nearby or more remote parts of the country – an example being vegetables for the needs of the townspeople, which may carry different infective stages of worms. As far as hookworm infections in the towns are concerned, their transmission frequency is considerably reduced for the simple reasons that a soil which is solid, dry and lacking vegetation scarcely allows the development of larvae, and because percutaneous invasion is prevented by the wearing of shoes.

In the countryside on the other hand, conditions for the transmission of soil-transmitted helminthiases are on average more favourable in the Philippines. As a result of the poorer sanitary conditions environmental contamination tends to be higher; the soil and vegetation conditions also favour the development of the eggs and larvae of worms. Considerable differences in infection-frequency are also to be expected in those sections of the population which are engaged in agriculture. They are affected by the form settlement takes, by the density of population, by crop types and by other factors. Soil-transmitted helminthiases are no more of the same frequency pattern in the country than they are in the town. This is decided by the local conditions prevailing at the time.

6 Conclusions – a Geomedical View

The epidemiological and geomedical situation of the occurrence and distribution of helminths in the Philippines is in part determined by the particular geographical location of the archipelago between the Asian continent and the Pacific region. The isolation of this border region has, for instance, led to the origin of a sub-species of the Japanese blood fluke, which evolved together with *Oncomelania hupensis quadrasi,* its intermediate host snail, and remains restricted to the Philippines. Largely restricted to the Ilocos region, intestinal capillariasis, too, may be regarded as specific to the Philippines despite the recent evidence of isolated cases in Thailand and Japan. This includes the complete absence of certain helminth species such as *Fasciolopsis buski* and *Opisthorchis sinensis,* which are very common in some regions of the Asian continent but which were unable to become endemically established in the Philippines, in spite of apparently suitable ecological conditions, although time and time again immigrants from the mainland have been found to be infected.

The geographical distribution of the various helminthiases is not without considerable differences, and presents a complex picture. Varying from island to island, occurrences point to a dependence upon geomedically effective factors, a geo-ecological analysis of which can explain the significance of the particular environmental conditions of the climate, the orographic location, the physical condition of the terrain, the utilization of the soil, the activities of man, and the habits of the ethnically diverse inhabitants as well.

The climate, and in particular the distribution of precipitation throughout the course of the year present the predominant causal factor. It is, for instance, the precipitation which decides whether *Oncomelania hupensis quadrasi,* the intermediate snail host of *Schistosoma japonicum,* meets with suitable living conditions or whether *Aedes poicilius,* the most important of the vectors of *Wuchereria bancrofti,* finds a sufficient number of breeding places or not. In both these cases nothing but high precipitation throughout the year can guarantee the presence or necessary density of intermediate hosts and vectors for the maintenance of endemicity. In the case of *Aedes poicilius,* the all-season rains even fulfill a two-fold function: they are the precondition for the cultivation of abaca (*Musa textilis*) and at the same time ensure that the abaca leaf axils, which serve as breeding places, contain sufficient water for the development of the larvae and pupae of this mosquito.

Perennial precipitation is also of significance for an evergreen vegetation and a high degree of ground-cover, both of which have a positive effect on the soil moisture, which in turn guarantees the undisturbed development of hookworm and *Strongyloides* larvae, themselves very sensitive to desiccation – indeed, their development on the ground can only take place in a humid environment. Only if there is sufficient humidity are these larvae able to crawl up those parts of plants, for example, from whence they can then reach the skin of man.

Apart from this, shade-giving vegetation offers protection against over-heating by direct solar radiation, as well as against over-high doses of ultra-violet rays, which do not agree with the eggs and larvae of soil-transmitted helminths. Those areas of the Philippines which have a dry season are more likely to reach soil temperatures and radiation intensities in excess of the tolerance limits of eggs and larvae of soil-transmitted helminths than those with precipitation throughout the year.

It is true that in this context the largely mountainous character of the tropical islands of the Philippines must be taken into account; due to the temperature gradient with increasing altitude, higher locations practically never experience such high temperatures as will affect the development of worm larvae and eggs. On the contrary, the relatively low temperatures there may even have a potentially damaging effect on their development. The differences in the frequency pattern of soil-transmitted helminthiases in the Philippines depend, however, on the limits of tolerance varying from one species to another.

The many-sided complexity of the distribution of helminthiases in the Philippines requires a discussion of the orographic conditions. The wide valleys with their large river systems only offer the intermediate snail hosts of *Schistosoma japonicum* the necessary conditions for settling if a certain velocity of flow in their waters is not exceeded. Such conditions tend only to be found in the plains, the topography of which has only a low gradient.

Another geofactor limiting the endemicity of helminthiases is the soil structure. The significance of the soil conditions comes into its own indirectly in the case of endemic wuchereriasis; for abaca, the leaf axils of which are the breeding places of *Aedes poicilius*, the *Wuchereria* vector, flourishes only in deep soils, the structure of which must be neither too fine nor too coarse, so that good water storage and firm rooting are ensured.

Thus the geofactors mentioned above determine whether particular helminthiases can occur in certain area or not; but it is the human factors that decide whether a potential endemic area becomes an actual one. Since the great majority of helminths involve species whose eggs or larvae are passed out with the stool, defecation habits – and thus the hygiene behaviour of man – are of special significance. However, these cannot be seen in isolation from the economic situation of the individual or the community in which he lives.

The existence or lack of toilet facilities may be the cause of the commonly-observed town-country gradient in helminthiases' frequency. In country areas defecation in the open is more common than in the towns, especially if the fields are some distance away from the houses. The less frequent occurrence of ancylostomiasis and strongyloidiasis in the towns, however, is in part a result of the firm, barren ground and the resulting unfavourable conditions for the worm larvae. In the case of ascariasis and trichuriasis, the town-country gradient can to a certain extent be evened-out by the fact that vegetables arriving at the urban markets from the country areas are highly contaminated by worm eggs and thus contribute considerably to the infection of the urban population.

In rural areas the density of population, settlement and residence can present an important factor for the degree of contamination; that applies particularly to places where defecation occurs in the immediate vicinity of houses or villages, or possibly – and chiefly by small children – in the farmyard. These areas, however, are places of direct infection only when it is a matter of monoxenous soil-transmitted helminths.

On the other hand, indirect infection does occur when the intermediate hosts of the two *Taenia* species become infected: cattle and carabaos when grazing, pigs as a result of coprophagy, something which is very common among these animals. For the majority of the heteroxenous helminths, above all the trematodes, it is necessary that stool containing eggs gets into the water or into its immediate vicinity if the cycle, i.e. the infection of intermediate hosts, is to continue.

On the other hand, the continuation of the cycle does not only depend on the presence of natural biotopes for the intermediate host. Decisive factors may include the interference by man, especially the prevailing land use. This applies, for instance, to rice cultivation and schistosomiasis, although rice fields only make suitable habitats for *Oncomelania hupensis quadrasi* if they are not utilized between harvest and the next planting, but are allowed to become the undisturbed milieu required for colonization by snails. Intensively cultivated and utilized rice-fields are, on the other hand, unsuited as snail biotopes. Another classic example for the creation of breeding places for vectors through man's activities is the cultivation of abaca, mentioned at an earlier stage, as well as the cultivation of the saba banana, in the leaf axils of which the larvae and pupae of *Aedes poicilius*, the vector of *Wuchereria bancrofti*, develop.

The behaviour of man and his habits and customs also decide the coming about of the different helminthic infections. In the case of the soil-transmitted helminths it is largely a matter of hygiene, schistosomiasis is dependent on behaviour in regard to contact with water, food-borne helminthiases relate to certain eating habits, whilst insect-borne helminthiases are affected by the distance of dwellings from the breeding places of the vectors. In the last resort it is these human factors which decide whether or not worm infections can indeed persist in an area.

In conclusion it remains to be stated that the distribution pattern of the human helminthiases in the Philippines is governed by the interplay of the geofactors and the human factors mentioned already. The elucidation of these inter-relationships for each of the helminth infections in the sense of a causal analysis has been the purpose of the work presented here.

Acknowledgements

This monographic treatment of the Helminthiases of Man in the Philippines would not have been possible without the helpfulness and support of Philippine colleagues. I am particularly obliged to Prof. Dr. Benjamin D. Cabrera of the Institute of Public Health, University of the Philippines in Manila, who made it possible for me to use the library of his institute. There it was above all Mrs R. R. Rosal, the Chief Librarian – who had herself already produced three annotated bibliographies on Helminthiases in the Philippines – who made her extensive collection of off-prints and copies, which concerned my theme, available to me.

In addition significant support for my project came from Dr. E. J. Zerrudo of the Schistosomiasis Control Council, from Dr. A. T. Santos of the Schistosomiasis Control and Research Service, Ministry of Health, as well as from Dr. B. L. Blas of the Schistosomiasis Control Pilot Project in Palo, Leyte, to whom I owe my thanks for their helpfulness and hospitality.

My stay in the Philippines was made possibly by a travel grant from the Deutsche Forschungsgemeinschaft.

I am particularly grateful to the Heidelberg Academy of Sciences for accepting this project.

Appendix A: REPUBLIC ACT NO. 299

AN ACT APPROPRIATING TWO HUNDRED THOUSAND PESOS FOR THE PREVENTION AND CONTROL OF SCHISTOSOMIASIS.

Be it enacted by the Senate and House of Representatives of the Philippines in Congress assembled:

SECTION 1. The sum of two hundred thousand pesos is hereby appropriated out of any funds in the National Treasury not otherwise appropriated, which sum shall be placed at the disposal of the Department of Health for the study of schistosomiasis, its cause, prevention and cure in coordination with the Institute of Science.

SEC. 2. This Act shall take effect upon its approval.

Approved, June 17, 1948.

Appendix B: REPUBLIC ACT NO. 2011

AN ACT APPROPRIATING ADDITIONAL FUND FOR EMERGENCY LABORERS AND FOREMEN TO BE EMPLOYED IN THE PREVENTION OR CONTROL OF SCHISTOSOMIASIS.

Be it enacted by the Senate and House of Representatives of the Philippines in Congress assembled:

SECTION 1. The sum of thirty-four thousand five hundred pesos is appropriated, out of any funds in the National Treasury not otherwise appropriated, as additional fund to be expended by the Department of Health for emergency laborers and foremen to be employed by it in the prevention or control of schistosomiasis during the fiscal year ending June thirty, nineteen hundred fifty-eight.

SEC. 2. This Act shall take effect upon its approval.

Approved, June 22, 1957.

Appendix C: REPUBLIC ACT NO. 4359

AN ACT TO CREATE THE NATIONAL SCHISTOSOMIASIS CONTROL COMMISSION, AUTHORIZING THE NECESSARY FUNDS THEREFOR.

Be it enacted by the Senate and House of Representatives of the Philippines in Congress assenbled:

SECTION 1. The National Schistosomiasis Control Commission is hereby created under the Department of Health to serve as a liaison agency between the goverment of the Philippines and the WHO/UNICEF, and as a coordinating body to ensure the concerted effort of various agencies concerned in planning and implementing a comprehensive program for schistosomiasis control in endemic and potentially endemic areas of the country.

SEC. 2. The membership of the Commission shall be composed of the following: The Secretary of Health as chairman, the Secretary of Education, the Secretary of Agriculture and Natural Resources, the Secretary of Public Works and Communications, the Chairman, National Economic Council, the Social Welfare Administrator, the Commissioner of Agricultural Productivity, the Director General of Program Implementation Agency, the National Irrigation Administrator, the Presidential Assistant on Community Development, the General Manager, National Waterworks & Sewerage Authority, the Budget Commissioner, and the President, Philippine Medical Association.

The chairman may designate the technical head of the schistosomiasis control program as executive secretary of the Commission. He may also secure the services on detail basis of officials and employees of other branches and agencies of the goverment to assist the Commission in the discharge of its duties and functions.

SEC. 3. The Commission shall create an executive committee and such other committees or sub-committees as may be necessary for the effective and efficient performance of its duties and proper exercise of its powers and responsibilities.

Under the executive committee there shall be created the secretariat and the necessary personnel provided therefor to carry out the administrative and general services functions of the National Schistosomiasis Control Commission.

SEC. 4. The members of the National Schistosomiasis Control Commission will serve in honorary capacity and shall receive no compensation or remuneration, except traveling expenses and per diems when attending sessions of the Commission, its committees and sub-committees in such amount and under such conditions as the Secretary of Health may prescribe.

SEC. 5. The National Schistosomiasis Control Commission shall have the following powers and duties:

(a) Formulate and carry out a comprehensive national schistosomiasis control program.

(b) Foster effective exchange of information and coordination of projects, programs and activities of various agencies connected with schistosomiasis control.

(c) Establish effective liaison with participating agencies.

(d) Submit recommendations to the President on administrative or legislative measures that will effectively carry out a national control program for the disease.

SEC. 6. The Commission, its committees and sub-committees shall meet regularly and may hold such special meetings as the exigencies of the undertaking may demand.

SEC. 7. To carry out the provisions of this Act, the sum of two hundred thousand pesos is hereby authorized to be appropriated, out of any funds in the National Treasury not otherwise appropriated, which sum shall be placed at the disposal of the national schistosomiasis control program in the Philippines.

SEC. 8. All Acts or parts of Acts, rules and regulations, and executive orders inconsistent with the provisions of this Act are hereby repealed or modified accordingly.

SEC. 9. This Act shall take effect upon its approval.

Approved, June 19, 1965.

Appendix D: PRESIDENTIAL DECREE NO. 893

RECONSTITUTING THE NATIONAL SCHISTOSOMIASIS CONTROL COMMISSION INTO THE SCHISTOSOMIASIS CONTROL COUNCIL AND FOR OTHER PURPOSES.

WHEREAS, the areas endemic and potentially endemic to schistosomiasis are scattered in the three major geographic divisions of the country and the disease generally affects the people of the rural areas;

WHEREAS, schistosomiasis affects not only the health of farmers and their families, but also, their domestic and work animals thus seriously imparing their productive capabilities;

WHEREAS, there is a need for a meaningful and effective coordination of projects and activities to combat and control the disease in order to protect and improve the quality of life of the people, especially those in the areas affected by the disease;

NOW, THEREFORE, I, FERDINAND E. MARCOS, President of the Philippines, by virtue of the powers vested in me by the Constitution, do hereby adopt, approve and make as part of the laws of the land the following:

SECTION 1. Creation of the Schistosomiasis Control Council. - The National Schistosomiasis Control Commission created under Republic Act Numbered four thousand three hundred fifty-nine is hereby abolished and in lieu thereof, there is hereby created the Schistosomiasis Control Council, hereinafter called the Council, which shall be attached to the Department of Health.

The powers, functions, balance of appropriations, records, assets, as well as the rights and obligations of the National

Schistosomiasis Control Commission are hereby transferred to, vested in, and assumed by the Council.

SEC. 2. Functions and Objectives of the Council. - The Council shall have the following functions and objectives:

a. Formulate an integrated national program for the control of Schistosomiasis in the Philippines;
b. Coordinate and integrate the implementation of the schistosomiasis control program, projects and activities of all agencies and instrumentalities of the government as well as those of the private sector involved in controlling the disease;
c. Formulate an effective information and education campaign scheme and other such complementary schemes necessary for the control of the disease;
d. Promote special studies and researches on related aspects of schistosomiasis control and treatment;
e. Represent the government in its dealings with the World Health Organization or with the United Nations International Children's Emergency Fund or any other international health agencies or entities or with foreign governments or their agencies in matters involving schistosomiasis control, subject to existing laws, rules and regulations; and
f. Undertake such other activity as may be deemed necessary for the attainment of the objectives of the Council.

SEC. 3. Organization of the Council. - To carry out its' foregoing functions and objectives. The Council shall have a Board of Directors, an Executive Committee, and a Technical Staff.

SEC. 4. The Board of Directors. - The Board of Directors which shall be the policy-making and governing body of the Council, shall be composed of seven (7) members, namely: the Secretary of Health as Chairman, the Secretary of Public Works, Transportation and Communications, the Secretary of Local Governments and Community Development, the Secretary of Agriculture, the Administrator of the National Irrigation Administration, the Secretary of Education and Culture and the President of the Philippine Medical Association as members. In case of the absence of the Chairman, the members shall choose among themselves a temporary Chairman who shall take his place and perform his functions and duties, as provided for under this decree.

The Board of Directors shall have the following functions and powers:

a. Approve and adopt the integrated national schistosomiasis control program envisioned in this Decree;
b. Consider and adopt periodic reports on the status of the aforementioned program;
c. Adopt broad policies and guidelines to carry out the goals and objectives of the Schistosomiasis Control Council;
d. Adopt such rules, regulations and procedures as maybe necessary for the attainment of the objectives of this Decree; and
e. Perform such other functions as may be provided by law.

SEC. 5. The Executive Committee. - The Executive Committee shall be headed by an Executive Director and composed of representatives each coming from and designated by the government agencies represented in the Board of Directors and three (3) other representatives each coming from the National Economic and Development Authority, the Department of Social Welfare and the Budget Commission. The said representatives who must not be the heads of the aforementioned agencies, can act in behalf of their respective agencies only on matters involving the formulation and execution of the integrated national schistosomiasis control program and related projects and activities envisioned in the Decree. The Chairman, upon the recommendation of the Board of Directors shall appoint the Executive Director and determine his qualifications and compensation.

The Executive Committee shall have the following powers and functions:

a. Review and recommend for approval and adoption by the Board of Directors the schistosomiasis control program prepared by the Technical Staff;
b. Review periodic reports prepared by the Technical Staff for the consideration and approval of the Board of Directors;
c. Provide guidelines for the operations and activities of the Technical Staff;
d. Act and decide on matters referred to it by the Technical Staff for its consideration; and
e. Perform such other functions as may be assigned to it by the Board of Directors and/or the Chairman.

SEC. 6. The Technical Staff. - The Council shall have a Technical Staff, likewise to be headed by the Executive Director of the Executive Committee. The Chairman, upon the recommendation of the Executive Committee, shall appoint the members of the technical staff. The technical positions of said staff shall be exempt from the rules and regulations of the Wage and Position Classification Office and from the examination and/or eligibility requirements of the Civil Service Commission.

The Technical Staff shall have the following functions and responsibilities:

a. Undertake the preparations of a coordinated schistosomiasis control program for the approval and adoption by the Board of Directors;
b. Determine the linkages for the effective coordination and implementation of schistosomiasis control projects;
c. Establish and maintain an up-to-date and comprehensive data system on the status of the disease specifying the areas affected, the number of people, the deaths caused by the disease and other relevant information;
d. Prepare periodic reports on the implementation of the program as well as the projects and activities relating thereto for the approval and adoption by the Board of Directos; and
e. Perform such other funtions as may be assigned to it by the Board of Directors and/or the Executive Committee.

For the purpose of the foregoing functions and responsibilities, the Council may likewise utilize the services of the Schistosomiasis Control and Research Service of the Department of Health and if deemed necessary, secure the detail to the Council of personnel from the said service unit.

The Council shall also hire such personnel, including consultants, as may be necessary for the performance of its functions, who shall be appointed by the Chairman.

SEC. 7. Meetings. - The Board of Directors and the Executive Committee shall hold meetings whenever necessary. For official meetings of either bodies, the presence of a majority of their respective members shall constitute the quorum to do business and all its decisions shall be concurred in by at least the majority of the quorum.

The members of the Board of Directors and the Executive Committee shall receive a reasonable amount of per diem as the Board may provide for every meeting actually attended by them.

SEC. 8. Essential Project Element. - All agroengineering projects, especially in endemic and potentially endemic areas identified by the Council from time to time, shall always have schistosoma control components.

SEC. 9. Reports. - Within three months after the end of every fiscal year, the Council shall submit to the President of the Philippines thru the National Economic and Development Authority its annual report embodying the status of its programs and projects as well as its problems and recommendations. It shall likewise submit periodic and other reports that may be required of it by the President or by the National Economic and Development Authority.

SEC. 1o. Role of Implementing Agencies. - All government agencies concerned shall be responsible for the implementation of the projects assigned to them. All other government agencies and instrumentalities, including government-owned and controlled corporations, as well as the private sector, shall support the projects of the Council for the control of the disease.

SEC. 11. Authority to Receive Gifts and Donations. - The Council is hereby authorized to receive gifts, donations or bequests and utilize the same for the implementation of its schistosomiasis control program and projects, in accordance with the terms of the gift, donation or bequest. Such gifts, donations or bequests shall be exempt from all taxes, duties dues or fees, whatsoever, imposed or hereafter to be imposed by the national government, its agencies instrumentalities or political subdivisions, and the same shall be fully deductible from the gross income of the donor or grantor for income tax purposes.

SEC. 12. Appropriations. - There is hereby appropriated the sum of one million pesos out of the funds in the National Treasury not otherwise appropriated for the operation of the Schistosomiasis Control Council and its projects for fiscal year 1975-76. Thereafter, such amounts as shall be needed for the continued operation of the Council and its projects shall be regularly included in subsequent annual General Appropriation Acts.

The government agencies represented in and composing the Council as indicated in Sections 4 and 5 hereof are hereinafter authorized to include in their respective regular annual appropriations such additional amount as may be deemed needed to be used exclusively in undertaking their activities involving the integrated national schistosomiasis control program provided for in this Decree.

SEC. 13. Repealing Clause. - Republic Act Numbered four thousand three hundred and fifty-nine is hereby repealed. Likewise, any provision of law, decree, rules and regulations inconsistent herewith is hereby repealed, amended or revised accordingly.

SEC. 14. Separability Clause. - If for any reason any section or provision of this Decree is declared to be unconstitutional or invalid, the other sections or provivions hereof which are not affected thereby shall continue in full force and effect.

SEC. 15. Effectivity. - This Decree shall take effect immediately.

Done in the City of Manila, this 16th day of February in the year of Our Lord, nineteen hundred and seventy-six.

FERDINAND E. MARCOS
President
Republic of the Philippines

Appendix E Schistosomiasis Control Council Organizational Chart

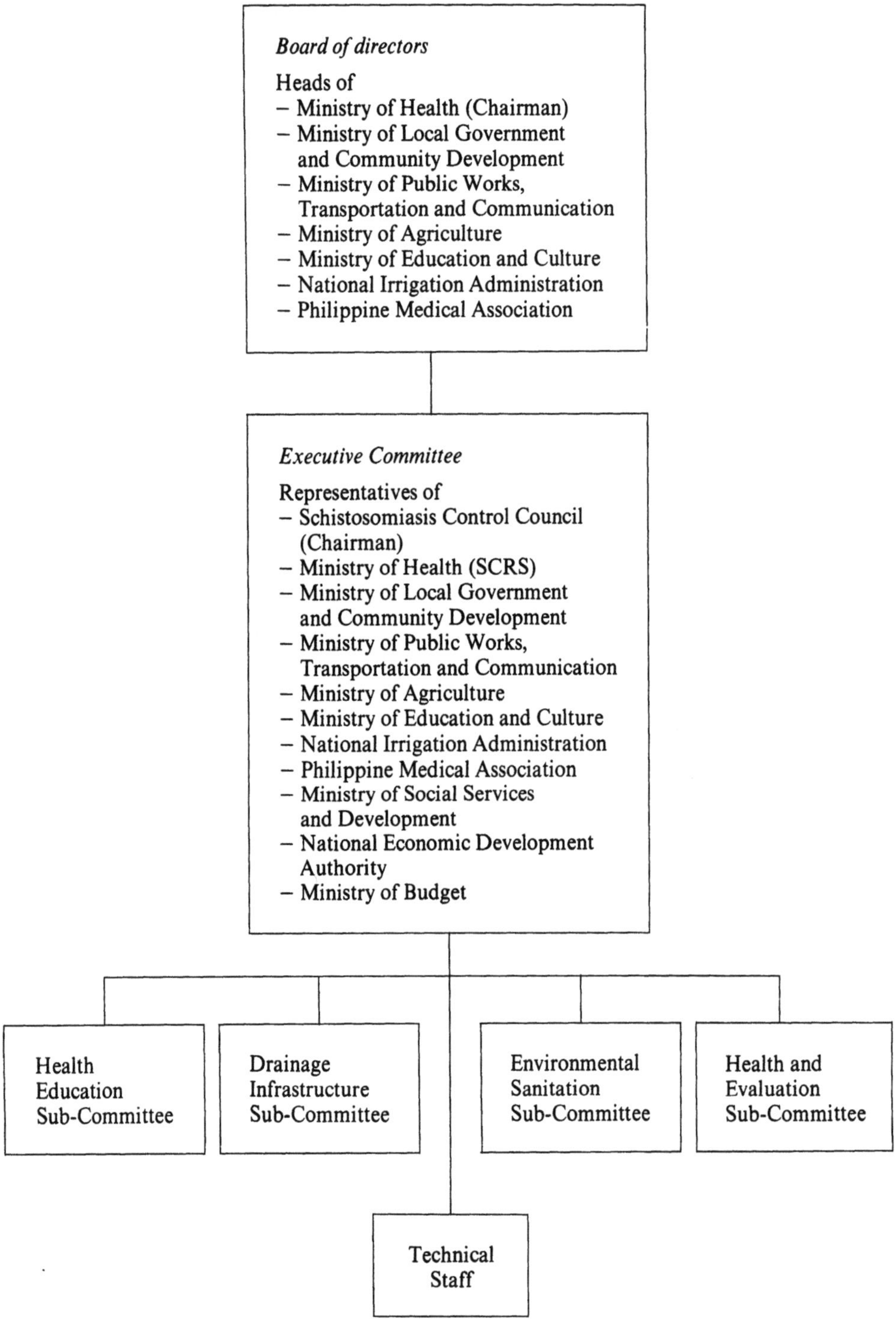

Appendix F: Habitats of Oncomelania quadrasi [Source: 609]

"Flood-plain forests and swamps. These apparently represent the most extensive original habitat of the snail. Because the land is desirable for growing lowland rice, these forests are gradually destroyed, and it is only under the pioneering conditions in certain parts of Mindanao that we can find the habitat in an undisturbed condition. The largest snail habitat yet discovered covers the entire Manat River swamp in Davao Province. It is approximately 20 km long by at least 2-3 km wide. At its northern end, a new road is under construction. The Manat River at this point is contained within banks, and the forest can be entered readily. Snails were found on the forest floor in places that seemed scarcely lower than the general ground level, and that contained no water at the time that tne place was observed. Elsewhere, as at a trail-crossing near the centre of the swamp, the river fingers out and has no defined channel, covering a large part of the forest with water. Snails were found all along this trail from the point of its entrance into the swamp onwards for at leats 2 km.

The sample description could be repeated with minor variations for many other places in Mindanao. As mentioned above, the land is desirable for rice-farming, and all such habitats seen are under agricultural encroachment.

Rice-fields. The above discussion brings up the subject of the fate of snails under the changing conditions described. In Mindanao, it is not common to find O.quadrasi in rice-fields, and in the localities where this happens, the fields are either newly cut out of swamps, or else have been abandoned for one year or more. This is in striking contrast to Leyte, where the largest snail habitat is the Cogon-Anahaway rice-field, which covers an estimated 12 km^2. The difference can only be accounted for on the basis of the primitive agriculture practised in eastern Leyte. The routine method in Mindanao seems to be to plough the land wherever absence of roots permits, and to perform at least some weeding during the growing season of the rice.

Thus, as rice-fields are pushed into a snail-inhabited swamp, the snails are able to continue living for a time, the length of which depends upon the intensity of cultivation. We are fortunate in having some record of the events in one case, that of the rice-fields at Managok, Malaybalay, Bukidnon Province. During 1952, a unit of the Division of Schistosomiasis of the Philippines Department of Health was stationed near by, and they reported that the rice-fields harboured snails. In September 1955, the place was resurveyed, and an intensive search revealed that snails remained in only one field but were abundant in the adjacent road ditch. The exception was a field that had been abandoned for two years. According to residents, the place was first farmed in 1948, and we may conclude that four years of the type of farming practised there are not sufficient to eliminate Oncomelania, but that after seven years, none remain. The reversion of one field after two fallow years shows that conditions are otherwise suitable for snails.

Streams. Throughout all the areas inhabited by O.quadrasi, there occur meandering, sluggish streams, densely populated with snails. The typical picture seen is a stream bed 3-15 m wide, at a level one metre or more below that of the surrounding country. The stream bed is flat, clogged with vegetation, and very soft. In spite of the vegetation, there is usually a moderate flow of water. The amount of flow and the absence of a defined channel constitute a paradox. The explanation seems to lie at least partly in the habits of the people and of their domestic animals. Several species of fish, particularly the mudfish (Ophicephalus sp.), inhabit these streams and are widely sought for food. The method used in catching them is to dike off or dam a part of the stream, and then to wade in and capture the fish by hand. The repeated temporary damming up of the streams prevents the water from making anything like a permanent channel. In addition, the streams are very widely used as wallowing sites for the water buffalo (carabao), the principal work-animal of the Philippines. The wallows are temporary, and many different sites may be used by the same animal. This wallowing has the same effect as the fishing - it forces the water away from any defined channel, softens the whole stream bed, and by making the flow more sluggish promotes the growth of semi-aquatic grasses and other plants, which in turn impede the water still further.

Small swamps. These areas are much more like the streams in appearance than they are like the large flood-plain swamps. The principal differences from the streams are their small size and the lesser flow of water through them. Like the streams, they are fished and used as carabao wallows. These small swamps, which we have called "pockets" in our records, are always located at the foot of rather high and steep banks, and the sources of their water are seepages and springs emerging below the banks.

In many places, but particularly in the Sorsogon area, spring outlets may support snails without forming a well-defined swampy area. Often the banks around such a spring, though rather steep, and even sandy, are continuously wet from seeping water. Such places may be very small. Omagom spring in Irosin, Sorsogon, forms an Oncomelania habitat whose maximum dimensions are 10 x 2 m. It is particularly these small swamps that have given such a vivid impression of discontinuity in snail distribution and have led us and others to refer to individual snail-inhabited areas as "colonies".

Road ditches and borrow pits. In the construction of roads in lowland areas, it is often necessary to do extensive filling in order to raise the road bed above ordinary high-water levels. For economic reasons, the source of the fill-dirt cannot be far removed, and the most common practice seems to be to take it from the immediately adjacent land. The result of this is that a rather large ditch is left on either side of the road. These ditches remain wet because of their low level, and in many places provide good habitats for O.quadrasi.

Discussion. Not every snail colony seen can be fitted exactly into the categories described above, as many of them are really combinations of types. The classification is intended mainly as an aid in description.

One interesting feature of these snail colonies is their permanence. Not only have all those that we have followed constantly since early 1953 continued to maintain themselves, but there is good evidence of much longer continuity. The swamp behind MacArthur (formerly Tarragona), Leyte, is discussed by Sullivan & Ferguson (1946) as a snail habitat, and there is every reason to suppose that the snails found by Tubangui in 1932 came from either Binog or Gacao streams - both flourishing colonies at present.

The indications are that the factors making for suitability do not ordinarily change over a period up to some tens of years. The example of the road ditches, however, is enough to warn us that it is possible to alter places so that snails can breed in them.

A consideration of all habitats of O.quadrasi reveals some striking differences. The extremes of conditions would appear to be the gravelbottomed springs and seeps in Sorsogon, the moist and not waterlogged forest floor in Davao, and the Cogon-Anahaway rice-field in Leyte. These areas are so different that any features common to all of them would appear more or less to define the requirements of the snail.

One characteristic appears to be that they never dry out for any length of time. Drying is prevented by constant seepage of underground water in Sorsogon, by the dense forest cover in Davao, and by the flat topography and clay subsoil in the rice-field. Therefore, the right amount of water proper to the slope and soil would appear to be the most important requirement for O.quadrasi.

As seen above, any further disturbance of the rice-field through more intensive cultivation would probably eliminate snails there, just as removal of the forest cover would permit the ground to dry enough to make that place unsuitable. We conclude, tentatively at least, that a relatively undisturbed situation is necessary for snails to thrive. These two features would thus appear to make the major differences between the places that are and those that are not suitable for snails, and the obvious implication is that control measures should be designed with them in mind".

Tables 1–40

Table 1. Filaria survey 1959-1963
(Source: 290)

Provinces	No. of barrios visited	Population examined	No. of houses canvassed	No. of blood smears taken	Positives	
					Number	%
1. Abra	142	26,073	5,079	4,337	0	0
2. Agusan	167	78,028	11,025	5,487	129	2.35
3. Albay	388	141,873	25,421	11,723	750	6.40
4. Bataan	33	15,903	2,849	1,435	1	0.07
5. Batangas	27	18,088	2,599	1,508	1	0.06
6. Bohol	48	609,900	1,377	5,414	15	0.27
7. Bulacan	18	6,524	917	149	0	0
8. Bukidnon	155	39,476	6,449	3,573	89	2.49
9. Cagayan	159	92,339	17,323	6,294	0	0
10. Camarines Norte	49	19,133	3,642	1,725	105	6.09
11. Camarines Sur	116	25,673	5,350	2,389	183	7.66
12. Catanduanes	33	12,109	2,164	2,252	77	3.43
13. Cavite	24	17,322	2,318	1,209	1	0.08
14. Cebu	54	1,030,380	1,379	3,766	1	0.03
15. Cotabato	230	71,062	13,709	6,498	135	2.08
16. Davao	328	117,520	20,695	8,423	173	2.05
17. Ilocos Norte	168	37,422	6,910	3,373	0	0
18. Ilocos Sur	118	35,047	6,484	4,909	0	0
19. Isabela	15	7,034	1,332	893	0	0
20. Laguna	18	4,990	891	938	0	0
21. Lanao del Norte	77	26,331	4,824	3,382	7	0.20
22. Lanao del Sur	27	13,555	218	3,617	69	1.90
23. La Union	99	22,227	4,005	2,970	0	0
24. Leyte del Norte	190	152,481	3,928	16,193	269	1.66
25. Leyte del Sur	56	32,364	1,373	5,555	105	1.89
26. Masbate	65	50,864	1,514	6,894	206	2.99
27. Misamis Occidental	79	28,174	5,639	4,364	6	0.14
28. Misamis Oriental	264	55,975	9,459	6,751	165	2.44
29. Mountain Province	249	44,959	8,478	3,790	5	0.13
30. Negros Oriental	27	570,330	562	1,958	1	0.05
31. Nueva Ecija	61	25,713	4,117	2,154	0	0
32. Nueva Vizcaya	37	11,742	7,183	1,697	0	0
33. Occidental Mindoro	50	16,307	2,470	3,751	9	0.24
34. Oriental Mindoro	117	46,305	9.795	4,373	65	1.49
35. Palawan	143	26,026	3,473	14,968	353	2.36
36. Pampanga	25	13,987	2,333	1,400	0	0
37. Pangasinan	252	105,330	17,970	6,511	0	0
38. Quezon	57	21,385	3,547	3,043	26	0.85
39. Rizal	14	6,264	1,029	776	1	0.13
40. Romblon	153	28,910	5,250	2,652	46	1.73
41. Samar	269	235,457	5,114	20.922	845	4.04
42. Sorsogon	370	136,139	13,770	77,650	6,525	8.40
43. Sulu	18	331,000	244	982	100	10.18
44. Surigao del Norte	134	35,676	6,269	4,406	226	5.13
45. Surigao del Sur	31	14,295	524	2,413	84	3.48
46. Tarlac	31	16,602	2,491	1,673	0	0
47. Zambales	104	38,965	7,255	2,991	0	0
48. Zamboanga del Norte	138	44,029	8,327	4,528	36	0.80
49. Zamboanga del Sur	76	28,562	5,361	4,053	14	0.35
Total	5,503	4,585,850	284,435	282,712	10,823	3.70

Table 2. Periodicity indices for Wuchereria bancrofti foci in the Philippines

Focus		Method of calculation	
		Sasa & Tanaka	Aikat & Das
Agusan del Norte		92.35	124.45
Quezon A		112.36	141.78
Quezon B		104.96	141.80
Muntinglupa		92.11	122.58
Bongao		100.42	131.81
Patient		85.79	107.81
Muntinglupa			
MFD max.	⩽25	95.75	122.32
	26-50	93.63	119.13
	51-100	81.90	104.44
	101-200	101.46	127.88
	>200	91.23	115.89

Table 3. Relative frequency (%) of Aedes larvae and pupae collected from abaca and banana axils

Species	abaca		banana	
	larvae	pupae	larvae	pupae
Aedes poicilius	22.6	24.5	80.5	75.1
Aedes ananae	53.9	36.4	4.0	2.4
Aedes meronephada	21.1	35.7	4.6	9.3
Aedes flavirostris	?	?	9.6	12.3
total number collected	54,285	1,447	44,587	1,589

Table 4. Infection rates of mosquitoes naturally infected by Wuchereria bancrofti in the Philippines

Species	Infection rate (%)			No. of larvae per positive			TM	Locality	Source
	L_1	L_2	L_3	L_1	L_2	L_3			
	12.1**		6.1	-	-	-	?	Sorsogon Prov.	74
Aedes poicilius	ca 17.5		17.5*	-	-	-	IO	Casiguran, Sorsogon	172
	13.0	2.5	3.1	11.1	7.5	5.4	I	Sorsogon Prov.	290
	4.1	1.7	1.5	9.3	4.9	3.4	O	Sorsogon Prov.	290
	4.2	1.1	1.1	12.9	8.0	4.4	H	Sorsogon Prov.	290
	2.7	0.9	0.9	10.1	10.4	5.1	H	Gubat, Sorsogon	85
	2.0	3.0	4.0	3.3	5.0	4.9	AH	Sablayan Is,Sorsogon	175
	5.7	2.6	2.6	4.6	3.5	3.1	H	Paraiso,Catanduanes	803
	0.0	0.0	0.0	0.0	0.0	0.0	A	Paraiso,Catanduanes	803
	4.3	0.9	1.5	-	-	-	IOH	Salcedo,E.Samar	830
	1.8	5.4	0.0	14.5	3.8	0.0	H	Jolo, Sulu	169
	2.5	0.0	0.9	24.3	0.0	4.0	A	Jolo, Sulu	169
Anopheles flavirostris	0.3	0.3	1.3	2.0	1.0	2.0	A	Calaccad,Kal.-Apayao	664
	4.6	1.9	1.2	6.7	5.7	3.5	A (H)	Bongao,Tawi-Tawi	127
	0.0	1.0	1.0	0.0	1.0	3.0	A	Odiong, Palawan	667
Culex quinque-fasciatus	12.7	0.0	0.0	-	-	-	IO	Irosin, Sorsogon	172
	13.9		0.0	-	-	-	IO	Casiguran, Sorsogon	172
	9.5**		1.4	-	-	-	?	Sorsogon Prov.	74
	21.1		0.0	-	-	-	A	Pilar, Sorsogon	74
	4.2	0.4	0.1	7.7	7.3	1.0	I	Sorsogon Prov.	290
	0.9	0.0	0.0	21.0	0.0	0.0	H	Sorsogon Prov.	290
	0.5	0.1	0.0	4.0	1.8	0.0	H	Gubat, Sorsogon	85
	0.9	0.0	0.9	1.0	0.0	1.0	I	Putiao, Sorsogon	448
	13.2	0.0	0.0	7.9	0.0	0.0	H	Paraiso,Catanduanes	803
	4.6	0.7	0.0	-	-	-	IOH	Salcedo, E.Samar	830
	2.6**		0.5	17.9**		10.3	IOAH	Palawan Prov.	290

Aedes ananae	2.7	0.0	0.0	1.3	0.0	0.0	I	Sorsogon Prov.	290
A.albopictus	3.3		0.0	-	-	-	?	Sorsogon Prov.	74
	3.8	0.0	0.0	6.8	0.0	0.0	I	Sorsogon Prov.	290
	0.5	0.0	0.0	4.8	0.0	0.0	O	Sorsogon Prov.	290
A.meronephada	0.8	0.0	0.0	30.0	0.0	0.0	H	Sorsogon Prov.	290
A.vexans nocturnus	0.0	0.0	0.1	0.0	0.0	1.0	IOAH	Palawan Prov.	290
A.niveus	0.0	0.0	2.1	0.0	0.0	2.0	H	Calaccad, Kal.-Apayao	664
Culex summorosus	0.6		0.0	-	-	-	?	Sorsogon Prov.	74
	8.5**		4.3	-	-	-	A	Pilar, Sorsogon	74
	5.1	0.0	0.0	7.8	0.0	0.0	I	Sorsogon Prov.	290
	0.1	0.0	0.0	5.0	0.0	0.0	H	Sorsogon Prov.	290
	0.1		0.0	10.0		0.0	IOAH	Palawan Prov.	290
C.bitaeniorhynchus	4.8	0.0	0.0	13.9	0.0	0.0	I	Sorsogon Prov.	290
	1.6		0.0	23.0		0.0	IOAH	Palawan Prov.	290
C.incognitus	2.7	0.0	0.0	12.5	0.0	0.0	I	Sorsogon Prov.	290
	0.2	0.0	0.0	78.0	0.0	0.0	H	Sorsogon Prov.	290
C.fuscocephalus	31.6		0.0	-	-	-	?	Sorsogon Prov.	74
	4.0	0.0	0.0	11.0	0.0	0.0	I	Sorsogon Prov.	290
C.fuscanus	11.1	0.0	0.0	2.0	0.0	0.0	O	Sorsogon Prov.	290
	5.0		0.0	17.0		0.0	IOAH	Palawan Prov.	290
C.gelidus	9.1		0.0	-	-	-	?	Sorsogon Prov.	74
	0.5	0.0	0.0	2.0	0.0	0.0	O	Sorsogon Prov.	290
C.nigropunctatus	1.2		1.2	6.0		4.0	IOAH	Palawan Prov.	290
Mansonia uniformis	3.6		0.0	-	-	-	IO	Casiguran, Sorsogon	172
	1.4		0.0	-	-	-	?	Sorsogon Prov.	74
	14.1	0.0	0.0	37.3	0.0	0.0	I	Sorsogon Prov.	290
	0.7	0.0	0.0	10.0	0.0	0.0	H	Sorsogon Prov.	290
M.longipalpis	12.5	0.0	0.0	15.0	0.0	0.0	I	Sorsogon Prov.	290
	6.1**		4.0	11.0**		5.7	IOAH	Palawan Prov.	290
M.crassipes	8.5**		8.5	53.7**		25.2	IOAH	Palawan Prov.	290
M.ochracea	6.3		0.0	52.0		0.0	IOAH	Palawan Prov.	290
Armigeres baisasi	0.0	0.0	0.7	0.0	0.0	4.0	O	Sorsogon Prov.	290
	0.4	0.0	0.0	5.0	0.0	0.0	H	Sorsogon Prov.	290
A.obturbans	6.7		0.0	-	-	-	?	Sorsogon Prov.	74

TM = Trapping method: A= Animal bait; H= Human bait; I= Indoor collections; O= Outdoor collections
* infective or near infective stages ** all stages ($L_1+L_2+L_3$)

Table 5. Monthly densities of Aedes poicilius larvae in Sorsogon Province, rainfall and number of raindays in Juban, Sorsogon (Sources: 75, 827)

Year and month		Larval density*		Rainfall** (mm)	Rainy days**
		Abaca	Banana		
1957	July	0.52	12.96	217.6	12
	August	2.16	12.55	235.6	13
	September	3.08	17.24	261.5	12
	October	1.25	9.89	407.8	14
	November	6.17	9.22	509.1	15
	December	5.97	9.59	534.8	18
1958	January	1.77	9.41	439.8	16
	February	1.76	3.05	245.3	11
	March	0.78	5.13	252.3	12
	April	1.08	3.65	174.5	10
	May	0.49	7.09	155.3	9
	June	0.19	6.20	157.9	9
		25.22	105.98	3591.5	151
$\bar{x}$		2.10	8.83	299.3	12.6

* number of larvae per plant
** 20 years' average

Table 6. Monthly density and infection rate of adult Aedes poicilius in Gubat, Sorsogon (Source: 85)

Month	Infection rate	Adult density	Absolute density of infected mosquitoes (x 100)	Relative density* of infected mosquitoes
July	3.44	0.31	1.07	35.5
August	17.04	0.05	0.85	28.2
September	5.66	0.23	1.30	43.2
October	1.71	0.67	1.15	38.2
November	4.48	0.29	1.30	43.2
December	5.31	1.07	5.68	188.7
January	3.33	1.44	4.80	159.5
February	2.29	1.52	3.48	115.6
March	4.37	1.30	5.68	188.7
April	4.76	1.38	6.57	218.3
May	2.38	1.32	3.14	104.3
June	2.94	0.37	1.09	36.2
Average	4.47	0.83	3.01	100.0

* monthly average (3.01) = 100 %

Table 7. Wuchereria bancrofti in the Philippines: prevalence, microfilarial density and transmission index (Sources: 158, 168-171, 175; average of data)

Age group (Yrs)	Prevalence		MFD/20µl		TI rel. (%)	
	♂♂	♀♀	♂♂	♀♀	♂♂	♀♀
1-5	0.46	0.44	2.5	2.8	0.06	0.10
6-10	2.14	3.82	26.5	41.6	2.18	3.54
11-15	6.12	3.78	19.1	13.5	2.82	1,36
16-20	8.61	8.24	35.7	28.9	8.56	4.58
21-25	14.98	6.64	50.3	20.6	14.14	3.00
26-30	16.47	7.35	33.0	20.9	8.68	2.10
31-35	14.84	3.83	33.1	25.0	8.44	1.32
36-40	12.64	8.13	48.7	19.6	8.86	1.86
41-45	19.15	5.78	32.9	15.0	5.72	1.66
46-50	14.42	7.18	27.6	34.0	2.94	2.28
51-55	26.55	6.51	15.7	22.5	2.58	1.72
56-60	19.22	5.57	52.8	10.7	5.14	0.88
>60	18.76	13.90	23.8	9.0	4.08	1.40
	8.25	4.85	35.1	26.1	74.20	25.80

MFD = microfilarial density = number of microfilariae per 20 µl blood
TI rel. = relative transmission index

Table 8. Absolute transmission indices for Wuchereria bancrofti in different endemic areas of the Philippines

Endemic area	Transmission index			Source
	♂♂	♀♀	Σ	
Mindanao	2,983	816	3,799	171
Jolo, Sulu	23,720	13,880	37,600	169
Sulu and Tawi-Tawi	8,943	2,263	11,206	158
Juban, Sorsogon	46,855	15,005	61,860	175
Quezon, Palawan	3,926	3,043	6,969	168
Brooke's Point,Palawan	33,236	11,121	44,357	168
Mindoro	20,823	11,731	32,554	285
	140,486	57,859	198,345	

Table 9. Abaca in the Philippines 1818-1978: quantity, area harvested and value of production
(Sources: 395 and 1980 Philippine Statistical Yearbook)

Year	Quantity of production per year			Area harvested			Value		
	metric tonnes	% of all crops	% of comm. crops	hectares	% of all crops	% of comm. crops	Pesos (mill.)	% of all crops	% of comm. crops
1818/25	159	-	-	-	-	-	-	-	-
1840/50	8,532	-	-	-	-	-	-	-	-
1854/60	23,729	-	-	-	-	-	-	-	-
1871/80	41,776	-	-	-	-	-	-	-	-
1891/1900	79,642	-	-	-	-	-	-	-	-
1903	-	-	-	217,810	-	-	-	-	-
1910	168,450	-	-	475,140	-	-	-	-	-
1918	-	-	-	512,510	-	-	-	-	-
1920	165,080	-	-	559,360	-	-	-	-	-
1925	-	-	-	477,110	-	-	-	-	-
1929	213,393	4.51	13.27	484,850	1.21	34.60	54.8	10.77	21.60
1930	195,500	4.03	12.09	496,080	1.22	34.60	37.7	8.01	17.10
1931	162,200	3.49	10.74	484,900	1.20	34.03	17.0	4.86	10.40
1932	130,400	2.70	7.85	467,000	1.15	33.10	8.2	2.81	5.25
1933	134,500	2.67	7.16	477,200	1.12	32.61	7.8	2.83	5.32
1934	180,300	3.23	8.19	458,600	1.06	31.21	10.9	3.22	5.64
1935	200,600	4.55	13.62	488,100	1.15	34.33	12.9	5.06	10.15
1936	150,000	3.23	8.28	478,900	1.06	32.44	19.6	5.74	10.62
1937	200,600	3.74	10.78	502,700	1.11	33.05	25.1	6.67	12.58
1938	167,900	3.06	8.52	508,300	1.15	33.79	22.7	5.73	10.58
1939	144,100	2.74	8.32	291,500	0.60	17.72	11.4	3.27	8.13
1946	30,400	1.01	9.64	272,400	0.70	20.93	8.3	0.83	7.68
1947	82,000	1.75	6.64	280,800	0.63	21.14	25.9	2.04	7.78
1948	99,400	1.95	6.83	283,200	0.61	20.61	47.7	3.60	11.16
1949	74,600	1.36	4.81	282,700	0.57	19.73	48.1	3.32	10.05
1950	82,200	1.40	5.10	291,500	0.57	19.89	52.6	3.51	10.89
1951	130,400	1.96	5.84	305,300	0.58	19.98	79.8	4.87	12.96
1952	114,600	1.65	5.83	275,100	0.49	18.07	68.7	4.42	14.67
1953	112,700	1.42	5.30	271,900	0.45	17.67	60.8	3.94	10.82

1954	106,200	1.26	4.29	238,600	0.39	15.34	38.9	2.55	6.40
1955	104,500	1.21	4.04	217,000	0.34	14.08	35.2	2.26	5.64
1956	120,300	1.36	4.79	216,800	0.32	14.10	35.4	2.26	6.00
1957	128,500	1.41	4.82	231,500	0.33	14.91	37.9	2.31	5.89
1958	124,600	1.31	3.99	192,800	0.28	12.65	38.7	2.15	4.90
1959	111,500	1.10	3.59	192,500	0.24	12.38	39.4	2.18	5.55
1960	94,500	0.91	3.05	175,200	0.23	11.05	58.8	2.87	6.70
1961	114,800	1.10	3.80	174,600	0.22	10.18	65.3	3.05	8.37
1962	116,300	1.02	3.30	182,600	0.23	9.90	61.3	2.50	6.07
1963	127,800	1.09	3.37	181,900	0.23	9.29	61.4	2.25	5.14
1964	134,300	1.10	3.45	210,500	0.26	10.09	78.9	2.51	6.04
1965	134,000	1.09	3.56	199,300	0.24	8.83	78.6	2.40	5.99
1966	135,300	1.11	3.76	198,000	0.24	8.86	65.4	1.79	4.44
1967	117,600	0.93	3.03	186,100	0.22	7.69	62.0	1.47	3.51
1968	103,400	0.78	2.62	170,700	0.19	7.10	52.0	1.01	2.55
1969	105,900	0.80	2.70	172,900	0.19	6.97	66.3	1.16	2.91
1970	122,400	0.81	2.70	173,000	0.19	6.81	105.7	1.30	3.16
1971	104,600	0.67	2.16	155,300	0.17	5.64	90.8	0.99	2.53
1972	110,100	0.72	2.41	145,200	0.15	5.15	102.6	1.00	2.90
1973	119,200	0.79	2.29	163,300	0.18	5.69	118.5	1.10	2.64
1974	135,900	0.77	2.48	170,100	0.17	5.68	374.7	2.11	5.07
1975	133,600	0.67	2.13	179,700	0.17	5.74	514.1	2.55	7.64
1976	136,500	0.60	1.88	244,300	0.21	7.09	308.2	1.52	5.28
1977	150,600	0.61	1.97	250,300	0.21	6.78	306.2	1.10	2.82
1978	129,800	0.50	1.68	243,800	0.21	6.52	240.1	0.89	2.76

Table 10. W.bancrofti prevalence by age groups in different areas of the Philippines. 1. Males

Area	Age groups (years)											
	1-5	6-10	11-15	16-20	21-25	26-30	31-35	36-40	41-45	>45	Total	Source
Bulusan, Sorsogon	1.8	3.3	9.3	11.0	20.6	29.5	17.1	16.7	26.2	25.2	13.0	74
Casiguran, Sorsogon	2.3	4.7	5.3	4.7	23.5	8.3	22.5	25.8	22.2	25.9	12.8	74
Gubat, Sorsogon	1.2	2.2	7.1	8.1	17.7	17.7	31.6	20.9	27.8	26.4	12.9	74
Irosin, Sorsogon	0.7	3.1	4.9	8.0	9.6	15.6	16.1	23.7	18.8	20.6	9.5	74
Prieto Diaz, Sorsogon	0.0	3.3	3.8	12.8	18.2	22.2	12.5	22.9	19.5	18.5	9.1	74
Barcelona, Sorsogon	0.0	4.8	6.5	4.9	14.1	24.5	23.5	17.5	36.1	30.3	14.9	74
Juban, Sorsogon	3.7	5.2	5.1	16.7	12.7	17.2	25.7	32.6	25.4	23.9	15.2	74
Juban, Sorsogon	1.0	1.1	14.8	16.0	30.4	33.3	25.0	14.8	53.8	42.9	14.8	175
Quezon, Palawan	0.0	3.1	7.0	6.0	12.5	14.9	10.3	10.8	0.0	11.1	8.1	168
Brooke's Point, Palawan	0.0	1.2	5.6	20.6	25.0	29.1	31.1	25.0	14.3	17.4	15.5	168
Quezon, Palawan	0.0	2.3	8.8	5.9	15.4	16.0	13.3	12.5	0.0	13.0	9.0	667
Siasi, Sulu	0.0	6.3	0.0	4.2	4.3	5.9	18.8	5.9	7.1	14.3	5.3	158
Bongao, Tawi-Tawi	0.9	0.5	3.5	1.2	7.9	10.8	14.6	20.5	17.6	3.9	4.6	158
Balimbing, Sulu	3.8	6.3	3.6	10.6	19.4	28.0	15.4	20.0	16.7	30.0	12.8	158
Jolo, Sulu	0.5	6.8	9.9	13.7	20.5	22.1	22.8	22.1	15.4	29.6	12.9	170
Mindanao	0.0	0.5	1.1	2.6	4.6	5.6	2.9	4.1	7.5	7.4	2.5	170
												171
Salcedo, Samar	2.1	1.8	1.8	4.6	0.9	12.6	12.7	11.0	11.2	15.5	6.5	830
Calaccad, Kalinga-Ap.	0.0	0.0	12.5	0.0	13.6	17.6	16.7	15.8	10.0	11.1	11.4	665
Average	1.00	3.14	6.14	8.42	15.05	18.38	18.48	17.92	18.31	20.39	10.60	
± s	1.26	2.14	3.86	5.74	7.76	8.26	7.38	7.29	12.96	9.78	3.96	

Table 11. W.bancrofti prevalence by age groups in different areas of the Philippines. 2. Females

Area	Age groups (years)											
	1-5	6-10	11-15	16-20	21-25	26-30	31-35	36-40	41-45	>45	Total	Source
Bulusan, Sorsogon	2.3	7.3	6.0	12.8	18.9	10.0	5.9	12.5	22.7	14.0	9.3	74
Casiguran, Sorsogon	5.1	5.2	2.0	10.8	14.3	0.0	5.9	0.0	23.5	17.6	7.8	74
Gubat, Sorsogon	0.0	2.5	4.5	9.7	16.3	28.2	0.0	24.4	22.7	13.7	8.6	74
Irosin, Sorsogon	2.4	4.1	3.1	4.0	6.5	9.7	16.7	5.4	10.4	6.2	5.6	74
Prieto Diaz, Sorsogon	0.0	3.2	2.1	2.8	11.3	6.5	9.8	8.0	8.3	16.9	5.5	74
Barcelona, Sorsogon	0.0	1.6	4.7	9.4	15.8	4.5	5.3	8.8	0.0	7.9	6.1	74
Juban, Sorsogon	1.1	9.5	7.4	13.5	12.1	8.3	16.7	3.7	12.1	13.7	9.6	74
Juban, Sorsogon	0.0	9.7	7.5	16.7	7.4	14.3	4.0	10.0	0.0	13.2	7.2	175
Quezon, Palawan	0.0	2.6	3.0	6.8	16.1	9.1	6.7	5.0	0.0	4.3	5.8	168
Brooke's Point, Palawan	0.0	4.5	9.6	6.0	16.3	11.6	14.3	13.3	13.6	17.0	9.5	168
Quezon, Palawan	0.0	4.5	5.0	11.1	22.2	13.0	11.1	7.1	0.0	7.1	8.4	667
Siasi, Sulu	0.0	0.0	0.0	6.3	0.0	0.0	0.0	0.0	0.0	0.0	0.8	158
Bongao, Tawi-Tawi	0.0	0.8	2.3	7.1	2.0	2.2	0.0	10.7	0.0	5.7	2.7	158
Balimbing, Tawi-Tawi	0.0	8.8	0.0	4.8	8.3	7.7	0.0	11.1	50.0	11.1	5.8	158
Jolo, Sulu	2.2	6.4	6.7	15.5	11.4	14.0	10.0	17.6	15.0	15.6	9.9	169
Mindanao	0.0	0.4	1.2	1.4	4.3	1.4	0.7	3.0	5.5	3.1	1.5	170
												171
Salcedo, Samar	1.6	2.6	3.2	1.3	6.0	6.2	5.4	5.2	9.7	6.5	4.3	830
Calaccad, Kalinga-Ap.	0.0	0.0	0.0	0.0	6.7	15.4	10.0	7.7	50.0	20.0	9.7	665
Average	0.82	4.09	3.79	7.78	10.88	9.01	6.81	8.53	13.53	10.76	6.56	
± s	1.41	3.18	2.83	4.98	6.12	6.81	5.62	6.05	15.62	5.78	2.85	

Table 12. Wuchereria bancrofti microfilarial densities in different areas of the Philippines
(Number of microfilariae per positive per 20 µl blood)
(Sources in brackets)

Sex	Age (yrs)	Mindanao (171)	Jolo, Sulu (169)	Sulu and Tawi-Tawi (158)	Juban, Sorsogon (175)	Quezon, Palawan (168)	Brooke's Point, Palawan (168)	Palawan (168)	Average
♂	1-5	0.0	3.5	3.0	6.0	0.0	0.0	0.0	2.5
	6-10	6.0	24.5	61.7	9.0	10.0	74.0	31.3	26.5
	11-15	8.4	32.1	13.0	20.7	17.3	23.8	21.4	19.1
	16-20	55.9	37.2	22.1	16.0	39.3	49.1	47.2	35.7
	21-25	28.6	32.8	66.9	93.3	5.8	40.8	30.1	50.3
	26-30	18.3	32.2	8.7	87.1	13.4	22.9	18.5	33.0
	31-35	3.0	22.6	50.0	27.8	8.7	77.5	62.3	33.1
	36-40	29.3	36.8	67.4	88.7	4.8	29.0	21.5	48.7
	41-45	18.6	62.9	34.6	32.6	0.0	20.2	15.8	32.9
	46-50	29.7	31.3	16.6	46.0	7.2	23.7	14.3	27.6
	51-55	16.7	25.7	0.0	30.7	0.0	6.5	5.6	15.7
	56-60	12.0	35.7	6.0	60.0	0.0	187.2	150.4	52.8
	>60	5.7	56.8	5.5	35.5	0.0	30.0	15.5	23.8
♀	1-5	0.0	13.9	0.0	0.0	0.0	0.0	0.0	2.8
	6-10	2.0	21.8	90.0	58.6	16.0	42.3	35.8	41.6
	11-15	5.2	37.9	1.0	19.3	4.0	4.0	4.0	13.5
	16-20	5.5	34.6	16.7	50.0	20.7	48.2	37.9	28.9
	21-25	3.3	42.5	25.0	15.5	8.6	21.4	16.5	20.6
	26-30	16.3	17.5	42.5	21.7	6.5	7.4	6.7	20.9
	31-35	4.0	17.4	0.0	71.0	14.0	39.0	32.8	25.0
	36-40	9.0	23.3	21.5	33.0	4.0	13.0	11.3	19.6
	41-45	11.8	24.0	26.0	0.0	0.0	13.3	13.3	15.0
	46-50	9.0	52.4	6.0	45.0	46.0	60.2	57.8	34.0
	51-55	62.0	17.7	0.0	26.0	0.0	7.0	7.0	22.5
	56-60	9.0	13.4	31.0	0.0	0.0	0.0	0.0	10.7
	>60	0.0	19.6	0.0	16.5	0.0	22.0	8.7	9.0

Table 13. Wuchereria bancrofti relative transmission indices in different areas of the Philippines (Sources in brackets)

Sex	Age (yrs)	Mindanao (171)	Jolo, Sulu (169)	Sulu and Tawi-Tawi (158)	Juban, Sorsogon (175)	Quezon, Palawan (168)	Brooke's Point, Palawan (168)	Palawan (168)	Average
♂	1-5	0.0	0.1	0.1	0.1	0.0	0.0	0.0	0.06
	6-10	0.6	3.2	5.6	0.1	3.2	1.5	1.4	2.18
	11-15	1.5	5.3	1.4	3.1	10.9	1.9	2.8	2.82
	16-20	18.6	6.7	3.4	2.1	16.8	11.3	12.0	8.56
	21-25	14.4	7.4	20.0	19.0	4.3	9.5	9.9	14.14
	26-30	10.1	7.1	2.6	17.5	10.7	5.6	6.1	8.68
	31-35	0.7	4.4	15.1	3.6	4.2	17.6	18.4	8.44
	36-40	8.4	5.8	20.8	5.6	2.0	4.3	3.7	8.86
	41-45	7.7	5.4	7.8	6.0	0.0	1.4	1.7	5.72
	46-50	5.0	3.7	1.4	3.3	4.2	1.0	1.3	2.94
	51-55	3.3	3.8	0.0	5.4	0.0	0.4	0.4	2.58
	56-60	2.9	3.7	0.8	3.2	0.0	19.2	15.1	5.14
	>60	5.3	6.5	0.8	6.8	0.0	1.2	1.0	4.08
♀	1-5	0.0	0.5	0.0	0.0	0.0	0.0	0.0	0.10
	6-10	0.1	2.6	6.1	6.3	4.1	3.0	2.6	3.54
	11-15	1.0	4.0	0.0	1.4	1.0	0.5	0.4	1.36
	16-20	1.0	7.2	3.2	6.9	10.3	3.3	4.6	4.58
	21-25	1.7	6.0	1.9	0.9	9.3	3.7	4.5	3.00
	26-30	2.4	2.7	2.3	2.0	3.5	0.8	1.1	2.10
	31-35	0.3	1.5	0.0	1.5	4.4	4.1	3.3	1.32
	36-40	1.8	2.8	2.1	1.4	0.7	1.0	1.2	1.86
	41-45	3.5	2.0	2.0	0.0	0.0	0.8	0.8	1.66
	46-50	1.6	2.5	0.3	1.3	10.5	4.9	5.7	2.28
	51-55	6.6	0.5	0.0	1.1	0.0	0.5	0.4	1.72
	56-60	1.3	0.8	2.3	0.0	0.0	0.0	0.0	0.88
	>60	0.0	3.7	0.0	1.6	0.0	2.5	1.7	1.40
♂+♀	1-5	0.0	0.5	0.1	0.1	0.0	0.0	0.0	0.14
	6-10	0.7	5.8	11.7	6.4	7.3	4.5	4.0	5.72
	11-15	2.5	9.3	1.4	4.5	11.9	2.4	3.2	4.18
	16-20	19.6	13.9	6.6	9.0	27.1	14.3	16.6	13.14
	21-25	16.1	13.4	21.9	19.9	13.6	13.2	14.4	17.14
	26-30	12.5	9.8	4.9	19.5	14.2	6.4	7.2	10.78
	31-35	1.0	5.9	15.1	5.1	8.6	21.7	21.7	9.76
	36-40	10.2	8.6	22.9	7.0	2.7	5.3	4.9	10.72
	41-45	11.2	7.4	9.8	6.0	0.0	2.2	2.5	7.38
	46-50	6.6	6.2	1.7	4.6	14.7	5.9	7.0	5.22
	51-55	9.9	4.3	0.0	6.5	0.0	0.9	0.8	4.30
	56-60	4.2	4.5	3.1	3.2	0.0	19.2	15.1	6.02
	>60	5.3	10.2	0.8	8.4	0.0	3.7	2.7	5.48

Table 14. Wuchereriasis in Palawan: prevalence, microfilarial density and percentage of positive barrios

Municipality	Latitude of Capital (N)	Prevalence	Microfilarial density*	Percentage of positive barrios
Brooke's Point	8°51'	12.9	39.5	53.3
Quezon	9°14'	7.2	12.5	66.7
Aborlan	9°27'	7.6	9.1	50.0
Puerto Princesa	9°44'	2.0	6.4	40.0
Roxas	10°20'	2.8	?	25.0
Taytay	10°50'	0.0	-	-
El Nido	11°10'	0.0	-	-

* average number of microfilariae per 20μl blood

Table 15. Population, land area and wuchereriasis in the Philippines (Sources: 158, 169, 171, 290 and 1980 Philippine Statistical Yearbook)

Province	Population 1960 (x 1000)	Prevalence	Estimated number of cases	Land area (km^2)	Population density per km^2	case density per km^2
Albay	515	6.40	32,960	2,553	201.7	12.91
Camarines Norte	188	6.09	11,449	2,113	89.0	5.42
Camarines Sur	820	7.66	62,812	5,267	155.6	11.93
Catanduanes	156	3.43	5,351	1,512	103.4	3.54
Sorsogon	348	8.40	29,232	2,141	162.4	13.65
Occidental Mindoro	84	0.24	202	5,880	14.3	0.03
Oriental Mindoro	229	1.49	3,412	4,365	52.5	0.78
Palawan	163	2.36	3,847	14,896	10.9	0.26
Masbate	336	2.99	10,046	4,048	83.4	2.48
Romblon	132	1.73	2,284	1,356	97.4	1.68
Bohol	592	0.27	1,598	4,117	143.8	0.39
Cebu	1,333	0.03	400	5,088	261.9	0.08
Leyte	963	1.66	15,986	6,268	153.7	2.55
Southern Leyte	210	1.89	3,969	1,735	120.8	2.29
Negros Oriental	538	0.05	269	5,402	99.6	0.05
Samar	869	4.04	35,108	13,429	64.7	2.61
Agusan del Norte	177	0.93	1,646	2,590	68.3	0.64
Agusan del Sur	94	3.37	3,168	8,966	10.5	0.35
Bukidnon	194	0.96	1,862	8,294	23.4	0.22
North Cotabato	279	0.13	363	6,702	41.6	0.05
Maguindanao	377	0.07	264	5,305	71.1	0.05
Sultan Kudarat	78	0.00	0	4,321	18.1	0.00
South Cotabato	295	0.86	2,537	7,469	39.5	0.34
Davao del Norte	263	0.64	1,683	8,130	32.4	0.21
Davao del Sur	498	3.75	18,675	6,378	78.1	2.93
Davao Oriental	133	2.18	2,899	5,165	25.8	0.56
Lanao del Norte	271	0.21	569	3,092	87.6	0.18
Lanao del Sur	378	0.05	189	3,873	97.6	0.05
Misamis Occidental	248	0.14	347	1,939	127.9	0.18
Misamis Oriental	344	1.67	5,745	3,570	96.4	1.61
Surigao del Norte	195	2.73	5,324	2,739	71.2	1.94
Surigao del Sur	165	3.39	5,594	4,552	36.3	1.23
Zamboanga del Norte	282	0.82	2,312	6,075	46.3	0.38
Zamboanga del Sur	586	0.35	2,051	8,595	68.2	0.24
Sulu	248	9.93	24,626	1,600	155.0	15.39
Tawi-Tawi	79	3.51	2,773	1,087	72.7	2.55

Table 16. Brugia malayi in Quezon, Palawan: prevalence, microfilarial density, population structure and transmission index (Source: 168)

Age group (yrs)	Number examined		Prevalence		MFD/20 µl		Population structure		Transmission index absolute		Transmission index rel. (%)	
	♂♂	♀♀	♂♂	♀♀	♂♂	♀♀	♂♂	♀♀	♂♂	♀♀	♂♂	♀♀
1-5	41	35	22.0	42.8	17.7	10.7	6.75	6.42	2628.45	2940.10	4.75	5.31
6-10	64	39	37.5	33.3	14.8	21.4	7.29	6.88	4045.95	4902.83	7.31	8.86
11-15	43	33	25.6	15.2	42.5	18.0	6.30	5.97	6854.40	1633.39	12.38	2.95
16-20	50	44	24.0	11.4	53.2	21.0	4.96	5.08	6332.93	1216.15	11.44	2.18
21-25	48	31	25.0	9.7	2.2	25.7	4.14	4.66	227.70	1161.69	0.41	2.10
26-30	74	44	25.7	6.8	42.3	49.3	3.74	4.08	4065.79	1367.78	7.35	2.47
31-35	29	15	37.9	13.3	32.4	79.0	3.23	3.24	3966.31	3404.27	7.17	6.15
36-40	37	20	32.4	10.0	40.3	34.5	2.66	2.58	3473.22	890.10	6.28	1.61
41-45	23	8	34.8	12.5	23.0	5.0	2.10	2.08	1680.84	130.00	3.04	0.23
46-50	16	11	18.8	9.1	20.3	2.0	1.63	1.74	622.07	31.67	1.12	0.06
51-55	10	8	30.0	12.5	75.3	4.0	1.36	1.54	3072.24	77.00	5.55	0.14
>55	10	4	30.0	25.0	2.3	4.0	3.50	3.83	241.50	383.00	0.44	0.69
, $\bar{x}$	445	292	28.5	17.8	28.7	21.1	47.66	48.10	37211.40	18137.98	67.23	32.77

MFD = microfilarial density = number of microfilariae per 20 µl blood per positive

Table 17. Schistosomiasis prevalence, daily egg output and transmission index in two communities in Leyte (Sources: 357,606)

Host and age group (years)	Population		Prevalence		Mean no. of hatchable eggs per day		Transmission index absolute		Transmission index relative (%)	
	(606)	(357)	(606)	(357)	(606)	(357)	(606)	(357)	(606)	(357)
Human										
<5	4331	121	18.2	10.7	148	225	116624	2925	2.35	0.35
5-9		136		51.5		594		41580		5.00
10-14	1852	107	59.9	76.6	2090	2282	2317810	187124	46.74	22.48
15-19	1648	76	66.6	81.6	429	424	471042	26288	9.50	3.16
20-29	2219	105	65.9	76.2	198	385	289476	30800	5.84	3.70
30-39	1621		57.3		183		170007		3.43	
40-49	1280	223	60.2	72.6	346	238	266766	38556	5.38	4.63
50-59	904		49.4		105		46935		0.95	
≥60	964		48.9		411		193581		3.90	
Total	14819	768	48.0	-	645	-	3872241	327273	78.09	39.32
Dog	1517	74	18.2	32.4	2333	2333	644127	55992	12.99	6.73
Cow	76	0	38.2	-	8780	-	254901	-	5.14	-
Pig	3193	176	13.3	45.5	153	153	64974	12240	1.31	1.47
Rat	148190	3075	22.7	64.0	2	222	67278	436896	1.36	52.49
Carabao	1318	58	1.5	0.0	2695	-	53280	-	1.07	-
Goat	204	1	1.4	0.0	680	-	1942	-	0.04	-

Table 18. Relative schistosomiasis transmission indices among humans in three villages in Leyte

Age group (yrs)	Village		
	Santol (24 % prev.)	Santa Rosa (39 % prev.)	Macanip (44 % prev.)
<10	3.3	23.3	10.2
10-14	39.2	28.2	30.5
15-19	8.3	10.6	27.4
20-29	11.6	11.8	11.9
30-39	28.2	13.5	6.8
40-49	3.1	7.5	9.0
50-59	5.7	1.9	2.0
⩾60	0.6	3.1	2.2

Table 19. Schistosoma japonicum prevalence in Oncomelania hupensis quadrasi in the Philippines*

Province and locality	Number examined	Number positive	Prevalence (%)	Source
SORSOGON				
Irosin	28	0	0.0	593
Irosin	-	-	0.0-26.0	605
Irosin	-	-	0.0- 7.7	621
Juban	-	-	0.0- 4.0	605
ORIENTAL MINDORO				
Naujan	-	-	3	795
Naujan	998	12	1.2	398
Naujan	-	-	1.0-80.0	605
Naujan	141	1	0.7	621
Naujan	162	2	1.2	621
Pola	624	5	0.8	398
Pola	-	-	10.0-18.7	605
Victoria	172	7	4.1	621
Victoria	78	1	1.3	621
NORTHERN SAMAR				
Bobon	-	-	1.3-42.0	605
Catarman	-	-	0.0-41.6	605
WESTERN SAMAR				
Calbiga	-	-	3	795

Table 19. (Contd.)

Province and locality	Number examined	Number positive	Prevalence (%)	Source
LEYTE				
Atogan	1026	142	13.8	87
Burauen	604	0	0.0	472
Burauen	-	-	1.2-33.0	605
Dagami	-	-	2-5	795
Dagami	-	-	0.8-21.6	605
Dagami	18728	1049	1.9-10.3	435
Dulag	-	-	5.0	605
Gacao	1078	-	0.0- 4.5	472
Julita	-	-	0.9-46.2	605
MacArthur	-	-	5.1-27.0	605
Palo	ca 400	-	0.0	781
Palo	965	-	1.3-21	472
Palo	4138	417	10.0	87
Palo	-	-	3.0-28.0	605
Palo	83059	3904	0.0- 7.0	609
Palo	-	-	0.0-39.0	609
Palo	-	-	0.0-27.9	609
Santa Fe	-	-	0.0-21.0	605
Tanauan	-	-	3.4-54.1	605
BOHOL				
Talibon	216	2	0.9	101
Trinidad	564	5	0.9	101
AGUSAN DEL NORTE				
Butuan	-	-	0.9- 9.3	605
Cabadbaran	-	-	0.0	605
Jabonga	-	-	0.0-10.0	605
DAVAO DEL SUR				
Calinan	35	0	0.0	610
Calinan	333	0	0.0	525
LANAO DEL NORTE				
Kapatagan	-	-	10.0-33.0	605
Lala	-	-	0.0-15.0	605
MISAMIS OCCIDENTAL				
Bonifacio	-	-	0.0-0.01	621
Bonifacio	-	-	0.0	605
Tangub	-	-	0.0	605
SURIGAO DEL NORTE				
Mainit	-	-	4	795
Mainit	-	-	0.0-33.0	605
ZAMBOANGA DEL SUR				
Aurora	-	-	0.0	605
Molave	-	-	0.0	605

*Prevalence in snails is highly variable, depending mainly upon distance from houses with Schistosoma-positive cases; there are also seasonal and other variations.

Table 20. Fresh water contact activities of 1935 household respondents in Leyte (Source: 769)

Activities	Water Body								All Water Bodies	
	River		Stream		Canal		Rice Field			
	No. (826)	% (42.7)	No. (360)	% (18.6)	No. (106)	% (5.5)	No. (261)	% (13.5)	No.	%
Washing clothes	447	23.1	183	9.5	12	0.6	0	-	642	34.1
Bathing	441	22.8	89	4.6	0	-	0	-	530	28.1
Farming	0	-	0	-	0	-	239	12.4	239	12.7
Crossing	46	2.4	48	2.5	71	3.7	1	0.1	166	8.8
Gathering vegetables	42	2.2	47	2.4	14	0.7	14	0.7	117	6.2
Fishing	64	3.3	31	1.6	6	0.3	0	-	101	5.4
Swimming	44	2.3	19	1.0	0	-	0	-	63	3.3
Defecating	13	0.7	11	0.6	3	0.2	10	0.5	37	2.0
All activities	1097	56.7	428	22.1	106	5.5	264	13.6	1885	100.0

Table 21. Percentage distribution of 653 exposures of people to water during one day in an inland village in Palo, Leyte (Source: 359)

Sex and age	Snail habitats		River		Total	
	Exposure (171)	Time (20.01 h)	Exposure (482)	Time (84.37 h)	Exposure (653)	Time (104.38 h)
Children						
Males	7.5	2.5	15.3	11.3	22.8	13.8
Females	6.7	1.6	25.9	29.9	32.6	31.6
Total	14.2	4.2	41.2	41.2	55.4	45.4
Adults						
Males	8.0	10.3	9.6	4.6	17.6	14.9
Females	4.0	4.7	23.0	35.0	27.0	39.7
Total	11.9	15.0	32.6	39.6	44.6	54.6
Total						
Males	15.5	12.8	25.0	15.9	40.4	28.7
Females	10.7	6.3	48.9	64.9	59.6	71.3
Total	26.2	19.2	73.8	80.8	100.0	100.0

(Figures in brackets indicate the total number and time of exposure)

Table 22. Results of soil and water analysis in O.h. quadrasi habitats and in similar places without O.h. quadrasi* (Source: 609)

Chemical	34 Snail Colonies		8 Areas without Snails	
	Range of values (p.p.m.)	Average (p.p.m.)	Range of values (p.p.m.)	Average (p.p.m.)
NO_3	0.8-20	5.9	0.8-4.0	1.65
P	12.5-100	90.8	75-100	93.3
K	45-185	90.4	50-1,110	49.37
Ca	250-5,000	1,286.8	50-1,000	781.25
NH_3	0.5-2.5	0.85	0.5-2.5	0.75
Mg	0.5-5.0	3.5	2.5-25	6.25
Mn	0.5-5.0	3.5	0.5-12.5	3.8
Al	0.5-50	23	12.5-25	23.4
NO_2	1-5	1.5	1-1	1
Fe	0.5-25	15	0.5-25	94.4
SO_4	50-250	85.2	50-250	81.25
Cl	25-100	51.0	25-100	43.75
pH (soil)	4.6-7.2	6.6	5.4-7.2	6.35
pH** (water)	6.2-7.8	7.2	6.2-7.8	7.4

* Soil analysis performed by LaMotte colorimeter method; water pH values determined with Beckman electronic pH meter.

**50 habitats, 35 negative areas.

Table 23. Human Schistosoma japonicum infections in the Philippines: population, prevalence, estimated number of cases and estimated snail area in the endemic areas (Source: 706)

Province	Population (1975) of endemic areas	Preva- lence	Estimated number of cases	Estimated snail area 1978 (ha)
1. Oriental Mindoro	118,149	7.94	9,383	2,690.08
2. Sorsogon	96,821	6.50	6,295	83.82
3. Bohol	55,137	38.45	21,198	19.12
4. Leyte	521,193	18.89	98,429	3,280.47
5. Northern Samar	247,836	12.90	31,962	4,474.93
6. Eastern Samar	156,692	13.03	20,415	2,942.63
7. Western Samar	341,241	13.12	44,770	4,563.48
8. Zamboanga del Norte	48,403	9.88	4,782	150.00
9. Zamboanga del Sur	148,379	24.20	35,901	1,418.31
10. Misamis Occidental	60,859	10.27	6,253	316.90
11. Lanao del Norte	50,968	25.12	12,801	3,727.54
12. Lanao del Sur	31,508	27.77	8,751	3.50
13. Agusan del Sur	200,819	18.12	36,382	16,436.11
14. Agusan del Norte	203,327	14.64	29,760	2,006.20
15. Bukidnon	211,964	21.32	45,201	25.49
16. Surigao del Norte	39,192	23.64	9,269	450.00
17. Surigao del Sur	77,594	33.70	26,150	1,061.33
18. North Cotabato	200,318	16.65	33,358	283.27
19. Sultan Kudarat	49,988	31.78	15,887	473.10
20. South Cotabato	17,147	9.98	1,712	150.00
21. Davao del Norte	508,497	17.63	89,640	1,298.64
22. Davao del Sur	570,265	7.12	40,620	403.05
All Provinces	3,956,297	15.90	628,919	46,257.97

Table 24. Sex differences in human Schistosoma japonicum infections in selected areas of the Philippines

Area	Prevalence		Case distribution		Source
	♂♂	♀♀	♂♂	♀♀	
Naujan, Or. Mindoro	23.7	14.1	62.7	37.3	795
Victoria, Or. Mindoro	26	15	63.4	36.6	198
Irosin, Sorsogon	3.6	2.0	64.3	35.7	181
Irosin, Sorsogon	54.0	30.2	64.1	35.9	838
Palo, Leyte	49.6	46.4	51.7	48.3	606
Carigara, Leyte	3.6	1.7	67.9	32.1	606
Tanauan, Leyte	49.3	43.2	53.3	46.7	606
Bagahupi, Leyte	53.6	30.0	64.1	35.9	606
Pastrana, Leyte	15.1	15.7	49.0	51.0	139
Santa Fe, Leyte	29.4	20.2	59.3	40.7	139
Jaro, Leyte	25.0	21.8	53.4	46.6	139
Barugo, Leyte	43	26	62.3	37.7	277
Basey, Samar	41.8	38.5	52.1	47.9	456
North Samar	20	9	69.0	31.0	245
North Bohol	7	4	63.6	36.4	197
Malaybalay, Bukidnon	20	12	62.5	37.5	200
Average	29.0	20.6	60.2	39.8	

Table 25. S. japonicum prevalence of the male population in different localities of the Philippines (Sources in brackets)

Age (yrs)	Sta. Rosa, Leyte (277)	Pastra-na, Leyte (139)	Sta. Fe, Leyte (139)	Jaro, Leyte (139)	Palo, Leyte (606)	Cari-gara, Leyte (606)	Tana-uan, Leyte (606)	Baga-hupi, Leyte (606)	Irosin, Sorso-gon (838)	San An-tonio, Samar (456)	Mala-yas, Mindoro (795)
1-4	4	0.0	12.8	17.3	20.7	0.3	23.7	37.5	11.9	0.0	9.5
5-9	17	12.5	27.5	19.6					40.3	14.3	
10-14	49	20.5	33.3	28.7	63.1	2.5	68.1	60.7	68.3	57.3	30.6
15-19	68	25.6	37.7	37.9					80.6	62.5	
20-24	44	15.5	33.3	32.2	71.3	10.7	53.5	69.2	81.2	72.7	30.9
25-29	57								74.4		
30-34	50	17.7	37.7	18.1	60.0	4.3	55.7	75.0		61.8	24.4
35-39									76.9		
40-44	51	7.8	25.0	30.8	65.2	9.2	62.4	62.5		71.4	19.5
45-49									64.7		
50-54	47	5.5	21.8	0.0	51.4	3.8	53.1	37.5		56.9	18.2
55-59									63.2		
60-64	42	11.1	36.3	0.0							
65-69									62.5		
≥70		0.0	0.0	0.0							
Total	43	15.1	29.4	25.0	49.6	3.6	49.3	53.6	54.0	41.8	23.7

Table 26. S.japonicum prevalence of the female population in different localities of the Philippines (Sources in brackets)

Age (yrs)	Sta. Rosa, Leyte (277)	Pastrana, Leyte (139)	Sta. Fe, Leyte (139)	Jaro, Leyte (139)	Palo, Leyte (606)	Carigara, Leyte (606)	Tanauan, Leyte (606)	Bagahupi, Leyte (606)	Irosin, Sorsogon (838)	San Antonio, Samar (456)	Malayas, Mindoro (795)
1-4	8	0.0	7.1	0.0	15.4	0.4	19.0	4.8	0.0	2.2	3.1
5-9	10	13.8	13.9	19.1					20.7	9.6	
10-14	18	20.6	31.1	25.6	62.5	1.9	61.8	39.1	47.8	36.5	24.1
15-19	46	14.2	27.7	15.7					52.4	47.8	
20-24	18	10.4	23.0	24.6	61.3	4.5	53.7	61.5	25.0	43.9	12.5
25-29	25								45.7		
30-34	38	22.4	27.6	21.1	55.3	2.1	44.7	22.2		65.4	17.1
35-39									59.4		
40-44	53	10.0	19.5	26.0	55.5	2.1	54.2	11.1		60.0	9.1
45-49									58.3		
50-54	37	10.8	3.1	28.9	47.5	1.5	40.6	60.0		50.0	10.0
55-59									23.1		
60-64	11	12.5	20.0	23.5							
65-69									16.7		
≥70		0.0	20.0	0.0							
Total	26	15.7	20.2	21.8	46.4	1.7	43.2	30.0	30.2	38.5	14.1

Table 27. Schistosoma japonicum infections in domestic animals and field rats (Rattus rattus mindanensis): average and range of data from different sources (cf. table 28) based on faeces examination

Final host	Number examined	Number positive	Prevalence (%)	
			Average	Range
dog	1330	198	14.9	7.7-75.0
pig	2996	296	9.9	1.9-69.2
carabao	2192	25	1.1	0.1-16.7
cattle	95	14	14.7	0.0-38.2
goat	141	3	2.1	0.0- 7.1
cat	19	0	0.0	-
rat	747	202	27.0	4.8-73.7

Table 28. Schistosoma japonicum infections in domestic animals and field rats (Rattus rattus mindanensis) in the Philippines

Final host	Number examined	Number positive	Prevalence (%)	Method	Locality	Source
dog	6	2	33.3	F	Naujan, Mindoro	795
	14	8	57.1	F	Naujan, Mindoro	398
	4	1	25.0	F	Misamis Occidental	598
	4	3	75.0	F	Cotabato	598
	40	4	10.0	F	Leyte	598
	35	14	40.0	F	Sorsogon	598
	4	3	75.0	F	Davao	598
	170	31	18.2	F	Palo, Leyte	606
	652	50	7.7	F	Dagami, Leyte	281
	28	5	17.9	F	22 Barangays, Leyte	281
	373	77	20.6	FC	Mayorga and La Paz, Leyte	297
pig	6	3	50.0	F	Naujan, Mindoro	795
	13	9	69.2	F	Naujan, Mindoro	398
	76	5	6.6	F	Misamis Occidental	598
	521	65	12.5	F	Leyte	598
	54	1	1.9	F	Sorsogon	598
	603	80	13.3	F	Palo, Leyte	606
	1124	47	4.2	F	Dagami, Leyte	281
	44	3	6.8	F	22 Barangays, Leyte	281
	555	83	15.0	FC	Mayorga and La Paz, Leyte	297
carabao	21	0	0.0	F	Naujan, Mindoro	795
	6	1	16.7	F	Naujan, Mindoro	398
	98	4	4.1	F	Lanao	598
	13	2	15.4	F	Davao	598
	137	2	1.5	F	Palo, Leyte	606
	1347	1	0.1	F	Dagami, Leyte	281
	478	6	1.3	F	22 Barangays, Leyte	281
	92	9	9.8	FC	Mayorga and La Paz, Leyte	297

Table 28. (Contd.)

Final host	Number examined	Number positive	Prevalence (%)	Method	Locality	Source
cattle	34	13	38.2	F	Palo, Leyte	606
	15	0	0.0	F	Dagami, Leyte	281
	46	1	2.2	FC	Mayorga and La Paz, Leyte	297
goat	69	1	1.4	F	Palo, Leyte	606
	21	0	0.0	F	Dagami, Leyte	281
	23	0	0.0	F	22 Barangays, Leyte	281
	28	2	7.1	FC	Mayorga and La Paz, Leyte	297
cat	19	0	0.0	FC	Mayorga and La Paz, Leyte	297
rat	43	11	25.6	LI	Palo, Leyte	471
	203	46	22.7	F	Palo, Leyte	606
	66	16	24.2		Palo, Leyte	425
	45	39	86.7	LLI	Palo, Leyte	581
	7	4	57.1	LI	Dagami, Leyte	471
	137	101	73.7	F	Dagami, Leyte	281
	199	163	81.9	T	Dagami, Leyte	433-435
	12	10	83.3	LI	Sta.Fe, Leyte	471
	126	83	65.1	L	Sta.Fe, Leyte	141
	126	57	45.2	I	Sta.Fe, Leyte	141
	126	6	4.8	F	Sta.Fe, Leyte	141
	167	122	73.0	L	Pastrana, Leyte	141
	166	90	54.2	I	Pastrana, Leyte	141
	166	35	21.1	F	Pastrana, Leyte	141
	264	201	76.1	L	Jaro, Leyte	141
	258	123	47.7	I	Jaro, Leyte	141
	93	7	7.5	F	Jaro, Leyte	141
	58	0	0.0	LI	Tacloban, Leyte	471
	41	37	90.2	LI	Gacao, Leyte	471
	4	2	50.0	LI	Alangalang, Leyte	471
	6	0	0.0	LI	San Antonio, Samar	471
	2	0	0.0	LI	Jinamoc Island, Samar	471
	3	2	66.7	F	Sorsogon	598
	19	5	26.3	F	Davao	598

Method: F = Faeces examination
FC = Faeces examination and circumoval precipitin test
I = Examination of intestine
L = Examination of liver
LI = Examination of liver and intestine
LLI= Examination of liver, lung and intestine
T = Examination of different tissues

Table 29. Estimations of schistosomiasis-affected area, population density and case density in the affected areas of the endemic provinces

Province	Area affected (km^2)	Population density per km^2	Case density per km^2
Sorsogon	375	258.2	16.8
Oriental Mindoro	675	175.0	13.9
Northern Samar	808	306.8	39.6
Eastern Samar	740	211.8	27.6
Western Samar	1,306	261.2	34.3
Leyte	1,738	299.8	56.6
Bohol	225	245.1	94.2
Surigao del Norte	152	258.0	61.0
Agusan del Norte	418	486.3	71.2
Agusan del Sur	2,029	99.0	17.9
Davao del Norte	2,274	223.7	39.4
Davao del Sur	1,276	446.8	31.8
Surigao del Sur	336	231.1	77.8
Cotabato	1,366	195.8	37.3
Bukidnon	1,217	174.2	37.1
Lanao del Sur	75	420.1	116.7
Lanao del Norte	204	250.3	62.7
Zamboanga del Sur	597	248.4	60.1
Misamis Occidental	211	288.7	29.6
Zamboanga del Norte	248	195.5	19.3
All Provinces	16,210	244.1	38.8

Table 30. Percentage of indicated net out-migration, by province, 1939-1960 (total of 3.056 million migrants and their descendants; Source: 831)

Province	Percentage of net out-migration	
Cebu	15.55	
Leyte	12.35	
Iloilo	9.60	
Bohol	7.86	
Antique	3.25	
Aklan	2.38	
Capiz	2.36	
Negros Oriental	2.30	
Negros Occidental	2.10	
Samar	1.86	
Visayas	59.61	59.61
Pangasinan	4.35	
Ilocos Sur	3.98	
Ilocos Norte	3.76	
Nueva Ecija	3.18	
Sorsogon	2.34	
Batangas	2.19	
Mountain	2.18	
La Union	1.91	
Albay	1.65	
Cagayan	1.62	
Abra	1.09	
Cavite	0.84	
Tarlac	0.68	
Pampanga	0.59	
Bulacan	0.25	
Laguna	0.04	
Luzon	30.65	30.65
Misamis Occidental	3.51	
Sulu	2.99	
Romblon	1.20	
Marinduque	0.78	
Surigao	0.73	
Catanduanes	0.34	
Batanes	0.19	
Other Islands	9.74	9.74

Table 31. Percentage of indicated net in-migration, by province, 1939-1960 (total of 3.056 million migrants and their descendants; Source: 831)

Province	Percentage of net in-migration	
Cotabato	17.11	
Davao	13.01	
Zamboanga del Sur	12.26	
Lanao del Sur	4.40	
Agusan	3.38	
Lanao del Norte	3.35	
Bukidnon	3.17	
Zamboanga del Norte	1.52	
Misamis Oriental	0.87	
Mindanao	59.07	59.07
Rizal	23.01	
Camarines Sur	5.45	
Manila	2.72	
Isabela	2.29	
Quezon	1.52	
Zambales	1.06	
Camarines Norte	0.71	
Nueva Vizcaya	0.17	
Bataan	0.02	
Luzon	36.95	36.95
Oriental Mindoro	2.37	
Masbate	0.88	
Occidental Mindoro	0.60	
Palawan	0.13	
Other Islands	3.98	3.98

Table 32. Five-year schistosomiasis control progranne (1980-1984): projects and cost (Source: 706)

Component	Measures	Total cost (Pesos)
	Case Finding	11,868,921.00
	Treatment	73,233,750.00
	Health Education	5,046.300.00
Health	Environmental Sanitation (water-sealed toilets)	17,670.092.00
	Snail Control (mollusciciding)	61,469,829.41
	Evaluation	330,000.00
	Total	169,618,892.41
	Survey, plans and designs	69,791,895.00
	Construction of drainage canals	2,791,675,800.00
	Construction of footbridges	186,121,400.00
Infrastructure	Water Supply	336,941,177.00
	Maintenance of drainage canals and deep wells	65,170,444.00
	Total	3,449,700,716.00
Grand total		3,619,319,608.41

Table 33. Developmental design of food-borne helminths

Type No.	Number of intermediate hosts	Number of free-living stages	Reservoir hosts	Paratenic hosts	Host function of man	Parasite
1	x*	0	+*	-	final and intermediate host	Trichinella spiralis
2	1	1	-	-	final host	Taenia saginata, T.solium
3	1	1	+**	-	final host	Capillaria philippinensis
4	1	1	-	+	abnormal host	Angiostrongylus cantonensis
5	1	3	+	-	final host	Fasciolopsis buski, Fasciola
6	2	1	+	+	final host	Diphyllobothrium spp.
7	2	1	-	+	abnormal host	Gnathostoma, "Spargana"
8	2	2	+	-	final host	Opisthorchis (Clonorchis), Heterophyes, Echinostoma, Metagonimus, Phaneropsolus, Dicrocoelium
9	2	2	+	+	final host	Paragonimus spp.

* due to the life cycle of Trichinella a considerable number of mammalian species simultaneously serve as intermediate and final hosts

** not yet found in nature

Table 34. Sources of infection and food-borne helminths

Source	Helminth genus		
	Trematodes	Cestodes	Nematodes
Mammals	Paragonimus	Taenia	Trichinella (Angiostrongylus)
Birds	-	"Spargana"	Gnathostoma
Reptiles, Amphibia	-	"Spargana"	Gnathostoma (Angiostrongylus)
Fish	Opisthorchis (Clonorchis) Heterophyes Metagonimus	Diphyllobothrium "Spargana"	Capillaria Gnathostoma (Angiostrongylus)
Insects	Dicrocoelium	(Hymenolepis)*	-
Crustaceans	Paragonimus	-	Angiostrongylus
Molluscs	Echinostoma	-	Angiostrongylus

*not really food-borne; infection happens by ingestion of fleas and other insects

Table 35. Human Taenia infections in the Philippines

Province and Location		T.saginata		T.solium		Taenia sp.		Source
	Number examined	Number positive	Prevalence (%)	Number positive	Prevalence (%)	Number positive	Prevalence (%)	
Philippines	263	-	-	-	-	2	0.8	863
Northern Luzon	7,417	-	-	-	-	-	1.0	248
Batanes,Sabtang	-	1	-	-	-	-	-	840
Cagayan,Tuguegarao	2,594	-	-	-	-	13	0.5	655
Isabela,Ilagan	802	-	-	-	-	11	1.4	655
Isabela,Ilagan	4,278	-	-	-	-	59	1.4	839
Ilocos Norte	445	-	-	-	-	14	3.1	772
Benguet,Baguio	119	-	-	-	-	15	12.6	209,210
Benguet,Baguio	183	4	2.2	5	2.7	-	-	209,210
Benguet,Antamok	1,925	-	-	-	-	10*	0.5	675
Pangasinan	550	-	-	-	-	1	0.2	109
Manila	136	-	-	1	0.7	-	-	189
Manila,Bilibid Prison	1,000	-	-	-	-	7	0.7	718
Manila,Bilibid Prison	4,106	-	-	-	-	30	0.7	337
Manila	-	27	-	2	-	-	-	336
Manila	6,000	-	-	-	-	5	0.1	336
Manila	385	-	-	-	-	1	0.3	343
Manila	36	-	-	-	-	1	2.8	346
Manila	-	-	-	-	-	10*	-	841
Manila,Bilibid Prison	7,843	-	-	-	-	57	0.7	842
Manila	500	1	0.2	-	-	-	-	252
Manila	500	1	-	-	-	2	-	714
Manila	730	-	-	-	-	4	0.5	329
Manila	-	2	-	1	-	-	-	644
Manila	3,862	1	-	-	-	-	-	186
Manila	27,993	-	-	-	-	2	-	504
Manila	20,902	-	-	-	-	14	-	504
Manila	28,117	-	-	-	-	2	-	505
Manila	30,778	-	-	-	-	2	-	505
Manila	-	6	-	-	-	-	-	506
Manila	-	13	-	2	-	1	-	182
Manila	-	-	-	3	-	-	-	213
Manila	5,565	-	-	-	-	1	-	5
Manila	-	-	-	1	-	-	-	125
Manila	-	1	-	-	-	-	-	817
Rizal,Las Pinas	6,018	-	-	-	-	4	0.1	655
Cavite,Canacao	932	3	0.3	-	-	-	-	734
Tayabas	332	-	-	-	-	13	3.9	628
Tayabas	90	-	-	-	-	9	10.0	628
Or.Mindoro,Victoria	1,058	-	-	-	-	-	1.0	198
Leyte	567	-	-	-				772
Leyte,Jaro,Pitogo	281	38	13.5	-	-	-	-	138,183
Leyte,Jaro,Buri	167	8	4.7	-	-	-	-	138,183
Leyte,Jaro,Pitogo,Buri	-	60	-	-	-	-	-	48,153
Leyte,Pastrana	1,222	4	0.3	-	-	-	-	139
Leyte,Santa Fe	980	4	0.4	-	-	-	-	139
Leyte,Jaro	1,664	33	2.0	-	-	-	-	139

Table 35 (Contd.)

Province and Location	Number examined	T.sagi-nata Number positive	Preva-lence (%)	T.solium Number positive	Preva-lence (%)	Taenia sp. Number positive	Preva-lence (%)	Source
Cebu	1,603	-	-	-	-	6	0.4	327
Cebu,Danao	51	-	-	-	-	3*	5.9	108
Cebu,Oslob	-	-	-	-	-	1*	-	491
Cebu,Argao	-	-	-	-	-	2	-	491
Bohol,Trinidad	972	-	-	-	-	1	0.1	197
Negros Oriental	1,249	-	-	-	-	3	0.2	731
Bukidnon,Malaybalay	831	-	-	-	-	-	1.0	200
Cotabato	50	-	-	1	2.0	-	-	807
Total	-	208	-	16	-	302	-	
Benguet Sanitarium Div.	81	-	-	3	3.7	-	-	113
Baguio Hospital	81	-	-	4	4.9	-	-	114
Bilibid Prison	5,928	26	0.4	-	-	-	-	114
Baguio Hospital	196	-	-	3	1.5	-	-	115
Baguio Hospital Outdoor Dept.	1,760	-	-	1	0.1	-	-	115
Baguio Hospital,Lab.	136	-	-	4	2.9	-	-	115
Bilibid Prison	3,556	1	-	-	-	7	0.2	115
Baguio Hospital	203	-	-	2	1.0	-	-	116
Baguio Hospital Outdoor Dept.	3,460	-	-	2	0.1	-	-	116
Baguio Hospital,Lab.	278	5	1.8	2	0.7	-	-	116
Iwahig Penal Colony	1,314	-	-	2	0.2	-	-	116
Baguio Hospital Outdoor Dept.	3,429	1	-	-	-	-	-	117
Manila,Phil.Gen.Hosp.	4,419	-	-	-	-	3	0.1	118
Bilibid Prison	4,887	35	0.7	1	-	-	-	118
Iwahig Penal Colony	1,114	-	-	3	0.3	-	-	118
Baguio Hospital	716	-	-	3	0.4	-	-	118
Baguio Hospital Outdoor Clinic	3,994	8	0.2	-	-	-	-	118
Bilibid Prison	2,416	-	-	-	-	2	0.1	120
Bilibid Prison	13,297	9	0.1	-	-	-	-	622
Mindanao and Sulu Div.	-	-	-	-	-	882	-	623
Bilibid Prison	16,476	-	-	-	-	10	0.1	624
San Ramon Penal Col.	592	-	-	-	-	6	1.0	625
Total	-	85	-	30	-	910	-	

*diagnosis not "Taenia" but "cestodes" or "tapeworms"

Table 36. Cysticercosis in slaughtered and inspected pigs, carabaos and cattle in the Philippines, 1970-1974 (Source: 48)

Province and/or City/Municipality	Pigs			Carabaos			Cattle		
	Number examined	Number positive	%	Number examined	Number positive	%	Number examined	Number positive	%
Nueva Vizcaya	73,659	0	0.00	619	0	0,00	10,454	0	0.00
Zambales, Santa Cruz	10,191	9	0.88	1,150	1	0.87	3,922	0	0.00
Olangapo City	30,601	10	0.33	759	0	0.00	1,306	0	0.00
Angeles City	100,484	67	0.67	15,002	0	0.00	1,352	0	0.00
Bulacan	238,959	0	0.00	20,537	0	0.00	24,233	0	0.00
Bataan	17,641	0	0.00	2,266	0	0.00	3,255	0	0.00
Caloocan City	163,365	215	1.32	13,314	0	0.00	11,261	0	0.00
Quezon City	284,998	318	1.12	0	-	-	9,631	0	0.00
Manila	2,542,002	1,602	0.63	186,631	0	0.00	114,866	0	0.00
National Abattoir	81,766	10	0.12	26,677	0	0.00	64,387	0	0.00
Sorsogon	36,287	1	0.03	654	0	0.00	3,474	0	0.00
Leyte, Jaro	2,577	26*	10.09	447	0	0.00	109	0	0.00
Cebu City	756,236	3,826	5.06	16,883	0	0.00	119,410	0	0.00
Aklan, Kalibo	9,218	0	0.00	5,582	0	0.00	6,643	0	0.00
Capiz	9,860	0	0.00	3,000	0	0.00	2,684	0	0.00
Iloilo	167,503	262	1.56	33,728	4	0.12	39,091	0	0.00
Surigao City	41,566	485	11.67	0	-	-	0	-	-
Butuan City	146,952	930	6.33	3,322	0	0.00	7,593	0	0.00
Cagayan de Oro City	134,647	217	1.61	1,007	0	0.00	37,754	0	0.00
Oroquieta City	65,756	317	4.82	9,699	3	0.31	23,311	13	0.56
General Santos City	44,731	7	0.16	810	0	0.00	10,232	0	0.00
Total	4,958,999	8,302	1.67	342,087	8	0.02	494,968	13	0.03

*condemned due to generalized cysticercosis
(Olangapo 1974 only; Angeles and Jaro 1972-1974 only; Bataan and Capiz 1971-1974 only)

Table 37. Human Echinostoma ilocanum infections in the Philippines

Number examined	Number positive	Prevalence (%)	Province of origin	Source
5000	5	0.1	Ilocos Sur (2), North-west Luzon (3)	338
4278	1	0.1	Isabela	839
-	5	-	Zambales	368,369
500	2	0.4	Ilocos Norte, Zambales	714
-	4	-	Ilocos Sur, Zambales (2), Pangasinan	792
30	8	26.7	Zambales	793
91	7	7.7	Ilocos Sur	793
-	1	-	Ilocos Sur	21
263	7	2.7	Nueva Ecija (4), Sulu,Laguna,Bataan	863
974	1	0.1	Ilocos	362
-	20	-	not given	130
831	1	0.1	Bukidnon	200
7417	-	11.	23 Barrios in Northern Luzon	248

839: after the author probably E.ilocanum
863: men of the Armed Forces stationed in the respective provinces
362: not Ilocano by origin, but had lived for 8 years in the Ilocos Region
200: after the author Echinostoma sp.

Table 38. Paragonimus infections of the first and second intermediate hosts in the Philippines

Host	PROVINCE Municipality	Number examined	Number positive	Preva-lence (%)	Source
1. Antemelania asperata, A.dactylus	SORSOGON	1986	2	0.1	791
	Casiguran	1000	7	0.7	864
		350	2	0.6	864
	LEYTE, Jaro	143	3	2.1	407
2. Sundathelphusa philippina	CAMARINES SUR	216	-	60	784
	SORSOGON				
	Bacon	47	9	19.1	179
	Barcelona	471	70	14.8	179
	Bulan	133	47	35.3	179
	Bulusan	216	12	5.5	179
	Casiguran	137	73	53.3	864
	Casiguran	947	434	45.8	179
	Castilla	119	2	1.6	179
	Donsol	104	0	0.0	179
	Gubat	123	5	4.0	179
	Irosin	282	16	5.6	179
	Juban	990	637	64.3	179
	Juban	35	33	94.3	876
	Magallanes	190	79	41.5	179
	Matnog	378	16	4.2	179
	Pilar	174	10	5.7	179
	Sorsogon	205	12	5.8	179
	Sta.Magdalena	68	17	25.0	179
	LEYTE				
	Jaro	65	64	98.5	862
	Jaro	792	422	53.3	137

Table 39. Human and animal Paragonimus infections in the Philippines

PROVINCE Municipality	Host	Number examined	Number positive	Prevalence (%)	Source
either not given or Bilibid Prison and Iwahig Penal Colony resp.	man	-	69	-	14,114,115,116,117, 122,301,403,558,628, 659,872
BATANES	man	-	1	-	726
BATAAN/ZAMBALES (Subic Bay area)	man	-	1	-	298
MANILA	man	-	1	-	337
	cat	32	1	3.1	558
CAVITE	man	-	1	-	337
CAMARINES NORTE	man	-	1	-	337
CAMARINES SUR					
Naga	man	-	1	-	363
ALBAY	man	-	3	-	185,337
SORSOGON	man	-	16	-	150,337,604
	rat	948	-	1.0	426
Casiguran	man	275	2	0.7	864
Casiguran	cat	2	2	-	864
Gubat	man	-	1	-	819
Irosin	man	-	1	-	593
ORIENTAL MINDORO	man	-	1	-	334
SAMAR	man	-	10	-	337,628
LEYTE	man	-	158	-	81,149,185,337,403, 542,529,590,591,659, 771,783,860
Baybay	man	-	1	-	192,183
Burauen	man	-	3	-	192,193,461
Dulag	man	-	1	-	192,193
Inopacan	man	-	1	-	611
Jaro	man	-	9	-	192,193,596
Jaro	man	97	4	4.1	862
Jaro	man	168	21	12.5	159
Jaro	man	1664	21	1.3	139
Jaro	cat	-	2	-	872
Jaro	cat	5	4	80.0	862
Jaro	cat	21	8	38.1	173,174
Jaro	dog	29	1	3.4	173,174
Jaro	rat	45	5	11.1	174
Jaro	rat	491	46	9.4	142

Table 39 (Contd.)

PROVINCE Municipality	Host	Number examined	Number positive	Prevalence (%)	Source
La Paz	man	-	1	-	334
Malitbog	man	-	1	-	461
Ormoc	man	-	2	-	334,192,193
Pastrana	man	1222	1	0.1	139
CEBU	man	-	1	-	604
BOHOL					
Trinidad	man	972	1*	0.1	197
NEGROS	man	-	1	-	604
MINDANAO	man	-	7	-	337,604
MISAMIS OCCIDENTAL					
Bonifacio	man	-	1	-	594
MISAMIS ORIENTAL	man	-	1	-	604

* Paragonimus-like eggs in feces

Table 40. Food-borne heterophyids and their second intermediate fish hosts (Source: 815)

Parasite species	Fish species
Stellantchasmus amplicaecalis	Anabas testudinaceus Mugil dussumieri
Procerovum calderoni	Ophicephalus striatus Glossogobius giurus Gerres filamentosus Creisson validus Amphacanthus javus Pelates quadrilineatus Eleutheronema tetradactyla Hemiramphus georgii Ambassis buruensis Mugil dussumieri Chanos chanos Butis amboinensis Platycephalus indicus Molienesia latipina
Haplorchis taichui	Ophicephalus striatus Puntius binotatus Puntius palata
Haplorchis yokogawai	Ophicephalus striatus Gerres sp. Amphacanthus javus Ambassis buruensis Hemiramphus georgii Mugil sp.
Carneophallus brevicaeca	Glossogobius giurus*

*paratenic host; second intermediate hosts are shrimps (Macrobrachium sp.)

References

Abbreviations of the Journals according to Index Medicus except the following

JPMA - J Philipp Med Assoc

JPMA - J Philipp Is Med Assoc

AMP - Acta Med Philipp

PJS - Philipp J Sci

SEA J - Southeast Asian J Trop Med Publ Hlth

Bibliographies

Argosino AY (1952) The Philippine Index Medicus. Vol. II, 1917-1944. XVI + 416 pp. National Science Development Board, Manila

Argosino AY (1961) The Philippine Index Medicus. Vol. 3, 1945-1951. XV + 176 pp. National Institute of Science and Technology, Manila

Pesigan TP, Ferrer NG (1953) A brief bibliography in parasitology in the Philippines. In: Golden Book, P.M.A. pp. 218-231

The Philippine Index Medicus, Vol. 4, 1952-1956. The Library, University of the Philippines, Quezon City

Philippine Index Medicus 1975-1979 (1985) Compiled by Dimalanta SG, Rosal RR. XV + 240 pp. SEAMIC Publication No. 41. Southeast Asian Medical Information Center, Tokyo

Rosal RR (assisted by Dugenia TP, Escasa CM) (1976) Public health surveys in the Philippines: a compilation of research reports. Research Guide No. 6. V + 107 pp. Inst.Publ.Hlth, U.P., Manila

Rosal RR (1978) Nematodes of medical importance in the Philippines, 1907-1977: an annotated bibliography. XVI + 192 pp. Institute of Public Health Library, University of the Philippines, Manila

Rosal RR, Manere NA (1974) Filariasis in the Philippines. A selected annotated bibliography. V + 55 pp. Inst.Publ.Hlth, U.P., Manila

Rosal RR, Manere NA (1974) Annotated bibliography on schistosomiasis in the Philippines (1951-1973). VII + 86 pp. Inst.Publ. Hlth, U.P., Manila

1 Abbott RT (1945) The Philippine intermediate snail host (Schistosomophora quadrasi) of schistosomiasis. Occasional Papers on Mollusks, Museum of Comparative Zoology, Harvard University, Cambridge, Massachusetts, 1:5-16
2 Abbott RT (1946) The egg and breeding habits of Oncomelania quadrasi Mlldff., the schistosomiasis snail of the Philippines. Occasional Papers on Mollusks, Museum of Comperative Zoology, Harvard University, Cambridge, Massachusetts, 1:41-48
3 Abbott RT (1948) Handbook of medically important mollusks of the Orient and the Western Pacific. Bulletin of the Museum of Comparative Zoology at Harvard College, 100:243-328
4 Abcede J (1973) Operation snail. World Health, Jan. 1973:28-33
5 Acosta AF, Silvestre J (1964) Preliminary results on the deworming program among Manila school children. Intern Med 2:13-21
6 Adiao AC, Mejia SS, Mundo F del (1961) Clinical observations on the use of pyrvinium pamoate in enterobiasis. JPMA 37:485-488
7 Africa CM (1927) Observations upon experimental feeding of various species of mosquitoes on filariated blood. Preliminary report. JPMA 7:330-336
8 Africa CM (1932) Studies on experimental creeping eruption in the Philippines. PJS 48:89-101
9 Africa CM (1934) Experimental infection of Philippine Cyclops with the coracidium of Diphyllobothrium mansoni Cobbold, 1882. Philipp J Publ Hlth 1:27-31
10 Africa CM (1937) Evidence of intramucosal invasion in the life cycle of Haplorchis yokogawai (Katsuta, 1932), Chen 1936 (Heterophyidae). JPMA 17:737-743
11 Africa CM (1938) An attempt to elucidate the filtration of eggs of certain heterophyid trematodes into the general circulation. Philipp J Anim Ind 5:187-201
12 Africa CM (1938) The menace of schistosomiasis in the Philippines (Editorial). JPMA 18:715-718
13 Africa CM (1938) Description of three trematodes of the genus Haplorchis (Heterophyidae) with notes on two other Philippine members of this genus. PJS 66:299-307
14 Africa et al. quoted from BD Cabrera (148)
15 Africa CM, Garcia EY (1934) A rat tapeworm (Raillietina garrisoni Tubangui, 1931) transmissible to man. With notes on Davainea madagascariensis Garrison, 1911. Philipp J Publ Hlth 1:44-51
16 Africa CM, Garcia EY (1935) The distribution of schistosomiasis japonica in the Philippines. Philipp J Publ Hlth 2:54-62
17 Africa CM, Garcia EY (1935) The occurrence of Bertiella in man, monkey and dog in the Philippines. PJS 56:1-11
18 Africa CM, Garcia EY (1935) Heterophyid trematodes of man and dog in the Philippines with descriptions of three new species. PJS 57:253-267
19 Africa CM, Garcia EY (1935) Two more new heterophyid trematodes from the Philippines. PJS 57:443-449
20 Africa CM, Garcia EY (1936) A new nematode parasite (Cheilospirura sp.) of the eye of man in the Philippines. JPMA 16:603-607
21 Africa CM, Garcia EY (1937) Plagiorchis sp. a new trematode parasite of the human intestine. Papers on Helminthology published in Commemoration of the 30 Year Jubileum of K.J. Skrjabin and of the 15th Anniversary of the All-Union Institute of Helminthology, Moscow 1937:9-10
22 Africa CM, Garcia EY, Layco J (1935) Periodic human microfilaria in the Philippines. JPMA 15:407-412
23 Africa CM, Garcia EY, Leon W de (1935) Intestinal heterophyidiasis with cardiac involvement; a contribution to the etiology of heart failure. An abstract. JPMA 15:358-361

24 Africa CM, Garcia EY, Leon W de (1935) Intestinal heterohyidiasis with cardiac involvement: a contribution to the etiology of heart failures. Philipp J Publ Hlth 2:1-35
25 Africa CM, Leon W de (1938) Observations on the mechanism of phagocytosis of various helminth ova. Livro Jubilar do Professor Lauro Travassos. Editado para commemorar o 25° anniversario de suas actividades scientificas (1913-1938), Rio de Janeiro 1938:1-9
26 Africa CM, Leon W de, Garcia EY (1935) Heterophyidiasis: II, ova in sclerosed mitral valves with other chronic lesions in the myocardium. JPMA 15:583-592
27 Africa CM, Leon W de, Garcia EY (1936) Heterophyidiasis: III, ova associated with a fatal hemorrhage in the right basal ganglia of the brain. JPMA 16:22-26
28 Africa CM, Leon W de, Garcia EY (1936) Heterophyidiasis, IV: Lesions found in the myocardium of eleven infested hearts including three cases with valvular involvement. Philipp J Publ Hlth 3:1-27
29 Africa CM, Leon W de, Garcia EY (1936) Somatic heterophyidiasis in fish-eating birds, II. Presence of adults and eggs in the bile ducts of the cattle egret. PJS 61:227-223
30 Africa CM, Leon W de, Garcia EY (1937) Heterophyidiasis, V, ova in the spinal cord of man. PJS 62:393-399
31 Africa CM, Leon W de, Garcia EY (1937) Heterophyidiasis: VI, two more cases of heart failure associated with the presence of eggs in sclerosed valves. JPMA 17:605-609
32 Africa CM, Leon W de, Garcia EY (1940) Visceral complications in intestinal heterophyidiasis of man. Rept Proc 3rd Internat Congr Microbiol, New York 1940:447-448
33 Africa CM, Leon W de, Garcia EY (1940) Visceral complications in intestinal heterophyidiasis of man. AMP, Monographic Series 1:1-132
34 Africa CM, Refuerzo PG, Garcia EY (1936) Observations on the life cycle of Gnathostoma spinigerum. PJS 59:513-523
35 Africa CM, Refuerzo PG, Garcia EY (1936) Further observations on the life cycle of Gnathostoma spinigerum. PJS 61:221-225
36 Africa CM, Sta. Cruz JZ (1927) Cysticercus cellulosae in man. JPMA 7:209-215
37 Aikat TK, DAS M (1977) A modified statistical method for analysis of periodicity of microfilaria. Indian J Med Res 65:58-64
38 Akagi K (1973) Enterobius vermicularis and enterobiasis. Progr Med Parasit Japan 5:229-279
39 Alicata JE (1966) The presence of Angiostrongylus cantonensis in islands of the Indian Ocean and probable role of the giant African snail, Achatina fulica, in dispersal of the parasite to the Pacific Islands. Canad J Zool 44:1041-1049
40 Alicata JE, Jindrak K (1970) Angiostrongylosis in the Pacific and Southeast Asia. Charles C Thomas, Springfield, Ill.
41 Andrews RW (1902) Tropical diseases as observed in the Philippines. Albany Med Ann 1902:19-23, 83-89
42 Annandale N (1924) The molluscan hosts of the human blood fluke in China and Japan, and species liable to be confused with them. Am J Hyg Monogr Ser 3:269
43 Antonio D, Penas MD, Pantangco E (1955) Renal echinococcosis JPMA 31:607-612
44 Arambulo PV (1967) Common food-borne diseases and their status as zoonoses in the Philippines. JPMA 43:937-955
45 Arambulo PV (1970) Country report on zoonoses in the Philippines. J Philipp Vet Med Assoc 2:41-44
46 Arambulo PV (1971) A summary of zoonoses in the Philippines. Am Vet Epidemiol Soc. Atlanta. 29 pp

47 Arambulo PV (1974) The natural nidality of zoonoses in the Philippines. Int J Zoonoses 1:58-74
48 Arambulo PV, Cabrera BD, Tongson MS (1976) Studies on the zoonotic cycle of Taenia saginata taeniasis and cysticercosis in the Philippines. Int J Zoonoses 3:77-104
49 Arambulo PV, Cada AB (1968) Zoonoses forum: IV. On food-transmitted zoonotic trematode diseases. A U Vet Dig 2:9-11, 22
50 Arambulo PV, Cada AB (1969) Zoonoses forum: V. On the cutaneous and visceral larva migrans. A U Vet Dig 3:18-20
51 Arambulo PV, Cada AB (1970) Some food-transmitted zoonotic trematode diseases in the Philippines. J Philipp Vet Med Assoc 2:84-90
52 Arambulo PV, Cada AB, Sarmiento RV, Cielo AB (1969) Cysticercus cellulosae in pigs slaughtered in the Manila abattoir. AMP 6:26-27
53 Arambulo PV, Cockrill WR, Borjal AC (1970) A preliminary note on the examination of carabaos for Cysticercus bovis. Philipp J Vet Med 9:113-115
54 Arambulo PV, Dolor RM, Tongson MS (1971) Serologic examination of pigs slaughtered in the Manila abattoir for trichinosis using bentonite flocculation test. Philipp J Vet Med 10:174-180
55 Arambulo PV, Hicarte V, Cada A, Sarmiento RV (1970) On the serologic screening for trichinosis of pigs slaughtered in the Manila abattoir using latex-trichina antigen. Philipp J Vet Med 9:104-107
56 Arambulo PV, Jueco NL, Sarmiento RV, Cada AB (1971) Zoonotic ancylostomes of dogs in Manila with note on the comparative morphology of A.braziliense (de Faria, 1910) and A.ceylanicum (Looss, 1911). Philipp J Anim Sci 8:7-13
57 Arambulo PV, Steele JH, Beran GW, Escudero SH, Carey JH (1978) Veterinary manpower in the Philippines - supply demand model and projections, and impact of animal diseases on human health. Int J Zoonoses 5:1-44
57a Arambulo PV, Yogore MG (1966) On the isolation of rat lung worm larvae from the giant African snail in the Philippines. AMP 3:56-57
58 Aranez JB (1962) The incidence of Fasciola hepatica and Fasciola gigantica among cattles and carabaos in Luzon. Philipp J Vet Med 1:33-46
59 Arcilla-Latonio A, Diaz de Rivera LM, Lacson P, Puzon CB (1973) Tropical syndromes with eosinophilia among Filipinos. JPMA 49:333-362
60 Arcilla-Latonio A, Miclat-Francisco F, Mendoza RA (1968) The giant African snail (Achatina fulica) - a new threat to public health. JPMA 44:418-444, 483-511
61 Areán VM (1971) Capillaria philippinensis. In: Marcial-Rojas RA (ed): Pathology of protozoal and helminthic diseases, pp 666-676. Baltimore
62 Ash LR (1963) Effect of temperature upon survival of first-stage larvae of Angiostrongylus cantonensis. Proc Int Congr Trop Med Mal, Rio de Janeiro, Sept. 1-11, 2:193-194
63 Ash LR (1971) A brief review of the biology of the genus Capillaria Zeder, 1800. Proc 7th SEAMEO-TROPMED Sem:255-258
64 Ashburn PM (1908) The board for the study of tropical diseases as they occur in the Philippine Islands. Milit Surg 23:298-304
65 Ashburn PM, Craig CF (1906) A new blood filaria of man: Filaria philippinensis. Am J Med Sci 132:435-443
66 Ashburn PM, Craig CF (1907) Study of tropical diseases in the Philippine Islands. J Am Med Assoc 48:691-693
67 Ashburn PM, Craig CF (1907) A study of tropical diseases as they occur in the Philippine Islands. NY Med J 85:1117-1119

68 Ashburn PM, Craig CF (1907) The work of the army board for the study of tropical diseases in the Philippine Islands. Milit Surg 21:38-49
69 Ashburn PM, Craig CF (1907) Tropical diseases in the Philippine Islands. Milit Surg 21:140-148
70 Ashburn PM, Craig CF (1907) Tropical diseases in the Philippine Islands. Milit Surg 21:336-348
71 Ashburn PM, Craig CF (1907) Observations upon Filaria philippinensis and its development in the mosquito. PJS B2:1-14
72 Ashburn PM, Craig CF (1907) Tropical diseases as they exist in the Philippine Islands. Milit Surg 21:220-225, 527-530
72a Avery JL (1945) Periodicity of microfilariae in the Philippines. J Parasitol 31(Suppl.):21
73 Avery JL (1946) The incidence of filariasis in the Central Philippines. J Parasitol 32:497-498
74 Baisas FE (1957) Notes on Philippine mosquitoes, XIX.The mosquito problem in the control of filariasis in Sorsogon Province. PJS 86:71-120
75 Baisas FE, Banez LFL, Leuenberger N (1960) Notes on Philippine mosquitoes,XXII. The axil-breeding species. PJS 89:183-199
76 Bang F, Ferguson MS, Hairston NG, Graham OH (1945) Hyperendemicity of schistosomiasis japonica on Leyte Island, P.I. Am J Trop Med 25:407
77 Bang FB, Hairston NG, Graham OH, Ferguson MS (1946) Studies on schistosomiasis japonica. II.Methods of surveying for schistosomiasis japonica. Am J Hyg 44:315-323
78 Banzon TC, Beltran AM (1951) Survey in Samar. JPMA 27:258-259
79 Banzon TC, Cross JH, Singson CN (1974) The capillariasis problem in the Philippines. Proc Joint Sci Meet Philipp Soc Parasitol SEAMEO-TROPMED, pp 12-25
80 Banzon TC, Singson CN, Cross JH (1976) Mebendazole treatment for intestinal nematodes in a Philippine barrio. A preliminary report. JPMA 52:239-243
81 Barrera (1938) pers. communication to Pesignan et al. (611)
82 Bartsch P (1936) Molluscan intermediate hosts of the Asiatic blood fluke Schistosoma japonicum, and species confused with them. Smithsonian Misc Coll 95, 5:1-60
83 Bartsch P (1946) Schistosomophora in China, with descriptions of two new species and a note on their Philippine relative. Smithsonian Misc Coll 104:1-7
84 Basaca-Sevilla V (1951) Survey in Leyte.JPMA 27:252-253
85 Basio DG (1969) Notes on the bionomics of the mosquito Aedes (Finlaya) poecilus (sic) Theobald, vector of Bancroftian filariasis in Sorsogon Province. Philipp J Publ Hlth 14:47-56
86 Baula PL (1968) The prevalent gastro-intestinal helminthic parasites in Bangued, Abra. Inst Publ Hlth Manila 1968:28 pp
87 Bauman PM, Bennett HJ, Ingalls JW (1948) The molluscan intermediate host and schistosomiasis japonica,II. Observations on the production and rate of emergence of cercariae of Schistosoma japonicum from the molluscan intermediate host, Oncomelania quadrasi. Am J Trop Med 28:567-575
88 Baylis HA, Lane C (1920) A revision of the nematode family Gnathostomidae. Proc Zool Soc London 3:245-310
89 Beck JW, Beverley-Burton M (1968) The pathology of Trichuris, Capillaria and Trichinella infections. Helm Abstr 37:1-26
90 Bequaert J (1934) The molluscan intermediate host of the blood fluke, Schistosoma japonicum Katsurada, in the Philippines. With a note on the genus Blanfordia by HA Pilsbry. J Parasitol 20: 280-284

91 Beran GW, Arambulo PV (1972) Zoonoses of cattle and carabaos in the Philippines. Philipp J Vet Med 11:45-62
92 Berger J (1969) The world's major fibre crops. Their cultivation and manuring. Centre d'Etude de l'Azote 6, Zurich, pp 233-235
93 Bhaibulaya M (1982) Angiostrongyliasis. In: CRC Handbook Series in Zoonoses, Sect C, Vol II:25-36
94 Bhaibulaya M, Benjapong W, Noeypatimanond S (1977) Infection of Capillaria philippinensis in man from Phetchabun Province, Northern Thailand: a report of the fifth case. J Med Assoc Thailand 60:507-509
95 Bhaibulaya M, Indra-ngarm S, Ananthapruti M (1979) Fresh-water fishes of Thailand as experimental intermediate hosts for Capillaria philippinensis. Int J Parasitol 9:105-108
96 Billings FT, Winkenwerder WL, Hunninen AV (1946) Studies on acute schistosomiasis japonica in the Philippine Islands. I. A clinical study of 337 cases with a preliminary report on the results of treatment with Fuadin in 110 cases. Bull Johns Hopkins Hosp 78:21-56
97 Blas BL (1972) The schistosomiasis control program in the Philippines. Proc 9th SEAMEO-TROPMED Sem, pp 215-217. Bangkok
98 Blas BL (1972) Treatment of schistosomiasis japonica in the Philippines: a review. Proc 9th SEAMEO-TROPMED Sem, pp 228-232. Bangkok
99 Blas BL (1976) Agro-engineering and sanitation improvement in the control of schistosomiasis japonica in the Philippines: a review. SEA J 7:341-345
100 Blas BL (1976) Ambilhar in the treatment of Schistosoma japonicum infection and as an egg suppressant for mass treatment. SEA J 7:348-350
101 Blas BL, Dazo BC (1968) Schistosomiasis survey and preliminary control work in the Province of Bohol. JPMA 44:80-87
102 Blas BL, Nosenas JS, Tanaka H, Matsudo O, Onodera N, Matsuda H, Hashiguchi J, Santos AT (1979) Longitudinal study on Schistosoma japonicum infections in the Philippines. I.Incidence and prevalence among school children in Dagami area, Leyte. Jpn J Exp Med 49:107-115
103 Blas BL, Santos AT, Banez E, Hambre R (1980) Snail control by means of mollusciciding. 10th Int Congr Trop Med Mal, Manila. Abstracts, p 353
104 Blatz I (1981) Der Darmhelminthenbefall des Haushundes und seine gesundheitliche Bedeutung für den Menschen. Untersuchung in Neckarsteinach (Hessisches Neckartal). Diss Med Fak Univ Heidelberg
105 Bonsdorff B von (1977) Diphyllobothriasis in man. London New York San Francisco
106 Bott R (1970) Die Süßwasserkrabben von Europa, Asien, Australien und ihre Stammesgeschichte. Abh Senckenb Naturforsch Ges 526:1-338
107 Bowman FB (1910) Observations on parasitic infections among Igorots. Bull Manila Med Soc 2:75-76
108 Brewer IW (1910) The animal parasites found in the intestines of native children in the Philippine Islands. NY Med J 91:1112-1113
109 Brody H (1947) Nematode infestation in Pangasinan, Luzon, Philippine Islands. Am J Clin Pathol 17:294-297
110 Bruce-Chwatt LJ, Garrett-Jones C, Weitz B (1966) Ten years' study (1955-64) of host selection by anopheline mosquitoes. Bull WHO 35:405-437
111 Bulatao-Jayme J, Villavieja GM, Alcalde PV, Valerio TE (1978) Food consumption, intestinal parasitism and anemia in the Visayas. JPMA 54:224-244

112 Burch JB (1964) Cytotaxonomy of the genus Oncomelania, intermediate host of Schistosoma japonicum. Am Malacol Union Ann Reps 31:28-29

113 Bureau of Health (1906) Annual Report of the Bureau of Health for the Philippine Islands. July1,1905 - June 30,1906. Manila. pp 94-96, 110-114

114 Bureau of Health (1908) Annual Report. Juli 1,1907 to June 30, 1908. Manila. pp 58-66,76,97,152-158

115 Bureau of Health (1909) Annual Report. July 1,1908 to June 30, 1909. Manila. pp 77-78,162-173

116 Bureau of Health (1910) Annual Report from July 1,1909 to June 30,1910. Manila. pp 84-85,145-155

117 Bureau of Health (1911) Annual Report for the Fiscal Year ended June 30,1911. Manila. pp 47-48,85-88,97-101

118 Bureau of Health (1913) Annual Report for the Fiscal Year ended June 30,1912. Manila. pp 83-84,161-175,186-191

119 Bureau of Health (1913) Annual Report for the Fiscal Year July 1, 1912 to June 30,1913. Manila. pp 126,257,265-267,270-274

120 Bureau of Health (1914) Report for the short fiscal year from July 1,1913 to December 31,1913. Manila. pp 63-66,206-208, 318-324,328-329

121 Bureau of Health (1915) Report for the Fiscal Year from January 1 to December 31,1914. Manila. pp 161-164,172-173

122 Bureau of Health (1940) Annual Report for the six months ending June 30,1939. Manila. p 21

123 Burley TM (1973) The Philippines. An economic and social geography. London

124 Burtner OW (1950) Schistosomiasis japonica in American soldiers in the Philippines. A clinical and laboratory study of 297 treated cases. JPMA 26:537-554

125 Cabrera BD (1965) Multiple infection with adult Taenia solium. A case report. AMP 1:147-150

126 Cabrera BD (1968) Anisakiasis - will it become a public health problem in the Philippines? AMP 4:214-215

127 Cabrera BD (1968) Determination of mosquito vector of filariasis in southern Sulu. JPMA 44:117-124

128 Cabrera BD (1969) Human intestinal capillariasis in the Republic of the Philippines. Philipp J Int Med 7:269-273

129 Cabrera BD (1969) Filariasis in the Philippine Islands. Jpn J Trop Med 10:45-49

130 Cabrera BD (1969) Paragonimiasis and other trematode infections in the Philippines. Proc 4th Southeast Asian Sem Parasitol Trop Med, pp 309-316

131 Cabrera BD (1970) Brugia malayi vector determination in Bunawan, Agusan, the third endemic focus for Malayan filariasis in the Philippines. SEA J 1:496-504

132 Cabrera BD (1971) A review of the parasitic diseases of the gastrointestinal system in the Philippines. Proc 7th SEAMEO-TROPMED Sem, pp 39-51

133 Cabrera BD (1971) Report of research work done by our group on Capillaria philippinensis. Proc 7th SEAMEO-TROPMED Sem, p 259

134 Cabrera BD (1971) Vector determination for Malayan filariasis in Bongao, Sulu: The second endemic focus for the disease in the Philippines. AMP 7:95-100

135 Cabrera BD (1972) The filaria control project of the Republic of the Philippines. Proc 9th SEAMEO-TROPMED Sem, pp 64-75

136 Cabrera BD (1972) Capillarias philippinensis, a rare nematode infection of the small intestines of man. Proc 4th Asian Pacific Congr Gastroenterol, pp 632-637

137 Cabrera BD (1973) Studies on Paragonimus and paragonimiasis in the Republic of the Philippines. I. Paragonimus metacercariae infection in Parathelphusa grapsoides. SEA K 4:55-62
138 Cabrera BD (1973) Treatment of taeniasis saginata with Bithionol (Bitin) in Jaro, Leyte. AMP 9:139-143
139 Cabrera BD (1975) Further evaluation of blood circumvol precipitin test (filter paper) and stool formalin-ether method in schistosomiasis japonica field surveys. SEA J 6:495-504
140 Cabrera BD (1975) Sundathelphusa philippina (Martens 1868) the new name of the crustacean intermediate host of mammalian lung fluke (Paragonimus) in the Philippines. SEA J 6:602
141 Cabrera BD (1976) Schistosomiasis japonica in field rats in Leyte, Philippines. SEA J 7:50-55
142 Cabrera BD (1977) Studies on Paragonimus and paragonimiasis in the Philippines. II. Prevalence of pulmonary paragonimiasis in wild rats Rattus norvegicus in Jaro, Leyte, Philippines. Int J Zoonoses 4:49-55
143 Cabrera BD (1978) Reinfection and infection rate studies of soil-transmitted helminthiasis in Juban, Sorsogon, Rupublic of the Philippines. AMP 14:8-19
144 Cabrera BD (1978) Reinfection and infection rate studies of soil-transmitted helminthiasis in Juban, Sorsogon, Republic of the Philippines. Proc 5th Conf APCO, pp 420-441
145 Cabrera BD (1980) Reinfection and infection rate studies of soil-transmitted helminthiasis in Juban, Sorsogon. Collected Papers on the Control of soil-transmitted Helminthiasis by the APCO Research Group. Vol I:181-191
146 Cabrera BD (1981) Prevalence of Strongyloides stercoralis infection in selected areas in the Philippines using a modified Harada-Mori culture technique. AMP 17:84-90
147 Cabrera BD (1981) A comparative study on the effect of mass treatment of entire community and selective treatment of children 0-14 years on the prevalence of soil-transmitted helminthiasis in two communities, Mindoro, Philippines. Proc 8th Conf APCO, pp 215-224
148 Cabrera BD (1984) Paragonimus in the Philippines: current status. Arzneim-Forsch 34(II):1188-1192
149 Cabrera BD et al. (1974) Quoted from 148
150 Cabrera BD et al. (1977-1978) Quoted from 148
151 Cabrera BD, Arambulo PV (1973) Human filariasis in the Philippines. AMP 9:160-173
152 Cabrera BD, Arambulo PV (1977) Malaria in the Republic of the Philippines. A review. Acta Trop 34:265-279
153 Cabrera BD, Arambulo PV (1977) Studies on the epidemiology and transmission of human taeniasis in the Philippines - a paradoxical public health problem. JPMA 53:105-130
154 Cabrera BD, Arambulo PV, Portillo GP (1975) Ascariasis control and/or eradication in a rural community in the Philippines. SEA J 6:510-518
155 Cabrera BD, Baisas FE, Santiago D (1970) Determination of the distribution and abundance of mosquitoes in selected geographic areas. AMP 7:40-81
156 Cabrera BD, Baltazar R, Caviles R, Domingo E, Gonzaga A, Lingao A, Campos PC (1969) Capillaria philippinensis eggs demonstrated in fish from Pudoc, Ilocos Sur. JPMA 45:275-280
157 Cabrera BD, Cruz AC (1980) Clinical trial of mebendazole on soil-transmitted helminthiasis in rural communities. AMP 16:55-57
158 Cabrera BD, Cruz I (1968) The second endemic focus for Malayan filariasis in the Republic of the Philippines. AMP 5:1-24

159 Cabrera BD, Fevidal PM (1974) Studies on Paragonimus and paragonimiasis in the Philippines. III. Prevalence and treatment of human paragonimiasis with bithionol in Jaro, Leyte, Philippines. SEA J 5:39-45
160 Cabrera BD, Garcia EG, Cruz TA, Jueco NL (1961) Studies on enterobiasis in the Philippines. II. The occurrence of Enterobius ova in the finger tips and finger nails of infected school children. JPMA 37:1046-1049
161 Cabrera BD, Garcia EG, Cruz TA, Jueco NL (1961) Studies on enterobiasis in the Philippines. VI. Reinfection of successfully treated cases of enterobiasis. JPMA 37:1061-1064
162 Cabrera BD, Garcia EG, Cruz TA, Salazar NP, Jueco NL (1961) Studies on enterobiasis in the Philippines. I. Frequency of enterobiasis among children in the City of Manila. JPMA 37:1033-1045
163 Cabrera BD, Jueco NL (1972) Filariasis survey among indigenous tribes of Palawan, Republic of the Philippines. SEA J 3:31-39
164 Cabrera BD, Lara ED, Fernandes J (1972) Intestinal parasitism among two indigenous tribes in Palawan, Republic of the Philippines. JPMA 48:153-157
165 Cabrera BD, Rozeboom LE (1964) Filariasis in Palawan, Philippine Islands. Nature 202:725-726
166 Cabrera BD, Rozeboom LE (1965) The periodicity characteristics of the filaria parasites of man in the Republic of the Philippines. Am J Epidemiol 81:192-199
167 Cabrera BD, Sy FS (1978) Oxantel-pyrantel in various regimens for the treatment of soil transmitted helminthiasis in rural and urban communities. Drugs 15(Suppl 1):78-85
168 Cabrera BD, Tamondong CT (1966) Bancroftian and Malayan filariasis in Palawan. Extent and distribution. AMP 3:20-36
169 Cabrera BD, Tamondong CT (1966) Filariasis survey in Jolo, Sulu. AMP 3:86-102
170 Cabrera BJ, Tamondong CT (1970) Filariasis studies in Mindanao Island. JPMA 46:74-98
171 Cabrera BD, Tamondong CT (1970) Filariasis studies in Mindanao, Republic of the Philippines: The third endemic focus for Malayan filariasis. AMP 6:102-131
172 Cabrera BD, Tubangui M (1951) Studies on filariasis in the Philippines. III. Aedes (Finlaya) poicilius (Theobald), the mosquito intermediate host of Wuchereria bancrofti in the Bicol Region. AMP 7:221-227
173 Cabrera BD, Vajrasthira S (1972) Occurrence of the lung fluke Paragonimus siamensis Miyazaki and Wykoff 1965, in the Republic of the Philippines. SEA J 3:446-448
174 Cabrera BD, Vajrasthira S (1973) Endemicity of Paragonimus siamensis Miyazaki and Wykoff 1965, the second species of lung fluke found in the Republic of the Philippines. SEA J 4:509-518
175 Cabrera BD, Valeza F (1972) Filariasis in Sablayan Island, Juban, Sorsogon. AMP 8:145-151
176 Cabrera BD, Valeza F (1977) The reinfection rate of soil-transmitted helminths in the pilot areas after treatment. Proc 4th Conf APCO, pp 237-254
177 Cabrera BD, Valeza F (1978) Distribution and density of mosquitoes in two endemic areas for Bancroftian filariasis in Sorsogon, Philippines. SEA J 9:398-405
178 Cabrera BD, Valeza F (1978) Distribution and density of mosquitoes in two endemic areas for Bancroftian filariasis in Sorsogon, Philippines. AMP 14:20-27
179 Cabrera BD, Valeza F (1979) Study of paragonimiasis in Sorsogon Province, Philippines. I. Survey of streams for infection rate in crab intermediate host. AMP 15:29-42

180 Cabrera BD, Valeza F (1980) The reinfection rate of soil-transmitted helminth in the pilot areas after treatment. Collected Papers on the Control of soil-transmitted Helminthiasis by the APCO Research Group. Vol I:141-149
181 Cabrera BD, Valeza F, Santos AT, Cruz I (1978) Current status of schistosomiasis japonica in Sorsogon Province, Republic of the Philippines. SEA J 9:86-92
182 Cabrera BD, Yogore MG (1961) Expulsion of Taenia saginata and Taenia solium by quinacrine (atabrine) treatment. JPMA 37:479-484
183 Cabrera MG, Cabrera BD, Ordinario E, Leon W de (1973) Comparison of circumoval precipitin test (COPT) filter paper method and formalin-ether concentration technique in schistosomiasis field survey. SEA J 4:359-366
184 Cabrera MG, Rosales AB, Leon W de (1974) A stool survey of second year medical students at the UERMMMC. UE J med 9:7-9
185 Cabrera (1927) pers. comn. to Pesigan et al. (611)
186 Calubaquib PB, Rolda H (1947) The incidence of intestinal parasitism among food handlers. JPMA 23:149-152
187 Calvert WJ (1902) A preliminary report on the blood in 2 cases of filariasis. Bull Johns Hopkins Hosp 13:23-24
188 Calvert WJ (1902) Eosinophilia in filariasis. Bull Johns Hopkins Hosp 13:133-136
189 Calvert WJ (1903) Record of parasitic infections in the Philippines. Boston Med Surg J 149:484-485
190 Camacho VM (1969) Control of filariasis in the Province of Sorsogon imperative to economic growth. Philipp J Publ Hlth 13:19-44
191 Campaign against hookworm in the Province of Cebu, Philippine Islands (1923). Publ Hlth Rep 38:2368-2370
192 Canizares M, Celis J (1951) Observations of paragonimiasis at the Quezon Institute. Dis Chest Chicago 19:668-676
193 Canizares M, Celis J (1951) Observaciones sobre la paragonimiasis en el Instituto Quezón. Rev Esp Tuberculosis 20:39-46
194 Canlas B, Cabrera BD, Dauz U (1967) Human intestinal capillariasis II. Pathological features. AMP 4:84-91
195 Canos AJ, Sayoc A, Gador F (1965) The schistosomiasis problem in Davao City. Institute of Hygiene, U P, Manila, 43 pp
196 Capistrano IM (1977) Notes on schistosomiasis control in the Philippines. Philipp J Publ Hlth 20:74-76
197 Carney WP, Banzon T, Veyra V de, Dano E, Cross JH (1980) Intestinal parasites of man in northern Bohol, Philippines, with emphasis on schistosomiasis. SEA J 11:473-479
198 Carney WP, Banzon T, Veyra V de, Papasin MC, Cross JH (1981) Intestinal parasites of man in Oriental Mindoro, Philippines, with emphasis on schistosomiasis. SEA J 12:13-18
199 Carney WP, Mercado AS, Pagaran IG, Vergel AG, Cross JH (1984) Intestinal parasites of man in Agusan del Norte, Mindanao, Philippines, with emphasis on schistosomiasis. SEA J (in press). Quoted from 248
200 Carney WP, Veyra V de, Cala EM, Cross JH (1981) Intestinal parasites of man in Bukidnon, Philippines, with emphasis on schistosomiasis. SEA J 12:24-29
201 Caro B (1923) Hookworm clinic in Bulacan. Mthly Bull Philipp Hlth Serv 3:305-316
202 Carroll D, Hunninen AV (1948) Studies on schistosomiasis japonica in the Philippine Islands. 3. A clinical study of 72 cases treated with tartar emetic. Bull Johns Hopkins Hosp 82:366-372

203 Case Study IV (1983) The integrated project on family planning, parasite control and nutrition in the Philippines. Proc 10th Asian Parasite Contr Fam Plan Conf, pp 146-169
204 Catto J (1905) Schistosoma cattoi, a new blood fluke in man. Br Med J 1905:1-13
205 Celis GR, Viloria EV (1970) A parasitological survey of Leveriza, Malate, Manila. AMP 6:95-101
206 Chamberlain WP (1901) Medical and sanitary conditions in the Philippines. Boston Med Surg J 145:507-513
207 Chamberlain WP (1910) A statistical study of uncinariasis among white men in the Philippines. PJS B5:249-266
208 Chamberlain WP (1912) The study of tropical diseases in the Philippine Islands. J Am Med Assoc 58:998-1002
209 Chamberlain WP, Bloombergh HD, Kilbourne ED (1910) Examinations of stools and blood among the Igorots at Baguio, Philippine Islands. PJS B5:505-514
210 Chamberlain WP, Bloombergh HD, Kilbourne ED (1910) Report of the board for the study of tropical diseases in the Philippine Islands. Quarter ending June 30th, 1910. Milit Surg 27:519-539
211 Chamberlain WP, Vedder EB, Barber JR (1912) Report of the U S Army Board for the study of tropical diseases as they exist in the Philippine Islands. Report for the quarter ending March 31, 1911. Milit Surg 30:306-331
212 Chanco PP (1940) Incidence of Enterobius vermicularis in the caecum of man at autopsy. Bull Natl Appl Sci 7:291-295
213 Chanco P (1960) Taeniasis solium: report of three indigenous cases in the Philippines. Sto Tomas J Med 15:267-276
214 Chanco PP (1968) Human gnathostomiasis. Fam Physician 6:1-3
215 Chanco PP (1975) Finger washings in enterobiasis: comparison of findings with those of the cellophane swab and fecal smears. Inst Publ Hlth, U P, Manila. 8pp
216 Chanco PP, Arevalo EC (1971) Digenin for the treatment of ascariasis and other intestinal nematode parasites of man. JPMA 47:440-448
217 Chanco PP, Refuerzo PG, Cabrera AC (1956) Human gnathostomiasis; report of the first case in a native Filipino. Sto Tomas J Med 11:136-139
218 Chanco PP, Reyes PR (1971) Incidence of intestinal parasitism among 609 medical technology students with the use of concentration technics. Sto Tomas J Med 26:347-362
219 Chanco PP, Soriano LJ (1939) The incidence of Enterobius vermicularis among Filipino school children: a preliminary report. AMP 1:81-89
220 Chanco PP, Villegas-Cinco A (1961) The unreliability of the routine fecal examination for parasitic ova and cysts. JPMA 37:653-600
221 Chandler A (1929) Hookworm disease. Quoted from Africa(8)
222 Chang T-H, Smith GW, Riesenman FR, Alston EF (1948) Cerebral granuloma due to schistosomiasis. J Am Med Assoc 136:230-238
223 Character of the sickness among troops in the Philippines (1899). J Am Med Assoc 32:447
224 Cheng TC (1982) Anisakiasis. CRC Handbook Series in Zoonoses, Sec C, Vol II:37-54
225 Chi LW, Wagner ED, Wold N (1971) Susceptibility of Oncomelania hybrid snails to various geographic strains of Schistosoma japonicum. Am J Trop Med Hyg 20:89-94
226 Chitwood MB, Velasquez C, Salazar NP (1966) Physiological changes in a species of Capillaria (Trichuroidea) causing a fatal case of human intestinal capillariasis. Proc 1st Int Congr Parasitol, Vol II:797-798

227 Chitwood MB, Velasquez C, Salazar NG (1968) Capillaria philippinensis sp.n.(Nematoda: Trichinellida), from the intestine of man in the Philippines. J Parasitol 54:368-371
228 Chiu JK (1967) Susceptibility of Oncomelania hupensis chiui to infection with Schistosoma japonicum. Malacologia 6:145-153
229 Chiu JK (1968) Cercaria production of geographic strains of Schistosoma japonicum in Oncomelania hupensis chiui. J Formosan Med Assoc 67:259-265
230 Chiu JK, Lu SC (1975) Parasitological studies at a new breeding area of Oncomelania hupensis chiui in Taipei County, Taiwan. Chinese J Microbiol 8:172-178
231 Chow CY (1969) Ecology of malaria vectors in the Pacific. Cah ORSTOM, Sér Ent Med Parasitol 7:93-97
232 Chow CY (1970) Bionomics of malaria vectors in the Western Pacific Region. SEA J 1:40-57
233 Codorniu y Nieto DA (1857) Topografía médica de las Islas Filipinas. Madrid
234 Coker-Vann MR, Subianto DB, Brown P, Diwan AR, Desowitz R, Garruto RM, Gibbs CJ, Gajdusek DC (1981) Elisa antibodies to cysticerci of Taenia solium in human populations in New Guinea, Oceania and Southeast Asia. SEA J 12:499-505
235 Cole CL (1907) Necator americanus in natives of the Philippine Islands. Milit Surg 21:298-312
236 Cole CL (1907) Necator americanus in natives of the Philippine Islands. PJS B2:333-342
237 Copeland EB (1927) Nomenclature of the abaca plant. PJS 33:141-153
238 Craig CF (1903) The occurrence of uncinariasis (ankylostomiasis) in soldiers of the United States Army. Am J Med Sci 126:798-816
239 Crook JR, Fulton SE, Supanwong K (1967) Studies on the animal parasite phases of Angiostrongylus cantonensis in Thailand (intermediate host studies; definitive host studies). SEATO Ann Prog Rep, SEATO Med Res Lab & SEATO Clin Res Center 1966/67, p 238
240 Cross JH (1969) Fasciolopsiasis in Southeast Asia and the Far East: a review. Proc 4th Southeast Asian Sem Parasitol Trop Med, pp 177-196. Bangkok
241 Cross JH (1976) Preliminary observations on the biology of the Indonesian strain of Schistosoma japonicum: experimental transmission in laboratory animals and oncomelanid snails. SEA J 7:202-207
242 Cross JH, Banzon T, Clarke MD, Basaca-Sevilla V, Watten RH, Dizon JJ (1972) Studies on the experimental transmission of Capillaria philippinensis in monkeys. Trans R Soc Trop Med Hyg 66:819-827
243 Cross JH, Banzon T, Murell KD, Watten RH, Dizon JJ (1970) A new epidemic diarrheal disease caused by the nematode Capillaria philippinensis. Industry and Tropical Health 7:124-131
244 Cross JH, Banzon T, Singson C (1978) Further studies on Capillaria philippinensis: development of the parasite in the Mongolian gerbil. J Parasitol 64:208-213
245 Cross JH, Banzon T, Wheeling CH, Cometa H, Lien JC, Clarke R, Petersen H, Sevilla J, Basaca-Sevilla V (1977) Biomedical survey in North Samar Province, Philippine Islands. SEA J 8:464-475
246 Cross JH, Basaca-Sevilla V (1983) Experimental transmission of Capillaria philippinensis to birds. Trans R Soc Trop Med Hyg 77:511-514
247 Cross JH, Basaca-Sevilla V (1983) Experimental infections of Capillaria philippinensis in multimammate rats (Mastomys natalensis). SEA J 14:164

248 Cross JH, Bhaibulaya M (1983) Intestinal capillariasis in the Philippines and Thailand. In: Croll NA, Cross JH (ed): Human ecology and infectious diseases, pp 103-136. New York London etc
249 Cross JH, Lo CT (1980) Susceptibility of new Taiwan foci of Oncomelania hupensis to geographic strains of Schistosoma japonicum. SEA J 11:374-377
250 Cross JH, Singson CN, Battad S, Basaca-Sevilla V (1980) Intestinal capillariasis: epidemiology, parasitology and treatment. R Soc Med Int Congr Symp Ser 24:81-87
251 Cross JH, Wheeling C, Banzon T, Basaca-Sevilla V, Sevilla J (1977) Amoebiasis and intestinal parasitic infections in a population on Cebu Island, the Philippines. Ann Trop Med Parasitol 71:435-441
252 Crowell BC, Hammack RW (1913) Intestinal parasites encountered in five hundred autopsies, with reports of cases. PJS B8:157-174
253 Cruz TA, Reyes WL (1968) Intestinal helminths of man in Bay, Laguna. J Mla Med Soc 6:241-250
254 Curry JJ (1901) US Army pathological laboratories in the Philippine Islands. Boston Med Surg J 144:175-177
255 Curry JJ (1901) The fevers of the Philippines. Boston Med Surg J 144:446-451
256 Curry JJ (1901) Dysenteric diseases of the Philippine Islands, with special reference to the Ameba coli as a causative agent in tropical dysentery. Boston Med Surg J 144:177-178
257 Cutler JG (1946) Schistosomiasis of the central nervous system. J Nerv Ment Dis 104:425-431
258 Daengsvang S (1980) A monograph on the genus Gnathostoma & gnathostomiasis in Thailand. Southeast Asian Medical Information Center Tokyo
259 Dakin WPH, Connellan JD (1947) Asiatic schistosomiasis: an outbreak in the Royal Australian Air Force. Med J Aust 1:257-265
260 Darling ST (1923) The occurrence of Ancylostoma braziliense, de Faria 1910 in the Philippine Islands. J Parasitol 9:234-235
261 Dauz U, Cabrera BD, Canlas B (1967) Human intestinal capillariasis I. Clinical features. AMP 4:72-83
262 Davao City Health (1977) Davao City health family planning, nutition and parasite control program. Proc 2nd Consult Meet Integr, Jap Org Int Coop Fam Planning, pp 109-130
263 Davis GM (1968) A systematic study of Oncomelania hupensis chiui (Gastropoda: Hydrobiidae). Malacologia 7:17-70
264 Davis GM (1969) Oncomelania and the transmission of Schistosoma japonicum: a brief review. Proc 4th Southeast Asian Sem Parasitol Trop Med, pp 93-99
265 Davis GM (1979) The origin and evolution of the Pomatiopsidae, with emphasis on the Mekong River Triculinae. Monogr Acad Nat Sci Phila 20; VIII+120 pp
266 Davis GM (1980) Snail hosts of Asian Schistosoma infecting man: evolution and coevolution. Malacological Review, Suppl 2:195-238
267 Dawes B (1968) The Trematoda with special reference to British an other European forms. XVI+644 pp. Cambridge
268 Dazo BC (1981) Control and research of parasitic diseases in the Western Pacific Region. Yonsei Repts Trop Med 12:43-53
269 Detels R, Gutman L, Jaramillo J, Zerrudo E, Banzon T, Valera J, Murrell KD, Cross J, Dizon JJ (1969) An epidemic of intestinal capillariasis in man. A study in a barrio in northern Luzon. Am J Trop Med Hyg 18:676-682
270 DeWitt WB (1952) Susceptibility of snail vectors to geographic strains of Schistosoma japonicum. J Parasitol 38 (Suppl):35
271 DeWitt WB (1953) Susceptibility of snail vectors to geographic strains of Schistosoma japonicum. J Parasitol 40:453-456

272 Dizon JJ (1981) Future plans of action. Proc 8th Asian Parasite Contr Fam Plan Conf, pp 173-176
273 Dizon JJ, Valeza FS (1975) Parasitism in the Philippines. Proc 2nd Conf Asian Parasite Control Org, pp 95-101
274 Dizon JJ, Watten RH, Associates (1969) Preliminary observations on a new disease in man - intestinal capillariasis. JPMA 45:5-20
275 Doherty DJ (1900) Medicine and disease in Philippines. J Am Med Assoc 34:1526-1531
276 Domingo EO, Lingao AL, Tiu E, Lao JY, Olvedo RM (1983) HBV exposure and HBsAg positivity rates in schistosomiasis japonica: study in a Philippine community endemic for both infections. SEA J 14:456-462
277 Domingo EO, Tiu E, Peters PA, Warren KS, Mahmoud AAF, Houser HB (1980) Morbidity in schistosomiasis japonica in relation to intensity of infection: study of a community in Leyte, Philippines. Am J Trop Med Hyg 29:858-867
278 Dominguez LS (1968) A study of hookworm infection among elementary public school children in selected provinces in various regions of the Philippines. MPH Thesis, Inst Hyg, UP, Manila; XIV+111 pp
279 Dominguez LS (1969) A study of hookworm infection among elementary public school children in selected provinces in various regions of the Philippines. JPMA 45:30-45
280 Dönges J (1980) Parasitologie. Mit besonderer Berücksichtigung humanpathogener Formen. Stuttgart
281 Dumag PU, Gajudo CE, Sena CY, Cardenas EC, Fementira EB (1981) Epidemiology of animal schistosomiasis in the Philippines. Philipp J Anim Ind 36:1-23
282 Dumag PU, Reyes PV, Kingscote AA (1970) Parasitic disease problems of cattle in the Philippines. Philipp J Anim Ind 27:91-112
283 Eastman WR (1910) A report on intestinal parasites in scout companies. Milit Surg 26:671-672
284 Edger BJ (1905) Some medical and sanitary experiences in a portion of the Philippines. Am Med 9:281-284
285 Enarson DA, Enarson PM (1982) Filariasis in an upland population in the Philippines. Trop Geogr Med 34:353-358
286 Enigk K (1979) Resistenz der Dauerformen von Endoparasiten der Haustiere. Berl Münch Tierärztl Wschr 92:491-497
287 Erhardt A (1935) Systematik und geographische Verbreitung der Gattung Opisthorchis Blanchard 1895, sowie Beiträge zur Chemotherapie und Pathologie der Opisthorchiasis. Z Parasitenk 8:188-225
288 Erhardt A, Germer WD, Hörning B (1962) Opisthorchiasis, hervorgerufen durch den Katzenleberegel Opisthorchis felineus (Riv). Parasitolog Schr Reihe, Heft 15, 171 pp
289 Esdorn I, Pirson H (1973) Die Nutzpflanzen der Tropen und Subtropen in der Weltwirtschaft. Stuttgart
290 Estrada JP, Basio DG (1965) Filariasis in the Philippines. JPMA 41:100-153
291 Family Planning-nutrition-parasite control integration pilot project (Philippines) (1977). 2nd progress report (1977). Proc 2nd Consult Meet Integr, pp 64-74
292 Fan PC, Khaw OK (1969) Schistosomiasis japonica in Taiwan: a review. Proc 4th Southeast Asian Sem Parasitol Trop Med, pp 17-47
293 Farooq M (1973) Review of national control programmes. In: Ansari N(ed): Epidemiology and control of schistosomiasis (bilharziasis), pp 388-421. Basel etc
294 Faustino LA (1933) Two fresh-water shells from the Philippine Islands. PJS 51:575-577

295 Feachem RG, Bradley DJ, Garelick H, Mara DD (1983) Sanitation and disease. Health aspects of excreta and wastewater management. Worldbank Studies in Water Supply and Sanitation 3
296 Ferguson MS, Bang FB (1960) Schistosomiasis. In: Preventive Medicine in World War II Vol 5:45-81. Washington
297 Fernandez TJ, Petilla T, Banez B (1982) An epidemiological study on Schistosoma japonicum in domestic animals in Leyte, Philippines. SEA J 13:575-579
298 Fischer GW, McGrew GL, Bass JW (1980) Pulmonary paragonimiasis in childhood. A cause of persistent pneumonia and hemoptysis. J Am Med Assoc 243:1360-1362
299 Fletcher M, Woodruff DS, LoVerde PT (1980) Genetic differentiation between Schistosoma mekongi and Schistosoma japonicum: an electrophoretic study. Malacol Rev Suppl 2:113-122
299a Flexner S, Barker LF (1900) Report of a special commission sent to the Philippines by the Johns Hopkins University to investigate the prevalent diseases of the Islands. Johns Hopkins Univ Circular 19:13-16
300 Foster CL (1906) Two cases of infection with Taenia nana in the Philippine Islands. J Am Med Assoc 47:685-686
301 Francisco (1939, 1946) pers. comn. to Pesigan et al. (611)
302 Frank W (1976) Parasitologie. Lehrbuch für Studierende der Human- und Veterinärmedizin, der Biologie und der Agrarbiologie. Stuttgart
303 Franke G (1967) Nutzpflanzen der Tropen und Subtropen. Bd II Obstliefernde Pflanzen. Zuckerliefernde Pflanzen. Getreide. Faserpflanzen. VII Faserbanane, pp 384-391. Leipzig
304 Franke W (1981) Nutzpflanzenkunde. Stuttgart
305 Fresh JW, Cross JH, Reyes V, Wahlen GE, Uylangco CV, Dizon JJ (1972) Necropsy findings in intestinal capillariasis. Am J Trop Med Hyg 21:169-173
306 Frohne W (1958) Report on assignment to the schistosomiasis control pilot project, Palo, Leyte, 16 November 1956 - 15 April 1958. WHO, Regional Office for the Western Pacific. Philippines 9, 16pp
307 Fülleborn F (1913) Beiträge zur Morphologie und Differentialdiagnose der Mikrofilarien. Arch Schiffs- u. Tropenhyg 17 Beih 1:1-72
308 Gajudo CE, Batolos JA, Serrano MS, Fementira EB (1982) Preliminary investigation on the role of animals in the transmission of Brugian filariasis. Philipp J Anim Ind 37:5-16
309 Garcia EG (1951) Survey in Sorsogon. JPMA 27:256-257
310 Garcia EG (1973) Sex ratio of Ascaris lumbricoides. AMP 9:152-154
311 Garcia EG (1974) Sex ratio of Ascaris lumbricoides recovered from autopsies. (Abstract) Proc 12th SEAMEO-TROPMED Sem, p 20
312 Garcia EG (1975) Studies on schistosomasis japonica and saponins. Socio-Tech Res Bull 1:113-115
313 Garcia EG (1976) The biology of Schistosoma japonicum, Philippine strain: a review. SEA J 7:190-196
314 Garcia EG (1979) Angiostrongylus cantonensis in the Philippines: a review. In: Cross JH(ed): Studies on angiostrongyliasis in Eastern Asia and Australia, pp 53-56. Taipei
315 Garcia EG, Cabrera BD, Cruz TA, Jueco NL (1961) Reinfection rates of successfully treated cases of ascariasis. JPMA 37:239-243
316 Garcia EY (1936) Study on the resistance of the metacercariae of Monorchotrema taihokui to different physical chemical agents. JPMA 16:561-565
317 Garcia EY (1950) Toxoplasmosis and sparganosis in native Filipinos. (Clinical reports). JPMA 26:225-230
318 Garcia EY (1952) Two rare tapeworms expelled by "Hetrazan". Medical Reporter 4:12-14

319 Garcia EY (1959) Schistosomiasis philippinensis and Schistosoma philippinensis, a new clinical entity and a new species of blood fluke respectively. MD J 8:225-231
320 Garcia EY (1960) Schistosoma philippinensis, new species, and pulmonary schistosomiasis. JPMA 36:24-31
321 Garcia EY (1964) Microfilaria philippinensis, a new species of human microfilaria in the Philippines. MD J 13:261-263
322 Garcia EY (1966) Microfilaria philippinensis, a new species of human microfilaria in the Philippines. Proc 1st Int Congr Parasitol 2:647-649
323 Garcia EY (1967) Schistosoma philippinensis, a new species of human schistosome in the Philippines. Helminthologia 8:147-154
324 Garcia EY, Africa CM (1935) Diphyllobothrium latum (Linnaeus, 1758) Lühe, 1910, in a native Filipino. PJS 57:451-457
325 Garcia EY, Garcia BS (1970) Cercarial enzyme-inhibition, a more effective control of schistosomiasis in the Philippines. J Mla Med Soc 8:21-25
326 Garcia EY, Refuerzo PG (1936) Two more species of the genus Stictodora Looss, 1899, in the Philippines, with description of a new species. PJS 60:137-139
327 Garcia F (1917) Common intestinal parasites. PJS B12:25-32
328 Garcia FA (1968) Intestinal capillariasis. Sto Tomas J Med 23:220-224
329 Garcia O, Vazquez-Colet A, Lacy GR (1924) Bacteriological examination of stools of food handlers in Manila. PJS 24:735-741
330 Garcia OP, Reyes AI (1962) Sparganosis in Filipinos. With a review of reported cases in the Philippines. JPMA 38:608-611
331 Garcia RG (1959) The Philippine schistosomiasis: a critical study of the snail vector (Oncomelania quadrasi) and the search for effective molluscicides. Unitas 32:681-709
332 Garcia RG (1960) The Philippine schistosomiasis: A critical study of the snail vector (Oncomelania quadrasi) and the search for effective molluscicides. Unitas 33:5-79
333 Garduno DM (1959) Eosinophilia, a clinical problem: report of 2 suspected cases of trichiniasis. Bull V Luna Gen Hosp 1:35-38
334 Garduno DM, Carpio D (1950) Paragonimiasis: a report of three cases with pulmonary and neurological symptoms. JPMA 26:253-260
335 Garduno DM, Icasiano CB (1954) A case of human cysticercosis (T. solium). JPMA 30:572-580
336 Garrison PE (1907) A preliminary report upon the specific identity of cestode parasites of man in the Philippine Islands, with a description of a new species of Taenia. PJS B2:537-550
337 Garrison PE (1908) The prevalence and distribution of the animal parasites of man in the Philippine Islands with a consideration of their possible influence upon public health. PJS B3:191-209
338 Garrison PE (2908) A new intestinal trematode of man (Fascioletta ilocana, gen. nov., sp. nov.). PJS B3:385-393
339 Garrison PE (1911) Davainea madagascariensis (Davaine) in the Philippine Islands. PJS B6:165-175
340 Garrison PE, Lago FP (1910) Specimens added to the helminthological collection, US Naval Medical School, June-August, 1910. US Naval Med Bull 4:511
341 Garrison PE, Leynes R (1909) The development of the miracidium of Paragonimus under various physical conditions. PJS B4:177-183
342 Garrison PE, Leynes R, Llamas R (1909) Medical survey of the town of Taytay. X. Animal parasites of the intestine. PJS B4:257-269
343 Garrison PE, Llamas R (1909) The intestinal worms of 385 Filipino women and children in Manila. PJS B4:185-186

344 Gazmen JMC (1974) A prevalence study of helminths at Palanan Elementary School, Palanan, Makati, Rizal. Inst Publ Hlth, UP, Manila, 12 pp
345 Gilman PK (1908) A report on the first one hundred autopsies at the Philippine Medical School. PJS B3:211-221
346 Gomez L (1911) A clinical study of hookworm infection in the Philippines. PJS B6:239-249
347 Gonzales A, Kumangai M, Rosales J, Silan RB, Tabo R (1967) A helminthic survey among the residents of Barrio San Jose, Bagong Buhay, Paco, Manila, April, 1967. Inst Hyg, UP, Manila, 19 pp
348 Greenleaf CR (1902) Health of American troops in the Philippine Islands. Med Rec 61:98
349 Grove DI, Valeza FS, Cabrera BD (1978) Bancroftian filariasis in a Philippine village: clinical, parasitological, immunological, and social aspects. Bull WHO 56:975-984
350 Guerrero LA, Guerrero RD (1972) Angiostrongylus cantonensis in commensal rats in Dumaguete City, Negros Oriental. AMP 8:33-35
351 Guill RE (1956) A review of intestinal parasites at Clark Field. J Philipp Fed Priv Med Pract 5:839-845
352 Gutierrez DD, Lozano AA, Pesigan TP (1946) Report of the first case of intestinal heterophyidiasis diagnosed in life in the Philippines. JPMA 22:287-292
353 Guzman RV (1933) Incidence of filarial infection among 980 male and female prisoners in Bilibid Prisons. Bull San Juan de Dios Hosp 7:187-193
354 Guzman T (1963) Survey of school pupils. Bureau of Public Schools, 1963 (unpublished) Quoted from Pesigan and Yogore (612)

355 Habs H (1952) Abdominaltyphus. In: Rodenwaldt E(ed): Welt-Seuchen-Atlas. World Atlas of epidemic diseases. Vol 1:1-5. Hamburg
356 Hairston NG (1959) Assignment report, Philippines-9 project. WHO unpublished document (1959) Quoted from Hairston (359)
357 Harston NG (1962) Population ecology and epidemiological problems. Ciba Foundation Symposium on Bilharziasis, pp 36-62. London
358 Hairston NG (1965) On the mathematical analysis of schistosome populations. Bull WHO 33:45-62
359 Hairston NG (1973) The dynamics of transmission. In: Ansari N (ed): Epidemiology and control of schistosomiasis (bilharziasis), pp 250-336. Basel etc
360 Hairston NG, Santos BC (1961) Ecological control of the snail host of Schistosoma japonicum in the Philippines. Bull WHO 25: 603-610
361 Hallock HM (1904) Tenia nana: report of two cases. J Am Med Assoc 42:891
362 Hamit HF, Cortese PT (1957) An extemporaneous study of intestinal parasites in Philippine scouts. Milit Med 121:84-90
363 Hasselmann CM (1941) Paragonimus Westermanii. Case report with microphotograph of operculum of the ovum sprung open. PJS 74:285-287
364 Hausherr K (1972) Die Entwicklung der Kulturlandschaft in den Lanao-Provinzen auf Mindanao (Philippinen) unter Berücksichtigung des Kulturkontaktes zwischen Islam und Christentum. Diss Phil Fak, Bonn
365 Heiser VG (1908) Prophylactic medicine as applied to the hygiene of Bilibid prison. PJS B3:69-71
366 Hesselbrock WB, Lippincott SW, Palmer ED, Henderson EW, Pauls FP (1947) A study of various methods of stool examination in the diagnosis of schistomiasis japonica. Am J Clin Pathol 17:197-204

367 Heyneman DD, Lim BL (1967) Angiostrongylus cantonensis: proof of direct transmission with its epidemiological implications. Science 158:1057-1058
368 Hilario JS, Wharton LD (1917) Echinostoma ilocanum (Garrison): a report of five cases and a contribution to the anatomy of the fluke. PJS 12:165-166
369 Hilario JS, Wharton LD (1917) Echinostoma ilocanum (Garrison): a report of five cases and a contribution to the anatomy of the fluke. PJS B12:203-211
370 Hinz E (1972) Eosinophile Meningoenzephalitis und Enzephalomyelitis beim Menschen in Südostasien. Med Klin 67:1236-1242
371 Hinz E (1984) Vegetation, Nutzpflanzen und Endemie. Erdkundliches Wissen 70:32-44
372 Hizon J (1928) Schistosomiasis - diagnosis and treatment. Bull San Juan de Dios Hosp 2:87-101
373 Hohorst W (1962) Die Rolle der Ameisen im Entwicklungsgang des Lanzettegels (Dicrocoelium dendriticum). Z Parasitenkd 22:105-106
374 Hohorst W, Graefe G (1961) Ameisen - obligatorische Zwischenwirte des Lanzettegels (Dicrocoelium dendriticum). Naturwiss. 48:229-230
375 Hosaka Y, Kajihara N, Horimi T, Minai, Hashiguchi J, Sakata Y, Banez EA, Blas BL (1980) An assessment of B-2 as a new molluscicide for the control Oncomelania snails, the intermediate host of Schistosoma japonicum. 10th Int Congr Trop Med Mal; Abstracts, pp 361-362
376 Hoyt RE (1908) Results of three hundred examinations of faeces with reference to the presence of amoebae. PJS B3:417-420
377 Hsü HF, Hsü SY Li (1957) On the intraspecific and interstrain variations of the male sexual glands of Schistosoma japonicum. J Parasitol 43:456-463
378 Hsü HF, Hsü SY Li (1958) On the size and shape of the eggs of the geographic strains of Schistosoma japonicum. Am J Trop Med Hyg 7:125-134
379 Hsü HF, Hsü SY Li (1958) The prepatent period of four geographic strains of Schistosoma japonicum. Trans R Soc Trop Med Hyg 52:363-367
380 Hsü HF, Hsü SY Li (1959) Characteristics of geographic strains of Schistosoma japonicum in the final hosts. Proc 5th Int Congr Trop Med Mal Vol II:58-66
381 Hsü HF, Hsü SY Li (1960) The infectivity of four geographic strains of Schistosoma japonicum in the rhesus monkey. J Parasitol 46:228
382 Hsü HF, Hsü SY Li (1960) Susceptibility of albino mice to different geographic strains of Schistosoma japonicum. Trans R Soc Trop Med Hyg 54:466-468
383 Hsü HF, Hsü SY Li (1960) Distribution of eggs of different geographic strains of Schistosoma japonicum in the viscera of infected hamsters and mice. Am J Trop Med Hyg 9:240-247
384 Hsü HF, Hsü SY Li (1962) Schistosoma japonicum in Formosa. A critical review. Exptl Parasitol 12:459-464
385 Hsü HF, Hsü SY Li (1967) The race complex of Oncomelania formosana in Taiwan, China. Z Tropenmed Parasit 18:417-432
386 Hsü HF, Hsü SY Li, Tsai CT (1960) Further studies on the prepatent periods of four geographic strains of Schistosoma japonicum. Libro Homenaje al Dr. E Caballero y Caballero, Jubileo 1930-1960, pp 153-160. Mexico
387 Hsü SY Li, Chu KY, Hsü HF (1963) Drug susceptibility of geographic strains of Schistosoma japonicum. Z Tropenmed Parasit 14:37-40
388 Hsü SY Li, Hsü HF (1960) Infectivity of the Philippine strains of Schistosoma japonicum in Oncomelania hupensis, Oncomelania formosana and Oncomelania nosophora. J Parasitol 46:793-796

389 Hsü SY Li, Hsü HF (1960) On the virulence of the geographic strains of Schistosoma japonicum. Am J Trop Med Hyg 9:195-198
390 Hsü SY Li, Hsü HF (1961) Behavior of different geographic strains of Schistosoma japonicum as determined by the circumoval test. Proc Soc Exp Biol Med 108:467-468
391 Hsü SY Li, Hsü HF, Chu KY (1962) Interbreeding of geographic strains of Schistosoma japonicum. Trans R Soc Trop Med Hyg 56:383-385
392 Huang SW (1967) Studies on Taenia species prevalent among the aborigines in Wulai District, Taiwan. Bull Inst Zool Acad Sin 6:29-34
393 Hubendick B (1956) Observations on the occurrence of Oncomelania quadrasi Möllendorff, the host-snail of Bilharzia in the Philippines. Proc 14th Int Congr Zool, pp 511-512
394 Huey-Ru Hwu, Banzon TC, Cross JH (1978) Counterimmunoelectrophoresis in the detection of antibodies to Schistosoma japonicum. Am J Trop Med Hyg 27:276-280
395 Huke RE (1963) Abaca production in the Philippines. Philipp Geogr J 7:18-35
396 Huke RE (1972) Shadows on the land: an economic geography of the Philippines. Manila Cebu Makati
397 Hunt AR (1945) Schistosomasis in naval personnel. US Nav Med Bull 45:407-419
398 Hunter GW, Dillahunt JA, Dalton HC (1950) The epidemiology of schistosomasis japonica in the Philippine Islands and Japan. I. Surveys for schistosomiasis japonica on Mindoro, P.I. Am J Trop Med 30:411-429
399 Hunter GW, Dillahunt JA, Faust EC, Modic JL, Connellan JD, Bennett HJ, Ingalls JW (1951) The diagnosis of schistosomiasis japonica. IV. Studies on schistosomiasis japonica in an Australian Air Force unit. Am J Trop Med 31:50-56
399a Iarotski LS, Davis A (1981) The schistosomiasis problem in the world: results of a WHO questionnaire survey. Bull WHO 59:115-127

400 Ignacio SG (1967) An investigation of the familial nature of enterobiasis. Inst Publ Hlth, UP, Manila, 24 pp
401 Ingalls JW (1946) The control of schistosomiasis japonica. III. Studies on the longevity of cercariae of Schistosoma japonicum in saline solutions. J Parasitol 32:521-524
402 Ingalls JW, Hunter GW, McMullen DB, Baumann DM (1949) The molluscan intermediate host and schistosomiasis japonica. II. Observations on conditions governing the hatching of the egg of Schistosoma japonicum. J Parasitol 35:147-151
403 Institute of Public Health (1952-1982) quoted from (148)
404 Ira-Concepcion F (1935) A new parasite in the Philippines. Bull San Juan de Dios Hosp 9:277-299
405 Ishii A, Cabrera BD, Suguri S, Kobayashi M, Go TG, Valeza F (1983) An epidemiological study of filariasis in Sorsogon province, Republic of the Philippines, with notes on experimental mosquito infection. J Trop Med Hyg 86:59-64
406 Ito J (1977) Studies on the fresh water cercariae in Leyte Island, Philippines. 3. Cercariae from Thiaridae. Jpn J Exp Med 47:223-248
407 Ito J, Yokogawa M, Araki K, Kobayashi M (1978) Studies on the morphology of larval and adult lung fluke in the Philippines, with a proposition of new name, Paragonimus philippinensis n.sp. Jpn J Parasitol 28:97-112
408 Iwanaga Y, Santos MJ, Blas BL (1977) The determination of the Oncomelania hupensis quadrasi population density using the banana leaf method in four municipalities of eastern Leyte, Philippines. Hiroshima J Med Sci 26:19-27

409 Jara ID, Paragas BP (1975) A survey on the prevalence of common helminthic infections in relation to excreta disposal practices in Barrio San Guillermo, Morong, Rizal. Inst Publ Hlth, UP, Manila, 9 pp
410 Jesus PI de (1933) Hydrodynamics of mosquito breeding places. Am J Hyg 17:502-514
411 Jesus PI de (1936) Physicochemical factors in anopheline ecology, I: studies on nitrogen. PJS 59:473-491
412 Jesus PI de (1937) Physicochemical factors in anopheline ecology, II: studies on turbidity, chloride, and iron. PJS 62:125-136
413 Jesus PI de (1940) Physicochemical factors affecting the breeding of Anopheles minimus in the Philippines. Rev Parasitol 4:157-161
414 Jesus PI de (1941) Physicochemical factors affecting the breeding of Anopheles minimus var. flavirostris. AMP 2:333-340
415 Jesus Z de (1932) Studies on cysticercosis in Philippine swine. UP Nat Appl Sci Bull 2:61-67
416 Jesus Z de (1932) Notes on an examination of native swine for trichinae. BAI Anim Gaz 2:11-12
417 Jesus Z de (1935) Lymnaea philippinensis, an intermediate host of Fasciola hepatica in the Philippines, with some observations on the bionomics of the parasite. PJS 58:299-315
418 Jesus Z de, Mallari AI (1957) Biology of Lymnaea philippinensis, an intermediate host of Fasciola hepatica and Fasciola gigantica in the Philippines. Philipp J Anim Ind 4:501-513
419 Johns Hopkins Center for Medical Research and Training, Report 1969-70. II. Parasite populations of man. A. Interaction of populations of man and hookworm, pp 5-24
420 Johns Hopkins Center for Medical Research and Training, Report 1964-65. IV. Parasite infections. A. Factors influencing distribution of hookworm, pp 21-24
421 JOICFP (1981) Overview of country projects. Proc 8th Asian Parasite Contr Fam Plan Conf, pp 64-79
422 Jones MF, Brady FJ (1947) Survival of Schistosoma japonicum cercariae at various temperatures in several types of water. Natl Inst Hlth Bull 189:131-136
423 Jueco NL (1968) A study of hookworm infection among school children from a coastal area and a mountainous area in Bay, Laguna. J Mla Med Soc 6:251-257
424 Jueco NL (1975) Raillietina (a rat tapeworm) infection in young children in the Philippines. AMP 11:49-50
425 Jueco NL (1975) Studies on zoonoses associated with rodents. Report submitted to CCB-TROPMED office as research fellowship grant from SEAMEO-TROPMED project
426 Jueco NL (1980) Unpublished tables on parasites of rats in the Philippines, 8 pp
427 Jueco NL (1982) Hymenolepis nana infection. In: CRC Handbook series in zoonoses, Sect C, Vol I:283-287
428 Jueco NL, Bobis TA, Ramirez LM (1971) Seasonal prevalence and density of Anisakis larvae in fishes (galunggong) sold in public markets in Manila. JPMA 47:467-476
429 Jueco NL, Cabrera BD (1971) Reinfection rates of successfully treated cases of ascariasis in Victoria, Laguna. Proc 7th Southeast Asian Reg Sem Trop Med Publ Hlth, p 293
430 Jueco NL, Cabrera BD (1971) Reinfection rates of successfully treated cases of ascariasis in Victoria, Laguna. JPMA 47:449-454
431 Jueco NL, Palaypay MA, Aceremo L (1973) Prevalence and intensity of intestinal parasitism on Talim Island, Binangonan, Rizal, Philippines. SEA J 4:582-587
432 Jusatz HJ (1966) The importance of biometeriological and geomedical aspects in human ecology. Int J Biometeor 10:323-334

433 Kamiya H, Tada Y, Matsuda H, Blas BL, Santos AT (1980) Annual occurrence of Schistosoma japonicum infections in the wild rat, Rattus rattus mindanensis, in Leyte, Philippines. Proc Philipp-Japan Joint Conf Schistosomiasis Res. Control, pp 65-69
434 Kamiya H, Tada Y, Matsuda H, Blas BL, Santos AT (1980) (Annual occurrence of Schistosoma japonicum infection in wild rat, Rattus rattus mindanensis, in Leyte, Philippines.) Jpn J Parasitol 29: (2 Suppl):71-80
435 Kamiya H, Tada Y, Matsuda H, Tanaka H, Blas BL, Nosenas JS, Santos AT (1980) Annual fluctuation of Schistosoma japonicum infection in field rats, Rattus rattus mindanensis, in Dagami, Leyte, Philippines. Jpn J Exp Med 50:375-382
436 Katagiri T (1982) Review of the six-year experience in the integrated projects of five countries (summary). Proc 9th Asian Parasite Contr Fam Plan Conf, pp 71-125
437 Katsurada F (1904) (Determination of the causative agent of a parasitic disease endemic in Yamanashi and several other prefectures.) Tokyo Iji 1371:1442-1446
438 Kindleberger CP (1908) The United States naval station, Olongapo, Philippine Islands. - Its location, climate and diseases. Milit Surg 22:308-316
439 Kolb A (1942) Die Philippinen. Leipzig
440 Kolb A (1974) Die Chinesen, Japaner und Inder auf den Philippinen. Schriften des Instituts für Asienkunde 38. Wiesbaden
441 Komiya K, Yasuraoka K (1966) The biology of hookworms. Progr Med Parasitol Japan 3:1-114
442 Kosin E, Depary A, Djohansjah A (1972) (Taeniasis auf der Insel Samosir.) J Med Fak Univ North Sumatra 2:5-11
443 Kumar V (1980) The digenetic trematodes, Fasciolopsis buski, Gastrodiscoides hominis and Artyfechinostomum malayanum, as zoonotic infections in South Asian countries. Ann Soc Belge Méd Trop 60:331-339
444 Kuntz RE (1955) Biology of the schistosome complexes. Am J Trop Med Hyg 4:383-413
445 Kuntz RE (1963) Intestinal parasites of man in Palawan, Republic of the Philippines. JPMA 39:590-600
446 Kuntz RE (1964) Intestinal parasites of man in Leyte. AMP 1:5-10
447 Kuntz RE, Lawless DK (1960) Intestinal protozoans and helminths in the indigenous population and employees associated with a Naval Air Station in the Philippines. Milit Med 125:561-566
448 Kurihara T, Ikemoto T, Eshita Y (1977) A brief survey of mosquito vectors of Bancroftian filariasis in Bicol region, the Philippines. Mosquito News 37:770-772

449 Lavadia P, Canlas BD, Durban VJ (1960) Hydatid cyst of the lung. (A case report in a Filipino.) JPMA 36:702-708
450 Leach CN, Haughwout FG, Ash JE (1923) The treatment of hookworm infestation with carbon tetrachloride: a clinical and laboratory study. PJS 23:455-513
451 Leavitt SS, Beck OH (1947) Schistosomiasis japonica. A report of its discovery in apparently healthy individuals. Am J Trop Med 27:347-356
452 Lee KM, Fan PC, Wu CC (1982) Further studies on the susceptibility of new Taiwan foci of Oncomelania hupensis to geographic strains of Schistosoma japonicum. SEA J 13:91-95
453 Leon DD de, Saulog TM (1965) Occurrence of Angiostrongylus cantonensis in the brown rat (Rattus norvegicus) in Manila, P.I. J Parasitol 51:947
454 Leon EF de (1974) Helminthic organisms found in flies collected from Barrio Palanan, Makati, Rizal. Inst Publ Hlth, UP, Manila, 7 pp

455 Leon W de, Leiva L (1925) Echinococcus cyst of the human lung. PJS 27:351-369
456 Lewert RM, Yogore MG, Blas BL (1979) Schistosomiasis japonica in Barrio San Antonio, Basey, Samar, the Philippines. I. Epidemiology and morbidity. Am J Trop Med Hyg 28:1010-1025
457 Lewert RM, Yogore MG, Blas BL (1980) Schistosomiasis japonica in Barrio San Antonio, Basey, Samar: Epidemiology and Morbidity. Proc Philipp-Japan Joint Conf Schistosomiasis Res Control, pp 52-55
458 Lewert RM, Yogore MG, Martin WR, Blas BL (1980) Schistosomiasis japonica in Barrio San Antonio, Basey, Samar in the Philippines. IV. "Atypical" precipitates in the circumoval precipitin test - a reaction of recent infection. Am J Trop Med Hyg 29:431-434
459 Leyva BT, Santos AT, Capistrano IM (1970) Schistosomiasis in a developing nation. Bull V. Luna Med Center Soc 12:33-38
460 Lim BL (1967) A review of recent studies on Angiostrongylus cantonensis in Malaysia. 1st Southeast Asian Reg Sem Trop Med, Abstr. of Papers, p 109
461 Limbo DM, Apostol RR (1961) Cerebral pulmonary paragonimiasis. (A short review of the literature, re-evaluation of aralen, emetin and glucosamin therapy). Surg Bull Philipp Gen Hosp 2:97-107
462 Lindsay SW, Denham DA, McGreevy PB (1984) The effect of humidity on the transmission of Brugia pahangi infective larvae to mammalian hosts by Aedes aegypti. Trans R Soc Trop Med Hyg 78:19-22
462a Lippincott SW, Ellerbrook LD, Rhees M (1947) Correlation of laboratory tests with the pathology of schistosomiasis japonica in American soldiers. Am J Path 23:894-895
462b Lippincott SW, Ellerbrook LD, Rhees M, Mason P (1947) A study of the distribution and fate of antimony when used as tartar emetic and Fuadin in the treatment of American soldiers with schistosomiasis japonica. J Clin Invest 26:370-378
463 Lippincott SW, Paddox FK, Rhees MC, Hesselbrock WB, Ellerbrook LD (1947) Tests of liver function in schistosomiasis japonica with particular reference to antimony treatment and with report of two autopsies. Arch Int Med 79:62-76
464 Lissner L (1928) Stuhluntersuchung von 500 philippinischen Schulkindern. Münch Med Wschr 75:1163
465 Loker ES (1983) A comparative study of the life-histories of mammalian schistosomes. Parasitology 87:343-369
466 Lopez CG, Alimurung NN, Fuente V de la (1953) Chronic eosinophilia (with case report). Philipp J Ped 2:22-28
467 Lopez-Rizal L, Padua RG (1926) Brief report of filaria surveys in the Philippines. JPMA 6:298-300
468 Low GC (1909) Filaria philippinensis. J Trop Med Hyg 12:256
469 Lucius R (1981) Untersuchungen zur Biologie, Pathologie und Ökologie von Dicrocoelium hospes Looss, 1907 (Trematodes, Dicrocoeliidae) Diss Fak II (Biologie), Univ Hohenheim, 106 pp
470 Lucius R, Romig T, Frank W (1980) Camponotus compressiscapus André (Hymenoptera, Formicidae) an experimental second intermediate host of Dicrocoelium hospes Looss, 1907 (Trematodes, Dicrocoeliidae). Z Parasitenkd 63:271-275
471 Magath TB, Mathieson DR (1945) Infection of wild rats on Leyte with Schistosoma japonicum. US Naval Med Bull 45:1195-1202
472 Magath TB, Mathieson DR (1946) Important factors in the epidemiology of schistosomiasis in Leyte. Am J Hyg 43:152-163
473 Magath TB, Mathieson DR (1946) Factors affecting the hatching of ova of Schistosoma japonicum. J Parasitol 32:64-68
474 Makiya K (1980) (Distribution and evaluation of control measures of the intermediate host snail Oncomelania quadrasi of Schisto-

soma japonicum in the Philippines. 1. Distribution of the snail, necessary sample size for density survey and significance test of snail density.) Jpn J Parasitol 29:293-304
475 Makiya K (1980) (Studies on the distribution pattern of, and evaluation of control measures against, the snail Oncomelania quadrasi, intermediate host of Schistosoma japonicum in the Philippines. 2. Evaluation of the control of the host snail by drainage.) Jpn J Parasitol 29:359-368
476 Makiya K, Tanaka H, Banez EA, Blas BL, Kumada N (1980) Distribution patterns of Oncomelania quadrasi, evaluation for snail control and determination of necessary sample size for the snail survey. Proc Philipp-Japan Joint Conf Schistosomiasis Res Control, pp 99-102
477 Makiya K, Tanaka H, Banez EA, Blas BL, Kumada N, Perez A (1980) Evaluation of the snail control projects in the pilot areas and the effect of drainage on the snail density. Proc Philipp-Japan Joint Conf Schistosomiasis Res Control, pp 103-104
478 Makiya K, Tanaka H, Banez EA, Blas BL, Kumada N, Santos AT (1981) Population studies on Oncomelania quadrasi, the snail intermediate host of Schistosoma japonicum, in the Philippines. 1. Distribution pattern of the snail in the field. Jpn J Exp Med 51:179-185
479 Makiya K, Tanaka H, Banez EA, Blas BL, Kumada N, Santos AT (1981) Population studies on Oncomelania quadrasi, the snail intermediate host of Schistosoma japonicum, in the Philippines. 2. Necessary sample size for snail density survey. Jpn J Exp Med 51:331-334
480 Makiya K, Tanaka H, Banez EA, Blas BL, Perez A, Kumada N, Santos AT (1982) Population studies on Oncomelania quadrasi, the snail intermediate host of Schistosoma japonicum, in the Philippines. 3. Data transformation for significance test of snail density. Jpn J Exp Med 52:33-38
481 Manalang C (1921) Critical tests of hookworm remedies on man. Am J Trop Med 7:57-60
482 Manalang C (1923) Hookworm campaign in Cebu Province, Philippines. Proc Pan Pacific Congr Vol 2:1431
483 Manalang C (1924) Ankylostomiasis, I. Laboratory diagnosis of hookworm infection. JPMA 4:421-425
484 Manalang C (1924) Ankylostomiasis, II. Carbon tetrachloride treatment of hookworm infection. JPMA 4:449-457
485 Manalang C (1924) Note on Ancylostoma braziliense as a human parasite in the Philippines. J Parasitol 11:90
486 Manalang C (1925) On the origin of a new species of Ankylostoma infesting man in the Philippines. Mthly Bull Philipp Hlth Serv 5:45-48
487 Manalang C (1925) Intestinal parasites with special reference to hookworm infestation and tetrachloride treatment. Mthly Bull Philipp Hlth Serv 5:49-55
488 Manalang C (1925) The isolation of hookworm larva from soil as an aid in sanitation work. Mthly Bull Philipp Hlth Serv 5:297-300
489 Manalang C (1925) Studies on ankylostomiasis in the Philippines. Far Eastern Assoc Trop Med Trans 6th Biennial Congr 1:351-368
490 Manalang C (1925) Hookworm infestation among Japanese immigrants and residents. Far Eastern Assoc Trop Med Trans 6th Biennial Congr 1:369-377
491 Manalang C (1925) A hookworm campaign in Cebu. PJS 27:483-493
492 Manalang C (1926) Ankylostomiasis, III. Hookworm counts and classification among hospital patients. JPMA 6:192-196
493 Manalang C (1926) Ankylostomiasis, IV Chenopodium treatment. J Philipp Is Med Assoc 6:226-229
494 Manalang C (1926) Ankylostomiasis, V. Thymol treatment. JPMA 6:228-229

495 Manalang C (1926) Ankylostomiasis: comparative efficiency of carbon tetrachloride, chenopodium and thymol in the treatment of hookworm infection. J Trop Med Hyg 29:101-103
496 Manalang C (1927) Critical tests of hookworm remedies on man. Am J Trop Med 7:57-60
497 Manalang C (1927) Ancylostomiasis: relation between number of ova per gram of formed stool and number of female worms harbored by the host. I and II. PJS 33:35-45, 47-65
498 Manalang C (1927) Effect of carbon tetrachloride, chenopodium, and thymol on the ova of expelled hookworms. PJS 32:507-511
499 Manalang C (1927) Observations on the development of Ascaris ova. PJS 33:249-255
500 Manalang C (1927) Comparative efficiency of carbon tetrachloride, chenopodium, and thymol. Mthly Bull Philipp Hlth Serv 7:721-722
501 Manalang C (1928) Trichuriasis: relation between the number of ova per gram of formed stool and the number of female worms harbored by the host. PJS 35:11-22
502 Manalang C (1928) Ascariasis: relation between the number of ova per gram of formed stool and the number of female worms harbored by the host. PJS 35:23-29
503 Mangmanee L, Aswapokee N, Vanasin B (1977) Intestinal capillariasis: report of the fourth case in Thailand. Siriraj Hosp Gaz 29:439-444
504 Manila Health Department (1952) Twelfth Annual Report of the Department of Health, City of Manila, for the period from July 1, 1951 to June 30, 1952, pp 135-136, 145-147
505 Manila Health Department (1953) Thirteenth Annual Report of the Department of Health, City of Manila, for the period from July 1, 1952 to June 30, 1953, pp 151-153
506 Manila Health Department (1954) Fourteenth Annual Report Fiscal Year 1953-1954. pp 128-129, 141-143
507 Manuel MF, Signey R (1966) Experimental infection of chicks with Sparganum Diesing, 1845. Philipp J Vet Med 5:93-99
508 Manuel MF, Cruz EC dela (1963) A study on the prevalence of Trichinella spiralis in native pigs in the Philippines. Philipp J Vet Med 2:235-241
509 Mao CP (1948) A review of the epidemiology of schistosomiasis japonica in China. Am J Trop Med 28:659-672
510 Mao CP, Li L, Wu CC (1949) Studies on the emergence of cercariae of Schistosoma japonicum from their Chinese snail host, Oncomelania hupensis. Am J Trop Med 24:937-944
511 Mao S-P, Shao B-R (1982) Schistosomiasis control in the People's Republic of China. Am J Trop Med Hyg 31:92-99
512 Marzan AM, Tantengco VO, Caviles AP, Villanueva L (1947) Nutritional anaemia in Filipino infants and preschoolers. SEA J 5:90-95
513 Maseras AA (1894) Estudio del cisticerco (Ladreria). La Correspondencia Medica de Filipinas 2:131
514 Mason PK, Daniels WB, Paddock FK, Gordon HH (1946) The latent phase of asiatic schistosomiasis. Arch Int Med 78:662-678
515 Mason PK, Paddock FK, Gordon HH, Daniels WB (1946) Schistosomiasis: diagnosis and treatment in American soldiers. N Engl J Med 235:179-182
516 Mateo DA (1967) Some observations on the development of Schistosoma japonicum in Oncomelania quadrasi. Leyte-Samar Studies 1:88-114
517 Mateo DA (1968) Observations on the developmental stages of Oncomelania quadrasi. Leyte-Samar Studies 2:156-176
518 Mateo DA, Dazo BC (1967) The morphology of Oncomelania quadrasi (Mollendorf), the intermediate host of Schistosoma japonicum in the Philippines. Leyte-Samar Studies 1:236-268

519 Matsusaki G (1963) Influence of low temperature, sunshine desiccation and excreta upon the eggs of Ancylostoma duodenale and Necator americanus. Part II. Sunshine. Yokohama Med Bull 14:123-126

520 Matsusaki G (1963) Influence of low temperature, sunshine, desiccation and human excreta upon the eggs of Ancylostoma duodenale and Necator americanus. Part III. Desiccation. Yokohama Med Bull 14:163-167

521 McCullough FS, Gayral PH, Duncan J, Christie JD (1980) Molluscicides in schistosomiasis control. Bull WHO 58:681-689

522 McDill JR (1905) Status of medical affairs in the Philippine Islands. J Am Med Assoc 45:503-507

523 McGreevy PB, Kolstrup N, Tao J, McGreevy MM, Marshall TF de C (1982) Ingestion and development of Wuchereria bancrofti in Culex quinquefasciatus, Anopheles gambiae and Aedes aegypti after feeding on humans with varying densities of microfilariae in Tanzania. Trans R Soc Trop Med Hyg 76:288-296

524 McMullen DB (1947) The control of schistosomiasis japonica. I. Observations on the habits, ecology and life cycle of Oncomelania quadrasi, the molluscan intermediate host of Schistosoma japonicum in the Philippine Islands. Am J Hyg 45:259-273

525 McMullen DB, Hubendick B, Pesigan TP, Bierstein P (1954) Observations made by the World Health Organization Schistosomiasis Team in the Philippines. JPMA 30:615-627

526 McMullin JJA (1917) A case of uncinariasis with severe anemia. US Naval Med Bull 11:380-383

527 Mead AR (1961) The giant African snail: a problem in economic malacology. Chicago

528 Medical Department (1901) Report of the Medical Department in the Philippines. Boston Med Surg J 144:120-122

529 Mehlhorn H, Piekarski G (1981) Grundriß der Parasitenkunde. Parasiten des Menschen und der Nutztiere. Stuttgart

529a Meier-Brook C (1983) Taxonomic studies on Gyraulus (Gastropoda, Planorbidae). Malacologia 24:1-113

530 Mejia-Talabis S (1963) Trichuriasis among Filipino children. Philipp J Ped 12:248-255

531 Mellnik SM (1981) Philippine war diary 1939-1945. Rev.ed., 346 pp. New York

532 Mendoza-Guazon MP (1916) A case of infestation with Dipylidium caninum. PJS B11:19-31

533 Mendoza-Guazon MP (1917) Study of the anatomicopathologic lesions in one thousand Filipino children under five years of age. PJS B12:51-84

534 Mendoza-Guazon MP (1922) Schistosomiasis in the Philippine Islands. PJS 21:535-567

535 Mendoza-Guazon MP (1922) The probable endemicity of Schistosomum japonicum in the Philippine Islands. JPMA 2:1-6

536 Mendoza-Guazon MP (1922) La probable endemicidad del Schistosomum japonicum en Filipinas. JPMA 2:31-36

537 Menges M (1981) Epidemiologie, Häufigkeit und geographische Verbreitung der Trematodiasen und Cestodiasen des Menschen in Indien. Diss Med Fak Univ Heidelberg

538 Mercado CM, Guioguio RV (1980) The role of the integrated project in community development and primary health care. Proc 7th Asian Parasite Contr Fam Plan Conf, pp 447-453

539 Micozzi MS, Ongchangco MN (1980) A case of pulmonary paragonimiasis in Mindanao, Philippines. SEA J 11:67-70

540 Millar CA, Putong PB (1951) Surveys in Surigao and Agusan. JPMA 27:259-261

541 Miller TA (1979) Hookworm infection in man. Adv Parasitol 17:315-384
542 Ministry of Health (1972-1978) quoted from B.D. Cabrera (148)
543 Mistica-Villanueva B, Belino ED, Hicarte VP (1966) The present status of trichinosis in the Philippines. Sto Tomas J Med 21:272-274
544 Miyazaki I (1978) Paragonimus filipinus sp.n. found in Leyte, the Republic of the Philippines (Trematoda: Troglotrematidae). Med Bull Fukuoka Univ 5:5-10
545 Miyazaki I (1978) Two types of the lung fluke which has been called Paragonimus westermani (Kerbert, 1878). Med Bull Fukuoka Univ 5:251-263
546 Miyazaki I, Habe S (1979) Paragonimus westermani filipinus Miyazaki, 1978, stat. n. occurring at Jaro, Leyte, the Philippines. Med Bull Fukuoka Univ 6:447-462
547 Miyazaki I, Kifune T, Habe S (1979) Lung flukes in the Philippines. Yonsei Rep Trop Med 10:83-84
548 Moloney NA, Webbe G (1983) The host-parasite relationship of Schistosoma japonicum in CBA mice. Parasitology 87:327-342
549 Monserrat C, Buenaventura A (1959) Pulmonary paragonimiasis in the Philippines. Report of a case discovered at autopsy with Paragonimum and Schistosoma lesions. PJS 88:363-371
550 Moose JW, Williams JE (1963) Infectivity of hybrid snails obtained from the reciprocal cross of Oncomelania nosophora and Oncomelania formosana to the Japanese strain of Schistosoma japonicum. J Parasitol 49:284
551 Moose JW, Williams JE (1963) Susceptibility of Oncomelania formosana from three different areas of Taiwan to infection with Formosan strain of Schistosoma japonicum. J Parasitol 49:702-703
552 Moose JW, Williams JE (1966) The susceptibility of geographical races of Oncomelania formosana to infection with human strains of Schistosoma japonicum. Proc 1st Int Congr Parasitol Vol II:700
553 Morishita K (1972) Studies on epidemiological aspects of ascariasis in Japan and basic knowledge concerning its control. Progr Med Parasitol Japan 4:1-153
554 Morishita K (1973) Studies on the biology of Ascaris in Japan; its morphology, ecology, biochemistry, development and infection. Progr Med Parasitol Japan 5:145-213
555 Morisita T (1973) Trichuris and trichuriasis. Progr Med Parasitol Japan 5:215-228
556 Muller R (1975) Worms and disease. A manual of medical helminthology. London
557 Mundo F del, Talabis SM (1963) A tetrachlorethylene-chenopodium preparation in trichuriasis. JPMA 39:710-715
558 Musgrave WE (1907) Paragonimiasis in the Philippine Islands. PJS B2:15-65
559 Musgrave WE (1914) Sanitary survey of the San Jose Estate and adjacent properties of Mindoro Island, Philippine Islands, with special reference to the epidemiology of malaria. 1. Introduction PJS B9:138-142
560 Musgrave WE, Vasquez A, Gutierrez B, Cox SL (1914) Sanitary survey of the San Jose Estate and adjacent properties of Mindoro Island, Philippine Islands, with special reference to the epidemiology of malaria. 6. The disease index. (B) Clinical examinations. PJS B9:173-181
561 Nanagas VT, Pascual AA, Canlas MS (1954) Fasciola hepatica in the common bile duct. Report of a case. JPMA 30:448-450
561a Nichols H (1909) Medical survey of the town of Taytay. XII. Filariasis, malaria, tuberculosis, typhoid fever, goitre, beriberi, venereal and skin diseases. PJS B4:279-286

562 Nishimura K, Yogore MG (1965) On the occurrence of the rat lungworm, Angiostrongylus cantonensis, in rats in the Philippines. J Parasitol 51:928

563 Nojima H, Santos AT, Blas BL, Kamiya H (1980) The emergence of Schistosoma japonicum cercariae from Oncomelania quadrasi in Leyte, Philippines. Proc Philipp-Japan Joint Conf Schistosomiasis Res Control, pp 77-79

564 Nojima H, Santos AT, Blas BL, Kamiya H (1980) The emergence of Schistosoma japonicum cercariae from Oncomelania quadrasi. J Parasitol 66:1010-1013

565 Nolasco JO, Gacoba C (1936) Helminthiasis in pupils of the Balala Elementary School. Mthly Bull Bur Hlth 16:277-280

566 Nolasco JO, Gacoba C (1937) Helminthiasis in children of preschool age in the non-leper district of Culion. Mthly Bull Bur Hlth 17:253-264

567 Nosenas JS (1980) Schistosomiasis control program: Health component national Irrigation Systems Improvement Project - I. Proc Philipp-Japan Joint Conf Schistosomiasis Res Control, pp 114-115

568 Nosenas JS, Matsuda H, Blas BL, Tanaka H, Santos AT (1975) Evaluation of the circumoval precipitin test using dried blood on filter paper as a diagnostic tool in epidemiological survey for schistosomiasis. Jpn J Exp Med 45:367-375

569 Nosenas JS, Santos AT, Blas BL (1980) Schistosomiasis control program in Leyte, the health component of the national irrigation systems improvement project - I. 10th Int Congr Med Mal Abstracts p 360

570 Nosenas JS, Tanaka H, Matsuda H , Santos AT (1975) Estimation of annual incidence of schistosomiasis japonica among school children in the Philippines. SEA J 6:359-365

571 Novilla MN, Flauta RF (1967) Observations on the anthelmintic value of thiabendazole against gastro-intestinal parasites in dogs. Philipp J Vet Med 6:135-144

572 Nwosu ABC (1978) Human neonatal infection with hookworm in an endemic area - a possible transmammary route. 4th Internat Congr Parasitol Short Comm C:10

573 Nwosu ABC (1981) Human neonatal infections with hookworm in an endemic area in southern Nigeria. A possible transmammary route. Trop Geogr Med 33:105-111

574 Ocampo-Paus R, Agor AB (1967) The incidence of Ascaris ova in vegetables commonly eaten raw in the Philippines. PJS 96:177-189

575 Odhner T (1911) Echinostomum ilocanum (Garrison), ein neuer Menschenparasit aus Ostasien. Zool Anz 38:65-68

576 Okabe K (1964) Biology and epidemiology of Schistosoma japonicum and schistosomasis. Progr Med Parasitol Japan 1:183-215

577 Olveda RM, Tiu E, Fevidal P, de Veyra F, Icatlo FC, Domingo EO (1983) Relationship of prevalence and intensity of infection to morbity in schistosomiasis japonica: a study of three communities in Leyte, Philippines. Am J Trop Med Hyg 32:1312-1321

578 Omori N (1966) On the role of Japanese mosquitoes, especially of Culex pipiens pallens in the transmission of Bancroftian filariasis. Progr Med Parasitol Japan 3:467-507

579 Ono Y (1974) (Epidemiology of schistosomiasis japonicum in Asia, and "Katayama-Ki" (1)) J Vet Med 620:793-797

580 Orange B (1966) A study on the raction of sera from imported and locally born swine to the Suessenguth Kline flocculation slide test. DVM Thesis, Coll Vet Med, UP, Diliman, Quezon City, 47 pp

581 Oshima S, Yasuraoka K, Irie Y, Blas BL, Nosenas JS, Santos AT (1978) Final localization of Schistosoma japonicum in the lungs of field rats, Rattus mindanensis in Leyte, Philippines. Jpn J Exp Med 48:503-509

582 Pabello PD (1965) A study of the prevalence of Trichinella spiralis in brown rats Rattus norvegicus found in Manila and suburbs. DVM Thesis Coll Vet Med, UP, Diliman, Rizal, 15 pp
583 Paguio AD (1959) Incidence of oxyuriasis in a district of Manila (Tondo) as determined with the use of the Scotch tape swab. Sto Tomas J Med 14:409-414
584 Paguio AD (1960) Cyst and ova of intestinal parasites found in vegetables usually eaten raw. Sto Tomas J Med 15:287-299
585 Paguio AD (1961) Cyst and ova of intestinal parasites found on paper money. Sto Tomas J Med 16:208-222
586 Pal R, Nair CP, Ramalingam S, Patil PV, Ram B (1960) On the bionomics of vectors of human filariasis in Ernakulam (Kerala), India. Indian J Malariol 14:595-604
587 Pambid ER (1967) The effect of sea water on the Enterobius ova. Inst Publ Hlth UP Manila, 11 pp
588 Pangilinan MV (1951) Survey in Panguil Bay area. JPMA 27:249-251
589 Pardo di Tavera M, Mirasol F (1956) Coexisting pulmonary tuberculosis and pulmonary paragonimiasis. Bull Quezon Inst 3:147-153
590 Paulino GB, Wittenberg J (1973) Intestinal capillariasis. A new cause of a malabsorption pattern. Am J Roentgenol Radium Ther Nucl Med 117:340-345
591 Pedroche (1936) pers.comn. to Pesigan et al (611)
592 Pesigan TP (1947) Results of a brief schistosomiasis survey around Lake Mainit, Mindanao. JPMA 23:23-32
593 Pesigan TP (1948) The endemicity of schistosomiasis japonica in Sorsogon, southeastern Luzon. JPMA 24:19-27
594 Pesigan TP (1948) Schistosomiasis reconnaissance in northwestern Mindanao. JPMA 24:495-505
595 Pesigan TP (1950) Studies on schistosomiasis: progress report on the Philippine campaign as of June 30, 1950. JPMA 26:339-348
596 Pesigan TP (1951) pers. comn. to Yogore (862)
597 Pesigan TP (1952) Studies on schistosomiasis: second progress report on mass surveys and other accomplishments as of December 31, 1951. JPMA 28:177-189
598 Pesigan TP (1953) The schistosomiasis problem in the Philippines: its public health and other aspects. Sto Tomas J Med 8:1-22
599 Pesigan TP (1953) Fifty years of progress in parasitology in the Philippines. Golden Book, Philipp Med Assoc, pp 177-183
600 Pesigan TP (1956) The schistosomiasis control pilot project. Philipp Hlth J 3:2-3
601 Pesigan TP (1959) Background information and guiding priciples in schistosomiasis control in the Philippines. Philipp J Publ Hlth 4:62-71
602 Pesigan TP (1961) The present status of the national schistosomiasis control program in the Philippines. Unpublished manuscript, 16 pp
603 Pesigan TP (1967) Screening of chemical compounds for molluscicidal activity against Oncomelania quadrasi. JPMA 43:861-870
604 Pesigan et al. (1948) Quoted from B.D. Cabrera (148)
605 Pesigan TP and staff (1951) Summary of all streams and swamps surveyed positive for Oncomelania quadrasi snails.JPMA 27:262-267
606 Pesigan TP, Farooq M, Hairston NG, Jauregui JJ, Garcia EG, Santos AT, Santos BC, Besa AA (1958) Studies on Schistosoma japonicum infection in the Philippines. I. General considerations and epidemiology. Bull WHO 18:345-455
607 Pesigan TP, Farooq M, Hairston NG, Jauregui JJ, Garcia EG, Santos AT, Santos BC, Besa AA (1958) Studies on Schistosoma japonicum infection in the Philippines. 3. Preliminary control experiments. Bull WHO 19:223-261

608 Pesigan TP, Hairston NG (1961) The effect of snail control on the prevalence of Schistosoma japonicum infection in the Philippines. Bull WHO 25:479-482
609 Pesigan TP, Hairston NG, Jauregui JJ, Garcia EG, Santos AT, Santos BC, Besa AA (1958) Studies on Schistosoma japonicum infection in the Philippines. 2. The molluscan host. Bull WHO 18:481-578
610 Pesigan TP, Pangilinan MV, Sarmiento AP (1949) Studies on schistosomiasis: further surveys in Mindanao. JPMA 25:417-433
611 Pesigan TP, Torres LF, Recio PM (1947) Paragonimiasis westermani: an unexpected case with cystic formation in the anterior abdominal wall. JPMA 23:293-298
612 Pesigan TP, Yogore MG (1965) Distribution and prevalence of infection caused by soil-transmitted helminths in the Philippines. JPMA 41:687-702
613 Pflüger W (1980) Experimental epidemiology of schistosomiasis. I. The prepatent period and cercarial production of Schistosoma mansoni in Biomphalaria snails at various constant temperatures. Z Parasitenkd 63:159-169
614 Pflüger W (1981) Experimental epidemiology of schistosomiasis. II. Prepatency of Schistosoma mansoni in Biomphalaria glabrata at diurnally fluctuating temperatures. Z. Parasitenkd 66:221-229
615 Pflüger W, Roushdy MZ, El Emam M (1984) The prepatent period and cercarial production of Schistosoma haematobium in Bulinus truncatus (Egyptian field strains) at different constant temperatures. Z Parasitenkd 70:95-103
616 Phalen JM, Nichols HJ (1908) Notes on the condition of the liver in schistosomiasis. PJS B3:223-229
617 Phalen JM, Nichols HJ (1908) Filariasis and elephantiasis in southern Luzon. PJS B3:293-304
618 Phalen JM, Nichols HJ (1908) Notes on the distribution of Filaria nocturna in the Philippine Islands. PJS B3:305-307
619 Phalen JM, Nichols HJ (1908) Tropical diseases in the Philippines. Milit Surg 23:462-468
620 Phalen JM, Nichols HJ (1909) The distribution of filaria in the Philippine Islands. PJS B4:127-140
621 Philippine Department of Health, Schistosomiasis Control Pilot Project. 2nd Quarterly Report FY 1967-1968
622 Philippine Health Service (1916) Report of the Philippine Health Service for the Fiscal Year from January 1 to December 31, 1915. Manila, pp 19-20, 31-32
623 Philippine Health Service (1917) Report for the Fiscal Year from January 1 to December 31, 1916. Manila, pp 20-21, 112, 119
624 Philippine Health Service (1919) Report for the Fiscal Year from January 1 to December 31, 1918. Manila, pp 14, 33-34
625 Philippine Health Service (1920) Report for the Fiscal Year from January 1 to December 31, 1919. Manila, pp 52-54, 408-410
625a Philippine Health Service (1926) Report for the Fiscal Year from January 1 to December 31, 1925. Manila, pp 101-102
626 Philipson WR (1975) Philippine crop occurrence in relation to soil and climatic factors. PhD Thesis Cornell Univ
627 Pichon G (1981) Migrations de microfilaires et de peuples océaniens. Une approche de la spécification chez les filaires de Bancroft et de Malaisie, par l'étude de la périodicité des microfilaires en fonction de la densité; contribution à la préhistoire du Pacifique. Ann Parasitol 56:107-120
628 Powell WA (1911) Report on Camp Wilhelm, Tayabas, P.I. Milit Surg 29:408-423
629 Pradatsundarasar AK, Pecharanond K, Chintanawongs C, Ungthavorn P (1973) The first case of intestinal capillariasis in Thailand. SEA J 4:131-134

630 Prakitrittranon W, Eua-ananta Y, Dhiensiri T, Bhaibulaya M (1980) Intestinal capillariasis from Maha Sarakham Province, Northeast Thailand: report of a case. SEA J 11:496-497
631 Progress Report. Integration of Family Planning & Parasite Control Pilot Project in the Philippines (1976) Proc 3rd Conf Asian Parasite Contr Org, pp 119-135
632 Progress Report of the Family Planning/Parasite Control/Nutrition Integration Pilot Project. Proc 6th Asian Parasite Contr Conf, pp 164-173
633 Purseglove JW (1975) Tropical crops. Monocotyledons. Second impression, in one volume. London
634 Quadras JF, Moellendorff OF von (1895) Diagnoses specierum novarum ex insulis Philippinis. Nachrichtsbl Dtsch Malakoz Ges 27:137-149
635 Quirolgico EO (1973) A comparative study of the methods of diagnosis of enterobiasis in private practice. Philipp J Ped 22:70-73
636 Rachford FW (1977) Oncomelania hupensis quadrasi from Mindoro (Victoria), Leyte (Palo), and Mindanao (Davao del Norte) of the Philippines: susceptibility to infection with Philippine isolates of Schistosoma japonicum. J Parasitol 63:1129-1130
637 Ramalingam S, Guptavanij P, Harinasuta C (1969) The vectors of Wuchereria bancrofti and Brugia malayi in Southeast Asia. In: Sandosham AA, Zaman V (eds): Proceedings of Seminar on Filariasis and Immunology of Parasitic Infections and Laboratory Meeting, pp 172-189
638 Ramas R (1951) Schistosomiasis research. An observation. JPMA 27:751-756
639 Ramos V (1928) Estudio de un caso de enfermedad de Katayama que se ha manifestado como una appendicitis cronica. Bull San Juan de Dios Hosp 2:127-136
640 Ratanachaiwongs S, Chintanawongs S, Piyachon C et al. (1976) Malabsorption due to intestinal capillariasis. Annual Conference of the Gastroenterology Association of Thailand. July 1976. Quoted from Mangmanee et al (503)
641 Recent Developments in the Integrated Project in the Philippines. (1980) Proc 7th Asian Parasite Contr Fam Plan Conf, pp 146-149
642 Reed AC (1915) Schistosomiasis japonica. Am J Trop Dis Prev Med 3:250-273
643 Refuerzo PG (1940) Some helminths of dogs and cats transmissable to man in the Philippines, with reference to their methods of transmission and prevention. Philipp J Anim Ind 7:493-524
644 Refuerzo PG, Albis FS (1947) Cysticercus bovis and Cysticercus cellulosae in animals with notes on human taeniasis in the Philippines. Philipp J Anim Ind 9:123-130
645 Refuerzo PG, Cabrera DJ (1941) Two species of Raillietina (Cestoidea: Davaineidae) from Filipino children. Philipp J Anim Ind 8:127-133
646 Refuerzo PG, Garcia EY (1936) The metazoan intestinal parasites of street dogs in Manila and environs. A preliminary report based on two hundred autopsies. Philipp J Publ Hlth 3:37-43
647 Refuerzo PG, Garcia EY (1937) Pygidiopsis marivillai, a new heterophyid trematode from the Philippines. PJS 64:359-363
648 Refuerzo PG, Garcia EY (1938) The crustacean intermediate hosts of Gnathostoma spinigerum in the Philippines and its pre- and intracrustacean development. Philipp J Anim Ind 5:351-362
649 Reimer WL (1983) Die Heringswürmer (Anisakis simplex und verwandte Arten). Merkblätter über angewandte Parasitenkunde und Schädlingsbekämpfung. Merkblatt Nr. 28, 16 pp. Beilage zu Angew Parasitol 24 (4), 1983
650 Rensch B (1932) Eine Hydrobiide von den Philippinen. PJS 49:551-552

651 Rensch B (1933) Bericht ueber "Eine neue Hydrobiide von den Philippinen." PJS 50:325
652 Reynes-Fortunato C (1973) The incidence of four endoparasitic worms among livestock slaughtered at the Cebu City abbattoir (Ext. Abstract-Mast Thesis) Philipp Scientist 10:113-117
653 Richards CS, Merritt JW (1967) Studies on Angiostrongylus cantonensis in molluscan intermediate hosts. J Parasitol 53:382-388
654 Riley WA (1919) A mouse oxyurid, Syphacia obvelata, as a parasite of man. J Parasitol 6:89-93
655 Rissler RS, Gomez L (1910) The prevalence of intestinal parasites in Rizal and Cavite Provinces and in Cagayan valley. PJS B5:267-276
656 Ritchie LS (1955) The biology and control of the amphibious snails that serve as intermediate hosts for Schistosoma japonicum. Am J Trop Med Hyg 4:426-441
657 Roa NO (1977) Progress report on the integrated family planning nutrition and parasite control project in Caramoan, Camarines Sur. Proc 2nd Consultative Meeting for Integration, pp 80-85
658 Roa NO (1978) Integrated program Caramoan, Camarines Sur, Philippines. Proc 5th Conf Asian Parasite Contr Org, pp 335-345
659 Roque et al quoted from BD Cabrera (148)
660 Rosario A del, Arenas JR (1960-1963) Experimental infection of Aedes aegypti, Aedes albopictus, and Culex fatigans with W. bancrofti and some notes on its development in the latter species. Bull Dept Hlth 32-35:2-6
661 Rozeboom LE, Cabrera BD (1956) Filariasis in the Philippine Islands. Am J Hyg 63:140-149
662 Rozeboom LE, Cabrera BD (1963) Filariasis in Mountain Province, Luzon, Philippine Islands. J Parasitol 49 (Suppl):47-48
663 Rozeboom LE, Cabrera BD (1963) Transmission of filariasis in the Philippine Islands by Anopheles minimus flavirostris Ludlow. Nature 200:915-916
664 Rozeboom LE, Cabrera BD (1964) Filariasis in Mountain Province, Luzon, Republic of the Philippines. J Med Ent 1:18-28
665 Rozeboom LE, Cabrera BD (1965) Filariasis in Mountain Province, Luzon. AMP 1:153-154
666 Rozeboom LE, Cabrera BD (1965) Filariasis caused by Brugia malayi in the Republic of the Philippines. Am J Epidemiol 81:200-215
667 Rozeboom LE, Cabrera BD (1965) Filariasis caused by Wuchereria bancrofti in Palawan, Republic of the Philippines. Am J Epidemiol 81:216-221
668 Rozeboom LE, Cabrera BD (1966) Filariasis in the Republic of the Philippines: epidemiology and possible origins. AMP 3:112-116
669 Ruff MD, Davis GM, Werner JK (1973) Schistosoma japonicum: disc electrophoretic patterns of the Japanese, Philippine and Formosan strains. Exp Parasitol 33:437-446
670 Russell PF (1934) Malaria and Culicidae in the Philippine Islands: history and critical bibliography, 1898 to 1933. Department of Agriculture and Commerce Technical Bulletin I:1-115. Manila
671 Russell PF, Baisas FE (1934) Habitats of Philippine Anopheles larvae. PJS 55:297-306
672 Salazar NP (1977) Commensal rodents and their parasites in Manila and suburbs. NRCP Res Bull 32:93-126
673 Salazar NP, Cabrera BD (1969) Angiostrongylus cantonensis in rodent and molluscan hosts in Manila and suburbs. AMP 6:20-25
674 Salazar NP, Sy FS, Elayda M (1977) Capillaria philippinensis: a case of intractable diarrhea. Unpublished Manuscript, 9 pp
675 Saleeby NM (1934) Health work of the Benguet Consolidated Mining Company. Philipp J Publ Hlth 1:52-64

676 Salita DC, Rosell D (1980) Economic geography of the Philippines. National Research Council of the Philippines, Manila
677 Sandground JH (1940) Plagiorchis javensis n.sp., a new trematode parasitic in man. Rev Med Trop Parasitol, Habana 6:207-211
678 Saniel VF (1951) Survey in Mindoro. JPMA 27:254-255
679 Sanpakit S, Suksungvol S, Bhaibulaya M (1974) Intestinal capillariasis from Saraburi Province, Thailand: report of the second case. J Med Assoc Thailand 57:458-460
680 Santillan JS, Gorospe AA (1968) Common diseases in a private college for girls. JPMA 44:300-304
681 Santos AT (1967) The schistosomiasis control program in the Philippines. JPMA 43:212-216
682 Santos AT (1967) The schistosomiasis control program in the Philippines. World Med J 14:23-24
683 Santos AT (1969) Schistosomiasis control in the Philippines: A review. Proc 4th Southeast Asian Sem Parasitol Trop Med, pp 1-4
684 Santos AT (1976) Prevalence and distribution of schistosomiasis in the Philippines: a review. SEA J 7:133-136
685 Santos AT (1976) Current chemotherapy of schistosomiasis japonica in the Philippines. SEA J 7:306
686 Santos AT (1976) Review of schistosomiasis control in the Philippines. SEA J 7:319-321
687 Santos AT, Blas BL, Eugenio LE, Portillo GD (1970) A preliminary report of early schistosomiasis japonica cases treated with sodium antimony dimethylcysteine tartrate (NaP). JPMA 46:254-258
688 Santos AT, Blas BL, Nosenas JS, Portillo GP (1970) Further trials with stibophen in the treatment of schistosomiasis japonica. JPMA 46:726-731
689 Santos AT, Blas BL, Nosenas JS, Portillo GP, Ortega OM (1980) Toxicity and efficacy on Embay 8440 (praziquantel) in early human Schistosoma japonicum infection in the Philippines. Proc Philipp-Japan Joint Conf Schistosomiasis Res Control, pp 83-84
690 Santos AT, Blas BL, Nosenas JS, Portillo GP, Ortega OM, Hayashi M, Boehme K (1979) Preliminary clinical trials with praziquantel in Schistosoma japonicum infections in the Philippines. Bull WHO 57:793-799
691 Santos AT, Blas BL, Portillo G, Nosenas J (1980) Toxicity and efficacy study of embay 8440 (praziquantel) in the treatment of Schistosoma japonicum infection in Leyte, Philippines. 10th Int Congr Trop Med Mal, Abstracts, pp 342-343
692 Santos AT, Blas BL, Portillo E, Nosenas JS, Poliquit O, Papasin M (1984) Phase III clinical trials with praziquantel in S. japonicum infections in the Philippines. Arzneim-Forsch 34(II), 9b:1221-1223
693 Santos AT, Blas BL, Portillo GP, Santos MJ (1969) A report of two schistosomiasis endemic municipalities in northwestern Leyte. JPMA 45:690-692
694 Santos AT, Blas BL, Redona SE, Santos MJ (1970) Schistosomiasis control in relation to improved agricultural practices. JPMA 46:732-738
695 Santos MJ, Yasuraoka K (1980) Molluscicidal effect of "gogo" bark against Oncomelania quadrasi in the laboratory. Proc Philipp-Japan Joint Conf Schistosomiasis Res Control, p 105
696 Sarmiento AG, Claridad CD (1975) Prevalence of enterobiasis among school children in Barrio San Guillermo, Morong, Rizal. Inst Publ Hlth, UP, Manila, 8 pp
697 Sarte LG (1977) Philippines' report. Proc 2nd Consultative Meeting for Integration, pp 75-79
698 Sasa M (1976) Human filariasis. A global survey of epidemiology and control. Baltimore London Tokyo, 819 pp

699 Sasa M, Tanaka H (1972) Studies on the methods for statistical analysis of the microfilarial periodicity survey data. SEA J 3:518-536
700 Sasa M, Tanaka H (1974) A statistical method for comparison and classification of the microfilarial periodicity. Jpn J Exp Med 44:321-346
701 Schad GA, Nawalinski TA, Kochar V (1983) Human ecology and the distribution and abundance of hookworm populations. In: Croll NA, Cross JH (ed) Human ecology and infectious diseases, pp 187-223
702 Schistosomiasis campaign (1941) Schistosomiasis campaign in Surigao, Leyte and Mindoro. Mthly Bull Bur Hlth 21:159-168
703 Schistosomiasis Control Council (1978) 1977 Annual Report, 24 pp
704 Schistosomiasis Control Council (1980) Annual Report 1978, 66 pp
705 Schistosomiasis Control Council (1980) Annual Report 1979, 157 pp
706 Schistosomiasis Control Council (1980) Integrated national program for the control of schistosomiasis in the Philippines (5-year-program), 34 pp + Append. A-N
707 Schneider A (1866) Monographie der Nematoden. Berlin
708 Schoening E (1973) Ecology, bionomics and control of Aedes (Finlaya) poicilius Theobald (Diptera: Culicidae). Philipp Scientist 10:35-51
709 Schwartz B (1922) Some cestodes from domestic animals in the Philippine Islands that are of economic and hygienic importance. Philipp Agric 11:113-116
710 Schwartz B (1924) Occurrence of Ancylostoma braziliense in Phillippine cats. J Parasitol 11:93
711 Schwartz B (1924) The occurrence of Fasciolopsis buski in the Philippine Islands. J Parasitol 11:93
712 Schwartz B (1925) Intestinal metazoan parasites collected from ruminants in the Philippine Islands. PJS 26:521-533
713 Schwartz B, Tubangui MA (1922) The prevalence of hookworm and other intestinal nematodes in adult Filipinos. J Parasitol 9:83-92
714 Schwartz B, Tubangui MA (1922) Uncommon intestinal parasites of man in the Philippine Islands. PJS 20:611-618
715 Sengbusch HG (1963) Enterobiasis in school children in the vicinity of Iloilo City, Republic of the Philippines. JPMA 39:802-812
716 Sengbusch HG, Sengbusch CH (1971) Studies on enterobiasis. II. The incidence of pinworm infection in a secondary school in Iloilo City, R.P. JPMA 47:455-462
717 Seo BS (1982) Report of the fourth APCO parasitologists' meeting. Proc 9th Asian Parasite Contr Fam Plan Conf, pp 43-58
718 Shattuck EC (1907) Clinical observations on uncinariasis. Am Med 13:683-686
719 Shibuya T, Cabrera BD, Tanaka H, Valeza FS, Instrella R (1980) Comparison of the blood film, Millipore filter and Nuclepore filter techniques for the detection of microfilaremia in a field survey in the Philippines. Jpn J Exp Med 50:463-468
720 Shibuya T, Tanaka H, Cabrera BD, Valesa FS, Instrella R, Go T, Kobayashi S (1981) Low density microfilaremia of Wuchereria bancrofti infections in pre- and post-treatment phases in the Philippines. Jpn J Exp Med 51:133-135
721 Shibuya T, Tanaka M, Valeza FS, Cabrera BD, Suguri S (1980) Detection of low density microfilaremia by different filtration methods in Sorsogon, Philippines. 10th Int Congr Trop Med Mal Abstracts, pp 106-107
722 Shimomura H, Santos MJ, Sakata Y, Blas BL, Tanaka H, Yasuraoka K (1979) (Dispersal of Oncomelania quadrasi, the intermediate host of Schistosoma japonicum, along a stream in Leyte, Philippines). Jpn J Parasitol 28:205-209
723 Simmons JS (1930) The US Army Medical Department Research Board. Milit Surg 67:479-489

724 Singson CN (1969) Human intestinal capillariasis. Philipp J Int Med 7:189-200

725 Sison ABM, Florentin AA (1951) Human gnathostomiasis: clinical study and report of a case. AMP 7:231-238

726 Sison AG (1913) pers. comn. to Willets (840)

727 Sison AG, Campos PC, Apostol R (1951) Eosinophilia in the spinal fluid. A report of four cases. AMP 8:95-107

728 Smith JW (1914) Some clinical observations on filariasis in Bilibid Prison. Read before the Manila Med Soc, April 5, 1914. Quoted from 821

729 Sosa RJ (1977) The effect of pH on the survival of hookworm filariform larva. Inst Publ Hlth, UP, Manila, 11 pp

730 Status Report of the Family Planning-Parasite Control-Nutrition Integration Pilot Project (Philippines) (1978). Proc 5th Conf Asian Parasite Contr Org, pp 144-163

731 Steele JH, Arambulo PV, Beran GW (1973) The epidemiology of zoonoses in the Philippines. Arch Environm Hlth 26:330-339

732 Steering Committee for the Integration of the Philippines (1977) Progress report of family planning-parasite control-nutrition integration pilot project Philippines. Proc 4th Conf Asian Parasite Contr Org, pp 126-143

733 Stiles CW (1905) The new Asiatic blood fluke (Schistosoma japonicum, Katsurada, 1904) in the Philippines. Am Med 10:854

734 Stitt ER (1911) A study of the intestinal parasites found in Cavite province. PJS B6:211-214

735 St.John JH, Simmons JS, Gardner LL (1929) Infestation of the lung by a nematode of the genus Cyathostoma. J Am Med Assoc 92:1816-1818

736 St.John JH, Simmons JS, Reynolds FHK (1931) A survey of mongrel dogs of Manila, Philippine Islands, for helminthic and blood protozoan parasites. PJS 44:309-311

737 Stransky E, Fagela CB, Tamayo CY (1962) On the incidence of enterobiasis among children admitted to the Philippine General Hospital. AMP 19:95-105

738 Stransky E, Felarca AB (1960) The first case of visceral larva migrans observed and diagnosed in the Philippines. JPMA 36:98-102

739 Stransky E, Joven IB (1964) On eosinophilic meningoencephalitis in the Philippines. Philipp J Ped 13:11-16

740 Stransky E, Joven IB (1964) On eosinophilic meningoencephilatis in the Philippines. MD J 13:14-19

741 Stransky E, Lorenzo AS (1960) On raillietiniasis in the Philippines. Acta Trop 17:80-83

742 Strong RP (1901) Animal parasites. Report of the Surgeon General, circulars on tropical diseases, chief surgeon's office, Manila, P.I., No.1:7-45

743 Strong RP (1909) Medical survey of the town of Taytay. I. Introduction. PJS B4:207-210

744 Strong RP (1909) Medical survey of the town of Taytay. XV. Summary and conclusions. PJS B4:289-301

745 Sullivan RR, Ferguson MS (1946) Studies on schistosomiasis japonica. III. An epidemiological study of schistosomiasis japonica. Am J Hyg 44:324-347

746 Sun S-C, Cross JH, Berg HS, Kau S-L, Singson C, Banzon T, Watten RH (1974) Ultrastructural studies of intestinal capillariasis. Capillaria philippinensis in human and gerbil hosts. SEA J 5:524-533

747 Sy FS, Ilagan NB (1978) Nutritional status of children with schistosomiasis japonica in a rural setting. JPMA 54:245-253

748 Tacal JV, Pabello PD (1965) Absence of Trichinella spiralis in brown rats (Rattus norvegicus) in the Philippines. J Parasitol 51:957

749 Tanaka H (1966) Genus Strongyloides. Progr Med Parasitol Japan 3:587-638
750 Tanaka H (1981) Periodicity of microfilariae of human filariasis analysed by a trigonometric method (Aikat and Das) Jpn J Exp Med 51:97-103
751 Tanaka H, Blas BL, Nosenas JS, Matsuda H, Ishige M, Kamiya H, Murata I, Hayashi Y, Santos AT (1983) Longitudinal study on Schistosoma japonicum infections in the Philippines. 3. Incidence and environmental modification at Dagami, Leyte. Jpn J Exp Med 53:87-94
752 Tanaka H, Nakai H, Omoto K, Shibuya T, Hirai M, Mercado AS, Pagaran IG, Eviota WS (1980) The high prevalence of Wuchereria bancrofti infections in indigenous tribes in northern Mindanao, Philippines. Jpn J Exp Med 50:85-89
753 Tanaka H, Santos AT (1976) Population research on Oncomelania quadrasi in Leyte, Philippines. SEA J 7:244-246
754 Tanaka H, Santos MJ, Banez EA, Pascua LM, Matsuda H, Santos AT (1978) Type of distribution, transformation of sampling data and population estimation of Oncomelania quadrasi in the Philippines. In: Research in Filariasis and Schistosomiasis 3:77-96
755 Tanaka H, Santos M, Matsuda H (1974) Snail sampling by the filter paper method and the distribution of Oncomelania quadrasi. Joint Scientific Meetings of the Philipp Soc Parasitol and SEAMEO TROPMED; Manila, Philipp., Proc p 26
756 Tanaka H, Santos MJ, Matsuda H, Hambre RS, Iwanaga Y, Shimomura H, Blas BL, Santos AT (1978) Distribution of Oncomelania quadrasi in waters in the Philippines. Jpn J Exp Med 48:193-202
757 Tanaka H, Santos MJ, Matsuda H, Santos AT (1976) Migration of Oncomelania quadrasi observed by the mark and release method in a wet bank of a stream in Leyte, Philippines. Jpn J Exp Med 46:31-35
758 Tanaka H, Santos MJ, Matsuda H, Yasuraoka K, Santos AT (1975) A quantitative sampling method for Oncomelania quadrasi by filter paper. Jpn J Exp Med 45:255-262
759 Tanaka H, Santos MJ, Matsuda H, Yasuraoka K, Santos AT (1980) Evaluation of quantitative sampling for Oncomelania quadrasi by filter paper. Proc Philipp-Japan Joint Conf Schistosomiasis Res Control, pp 94-96
760 Tantengco VO, Marzan AM, Castro CR de (1972) Helminthic infections and blood parameter studies in Filipino school children. SEA J 3:580-586
761 Tantengco VO, Marzan AM, Rapanot N, Villanueva L, Castro CR de (1973) Nutritional anaemia in Filipino school children. SEA J 4:524-533
762 Technical Guide for Schistosomiasis Control in the Philippines (1976). Schistosomiasis Control and Research Project, Department of Health, Philippines. Japan International Cooperation Agency (JICA)
763 Technical Guide for Schistosomiasis Control in the Philippines (1980). 2nd ed., Schistosomiasis Control and Research Project, Ministry of Health, Philippines. Japan International Cooperation Agency
764 Tenney ES (1912) Blood and stool examinations in a company of Philippine Scouts. Bost Med Surg J 167:4-5
765 Tenney ES (1913/14) Some observations on the prevalence of intestinal parasites in the Philippine Islands. Am J Trop Dis Prev Med 6:44-48
766 Tesana S, Bhuripanyo K, Sanpitak P, Sithithaworn P (1983) Intestinal capillariasis from Udon Thani Province, northeastern part of Thailand: report of an autopsy case. J Med Assoc Thailand 66:128-131
767 Tharavanij S (1982) Pathogenesis of diarrhoea caused by parasites. SEA J 13:331-338

768 Thomas JB (1905) Notes on diseases encountered in Baguio, Benguet, P.I., and the adjacent highlands of Central Luzon, including revised excerpts from the advance sheets of a report to the secretary of the interior. Am Med 10:28-32
769 Tiglao TV (1982) Health knowledge, attitudes and practices related to schistosomiasis in Leyte. Tropical Medicine 24:103-114
770 Tiglao TV, Camacho AC (1983) Water contact behaviour among humans in Leyte, Philippines. SEA J 14:18-24
771 Tillman AJB, Philipps HS (1948) Pulmonary paragonimiasis. Am J Med 5:167-187
772 Tirona JP (1927) Intestinal parasites in the Philippines (Preliminary report). Rev Fil Med Farm 18:217-220
773 Tombo JM (1962) On the role of excreta and sewage disposal on the prevalence of selected intestinal parasites in two communities. College of Medicine, UP, Manila, 20 pp
774 Tongson MS, Leon EB de (1970) The incidence of Toxocara canis and other gastro-intestinal parasites among dogs in Manila and suburbs. Philipp J Vet Med 8:1-12
775 Tongson MS, San Pablo FG (1979) A study of the prevalence of gastro-intestinal worms of cats in Metropolitan Manila. Philipp J Vet Med 18:1-15
776 Torres LF (1936) A case of Schistosoma japonicum infestation in a boy of ten. Bull San Juan de Dios Hosp 10:86-89
777 Tubangui MA (1924) Two larval parasites from the Philippine palm civet (Paradoxurus philippinensis) PJS 24:749-753
778 Tubangui MA (1925) Metazoan parasites of Philippine domesticated animals. PJS 28:11-37
779 Tubangui MA (1931) Worm parasites of the brown rat (Mus norvegicus) in the Philippine Islands, with special reference to those forms that may be transmitted to human beings. PJS 46:537-591
780 Tubangui MA (1931) Trematode parasites of Philippine vertebrates, II: two echinostome flukes from rats. PJS 44:273-284
781 Tubangui MA (1932) The molluscan intermediate host in the Philippines of the oriental blood fluke Schistosoma japonicum Katsurada. PJS 49:295-304
782 Tabangui MA (1938) Pseudophyllidean cestodes occurring in the Philippines. Livro Jubilar do Professor Lauro Travassos. Editado para commemorar o 25° anniversario de suas actividades scientificas (1913-1938). Rio de Janeiro, 1938, pp 489-494
783 Tubangui MA (1939) pers.comn. to Pesigan et al. (611)
784 Tubangui MA (1946) Preliminary notes on the crustacean vector of the mammalian lung fluke (Paragonimus) in the Philippines. J Parasitol 32:150-151. Correction: J Parasitol 33:284
785 Tubangui MA (1947) A summary of the parasitic worms reported from the Philippines. PJS 76:225-304
786 Tubangui MA (1948) Schistosomiasis japonica and other helminthic diseases in the Philippines. 4th Int Congr Trop Med Mal, Vol 2: 1034-1042
787 Tubangui MA, Africa CM (1938) The systematic position of some trematodes reported from the Philippines. PJS 67:117-127
788 Tubangui MA, Basaca M, Pasco AM (1934) Human infestations with Ascaris and Trichuris in different parts of the Philippine Islands. PJS 55:91-112
789 Tubangui MA, Basaca M, Pasco AM, Rosario F del (1935) Observations on the geographical distribution of hookworm parasites and hookworm disease in the Philippines. PJS 58:447-469
790 Tubangui M, Cabrera BD (1948) Studies on filariasis in the Philippines. I. Results of a survey in the province of Sorsogon and in the New Bilid Prison at Muntinlupa, Rizal. AMP 5:50-56
791 Tubangui MA, Cabrera BD, Yogore MG (1950) Studies on the life cycle of the human lung fluke (Paragonimus) in the Philippines: a preliminary report. AMP 6:371-372

792 Tubangui MA, Francisco SA (1925) Report on the hookworm campaign carried among the students of the Los Banos Colleges during the collegiate year 1924-1925. Annual Report, Dean, College of Agriculture, University of the Philippines for the year 1924-1925, pp 114-118. Quoted from Arambulo (46)
793 Tubangui MA, Pasco AM (1933) The life history of the human intestinal fluke, Euparyphium ilocanum (Garrison, 1908). PJS 51:581-606
794 Tubangui MA, Pasco AM (1940) Studies on the geographical distribution, incidence and control of schistosomiasis japonica in the Philippines. Nat Res Council Philipp Bull 25:83-84
795 Tubangui MA, Pasco AM (1941) Studies on the geographical distribution, incidence and control of schistosomiasis japonica in the Philippines. PJS 74:301-329
796 Tudor RC, Blair E (1971) Gnathostoma spinigerum: an unusual cause of ocular nematodiasis in the Western hemisphere. Am J Ophthalmol 72:185-190
797 Turner LH, Edeson JFB (1957) Studies on filariasis in Malaya: the periodicity of the microfilariae of Wuchereria malayi. Ann Trop Med Parasitol 51:271-277
798 Tuttle AD (1910) The Gastrodiscus hominis in the Philippines. Report of a case, with introductory helminthological comment. Milit Surg 26:673-677
799 Urbino CM (1936) The density of "Anopheles minimus" var. "flavirostris," Ludl. by trapping with human baits and its relation with the factors that influenced its variations. Mthly Bull Bur Hlth 16:471-481
800 Uylangco CV, Reyes VM, Calilong VP, Camacho RC (1968) Capillariasis. JPMA 44:743-757
801 Valeza FS (1977) Progress report on the integrated family planning, nutrition and parasite control project in San Pedro, Laguna. Proc 2nd Consultative Meeting for Integration, pp 86-102. Japanese Organization for International Cooperation in Family Planning
802 Valeza FS (1977) Report on the activities and accomplishments of the integrated family planning, nutrition and parasite control project of the Department of Health. Proc 2nd Consultative Meeting for Integration, pp 103-108. Japanese Organization for International Cooperation in Family Planning
803 Valeza FS, Grove DI (1979) Bancroftian filariasis in a Philippine village: entomological findings. SEA J 10:51-61
804 Vazquez-Colet A, Africa CM (1938) Determination of the piscine intermediate hosts of Philippine heterophyid trematodes by feeding experiments. PJS 65:293-302
805 Vazquez-Colet A, Africa CM (1939) Determination of the piscine intermediate hosts of Philippine heterophyid trematodes by feeding experiments. Progress report. PJS 70:201-215
806 Vazquez-Colet A, Africa CM (1940) Morphological studies on various Philippine heterophyid metacercariae with notes on the incidence, site, and degree of metacercarial infection in three species of marine fish. PJS 72:395-419
807 Vedder EB (1906) An examination of the stools of 100 healthy individuals, with especial reference to the presence of Entamoebae coli. J Am Med Assoc 40:870-872
808 Velasquez CC (1954) Studies on resistance and dissemination of Ascaris ova for the control of Ascaris infection in rural districts. Nat Appl Sci Bull 14:159-244
809 Velasquez CC (1968) Parasitism and foods. J Philipp Pharm Assoc 54:35-46
810 Velasquez CC (1968) List of Philippine marine fishes harboring Anisakinae (Heterocheilidae: Nematoda). Paper presented in a

seminar sponsered by the University of the Philippines Zoological Society, Quezon City 1968. Quoted from Arambulo (46)
811 Velasquez CC (1973) Observations on some Heterophyidae (Trematoda: Digenea) encysted in Philippine fishes. J Parasitol 59:77-84
812 Velasquez CC (1973) Intramolluscan stages of Haplorchis taichui (Nishigori) in Melania juncea Lea in the Philippines. J Parasitol 59:281
813 Velasquez CC (1973) Life cycle of Procerovum calderoni (Africa and Garcia, 1953) Price, 1940 (Trematoda: Digenea: Heterophyidae). J Parasitol 59:813-816
814 Velasquez CC (1974) Ecological aspects of fish parasitology in the Philippines. Professorial Chair Lecture Series, Monograph 3:14-32
815 Velasquez CC (1978) Some parasitic helminths as environmental hazards in the Philippines. Professorial Chair Lectures Monograph No 53, 23 pp
816 Velasquez CC, Cabrera BC (1968) Ancylostoma ceylanicum (Looss, 1911) in a Filipino woman. J Parasitol 54:430-431
817 Velasquez CC, Chanco P (1969) An abnormal Taenia saginata (Goeze, 1782) from the Philippines. J Parasitol 55:199
818 Veyra EA de (1967) An unusual case of parasitic protein-losing enteropathy. Intern Med 5:311-319
819 Vitug W, Cruz JR (1941) A case of endemic hemoptysis (pulmonary paragonimiasis). With parasitological discussion by CM Africa and radiological discussion by PS Chikiamco. Bull Quezon Inst 1:389-398
820 Walker D (1953) A selected bibliography on malaria in the Philippines. Bulletin of the Institute of Science and Technology. Bull No.1, 41 pp, Manila
821 Walker EL (1914) The morphology of the adults of the filaria found in the Philippine Islands. PJS B9:483-491
822 Walker EL, Guzman AM, Concepcion I (1914) Sanitary survey of the San Jose Estate and adjacent properties of Mindoro Island, Philippine Islands, with especial reference to the epidemiology of malaria. 6. The disease index. (A) Laboratory examinations. PJS B9:167-173
823 Wang YC, Fan PC (1972) Susceptibility of the Ilan race of Oncomelania hupensis formosana to the anthropophilic Philippine strain of Schistosoma japonicum. J Parasitol 58:1204
824 Warren KS, Berry EG (1972) Induction of hepatosplenic disease by single pairs of the Philippine, Formosan, Japanese, and Chinese strains of Schistosoma japonicum. J Infect Dis 126:482-491
825 Watten RH (1971) Clinical studies on intestinal capillariasis. In: Cross JH (ed): Proc 7th Southeast Asian Regional Seminar on Tropical Medicine and Public Health, p 261
826 Watten RH, Beckner WM, Cross JH, Gunning J-J, Jarimillo J (1972) Clinical studies of capillariasis philippinensis. Trans R Soc Trop Med Hyg 66:828-834
827 Weather Bureau (1973) Normals of climat and rain stations of the Philippines as of 1970. Prepared by: Climatic Statistics Section, Climatological Division, Weather Bureau, Philippines.
828 Wehr EE (1930) Helminth parasites of mammals in the Philippines. PJS 41:261-272
829 Weinberg HB, Tillinghast AJ (1946) The pulmonary manifestations of schistosomiasis caused by Schistosoma japonicum. Am J Trop Med 26:801-809
830 Wenceslao JM, Oban E, Cabrera BD (1972) Eastern Samar, the fourth endemic focus for Malayan filariasis in the Republic of the Philippines. SEA J 3:552-561
831 Wernstedt FL, Spencer JE (1967) The Philippine Island World. A physical, cultural and regional geography. Berkeley and Los Angeles
832 Westerlund NC, Chamberlain M (1969) Further observations on Angiostrongylus cantonensis in the Philippines. AMP 6:3-11

833 Whalen GE, Rosenberg EB, Strickland GT, Gutman RA, Cross JH, Watten RH, Uylangco C, Dizon JJ (1969) Intestinal capillariasis. A new disease in man. Lancet 1:13-16
834 Wharton LD (1917) The intestinal worms of dogs in the Philippine Islands. J Parasitol 4:80-82
835 Wharton LD (1921) Opisthorchis wardi, a new species of liver fluke from the cat in the Philippine Islands. PJS 19:243-245
836 Wharton RH (1962) The biology of Mansonia mosquitoes in relation to the transmission of filariasis in Malayia. Bull No. 11, Inst Med Res Malaya. Quoted from Ramalingam et al (637)
837 Whitmore ER (1908) A case of infection with Schistosoma japonicum in a Filipino. Arch Int Med 1:430-432
838 WHO Workshop (1980) Quatitative aspects of the epidemiology of Schistosoma japonicum infection in a rural community of Luzon, Philippines. Bull WHO 58:629-638
839 Willets DG (1911) A statistical study of intestinal parasites in tobacco haciendas of the Cagayan Valley, Philippine Islands. PJS B6:77-92
840 Willets DG (1913) General conditions affecting the public health and diseases prevalent in the Batanes Islands, P.I. PJS B8:49-57
841 Willets DG (1914) Intestinal parasitism, particularly entamoebiasis, in patients of the Philippine General Hospital, Manila, P.I. PJS B9:81-92
842 Willets DG (1914) Intestinal helminthiasis in the Philippine Islands as indicated by examinations of prisoners upon admission to Bilibid Prison, Manila, P.I. PJS B9:233-240
843 Winkenwerder WL, Hunninen AV, Harrison T, Billings FT, Carroll DG, Maier J (1946) Studies on schistosomiasis japonica. 2. Analysis of 364 cases of acute schistosomiasis with report of results of treatment with Fuadin in 184 cases. Bull Johns Hopkins Hosp 79:406-435
844 Woolley PG (1906) The occurrence of Schistosoma japonicum vel cattoi in the Philippine Islands. PJS 1:83-90
845 Worcester DC (1930) The Philippines past and present. New York
846 Wright WH (1950) Bilharziasis as a public-health problem in the Pacific. Bull WHO 2:581-595
847 Wright WH (1973) Geographical distribution of schistosomes and their intermediate hosts. In: Ansari N(ed): Epidemiology and control of schistosomiasis (bilharziasis), pp 32-249. Basel etc.
848 Wright WH, McMullen DB, Faust EC, Bauman PM (1946) The distribution of schistosomiasis japonica in Mindanao, Philippine Islands. J Parasitol 32(Suppl):12
849 Wright WH, McMullen DB, Faust EC, Bauman PM (1947) The epidemiology of schistosomiasis japonica in the Philippine Islands and Japan. II. Surveys for schistosomiasis japonica on Mindanao, Philippine Islands. Am J Hyg 45:164-184
850 Wyckoff HA, French WO (1925) Intestinal parasites among Filipino food handlers. J Am Med Assoc 84:1029-1030
851 Yagi S (1982) Overview of the annual project activities in seven countries. Proc 9th Asian Parasite Contr Fam Plan Conf, pp 62-70
852 Yanagisawa R (1966) The epidemiology of hookworm disease. Progr Med Parasitol Japan 3:283-366
853 Yasuraoka K (1969) Snail vectors in schistosomiasis japonica. In: Harinasuta C (ed): Proc 4th Southeast Asian Sem Parasitol Trop Med, pp 83-90
854 Yasuraoka K (1979) Schistosomiasis and water resource development. SEA J 10:630-633
855 Yasuraoka K, Hashiguchi J, Banez EA (1980) Laboratory assessment of the mulluscicidal activity of the plant Croton tiglium against Oncomelania snails. Proc Philipp-Japan Joint Conf Schistosomiasis Res Control, pp 106-108

856 Yasuraoka K, Hashiguchi J, Blas BL (1980) Laboratory assessment of the molluscicidal activity of the plant Jatropha curcas against Oncomelania snail. Proc Philipp-Japan Joint Conf Schistosomiasis Res Control, pp 110-112

857 Yasuraoka K, Hashiguchi J, Kawanaka M, Ishigi M, Santos MJ, Banez EA, Blas BL, Santos AT (1980) Molluscicides of plant origin in the Philippines. 10th Int Congr Trop Med Mal Abstracts, pp 362-363

858 Yasuraoka K, Irie Y, Takamura K, Shimomura H, Hashiguchi J, Santos MJ, Santos AT (1977) Laboratory and field assessment of the molluscicidal activity of gogo (Entada phaseoloides) against the amphibious snail intermediate host of Schistosoma japonicum. Jpn J Exp Med 47:483-487

859 Yasuraoka K, Santos AT, Santos MJ, Takamura K, Hosaka Y, Ito Y, Shimomura H (1976) Laboratory trials of two new molluscicides, the bark of "gogo" (Entada phasholoides) and polynactin, against Oncomelania snails: a preliminary report. SEA J 7:346-347

859a Yii C-Y, Chen C-Y, Fresh JW, Chen T, Cross JH (1968) Human angiostrongyliasis involving the lungs. Chin J Microbiol 1:148. Quoted from 93

860 Yogore MG (1955) Studies on paragonimiasis: seventy-four cases of paragonimiasis encountered in the Tacloban Tubercolosis Pavillon, 1949-1955 (unpublished)

861 Yogore MG (1957) Studies in paragonimiasis, I. The molluscan and crustacean hosts of Paragonimus in the Philippines. PJS 86:37-45

862 Yogore MG (1958) Studies on paragonimiasis, V. A survey in Jaro, Leyte, Philippines. PJS 87:217-253

863 Yogore MG, Cabrera BD (1957) A report on intestinal parasitism. Appendix 2, Nutrition Survey of the Armed Forces by WJ Darby et al, pp 116-126

864 Yogore MG, Cabrera BD, Noble GA (1958) Studies on paragonimiasis, IV. A survey in Casiguran, Sorsogon, Philippines. PJS 87:197-215

865 Yogore MG, Juliano SJ (1951) Report of a case of human gnathostomiasis. AMP 7:239-247

866 Yogore MG, Lewert RM, Blas BL (1978) Schistosomiasis japonica in Barrio San Antonio, Basey, Samar in the Philippines: II. Quantitative fecal examination and circumoval precipitin test. SEA J 9:344-355

867 Yogore MG, Lewert RM, Blas BL (1979) Schistosomiasis japonica in Barrio San Antonio, Basey Samar, in the Philippines: III. The plasma circumoval precipitin test-procedure and use in epidemiologic studies. SEA J 10:23-31

868 Yogore MG, Lewert RM, Blas BL (1980) Schistosomiasis japonica in Barrio San Antonio, Basey, Samar: Studies with the circumoval precipitin (COP) test. Proc Philipp-Japan Joint Conf Schistosomiasis Res Control, pp 49-50

869 Yogore MG, Lewert RM, Blas BL (1980) Schistosomiasis japonica in Barrio San Antonio, Basey, Samar in the Philippines vs ELISA compared with quantitative stool examination and the COP test. 10th Int Congr Trop Med Mal Abstracts, pp 312-313

870 Yogore MG, Lewert RM, Blas BL (1981) Schistosomiasis in Barrio San Antonio, Basey, Samar, in the Philippines. V. The enzyme-linked immunsorbent assay (ELISA) compared with quantitative stool examination and the circumoval precipitin (COP) test. Am J Trop Med Hyg 30:1252-1262

871 Yogore MG, Lewert RM, Blas BL (1983) Seroepidemiology of schistosomiasis japonica by ELISA in the Philippines. I. Underestimation by stool examination of the prevalence of infection in school children. Am J Trop Med Hyg 32:1322-1334

872 Yogore MG, Lewert RM, Madraso ED (1965) Immunodiffusion studies on paragonimiasis. Am J Trop Med Hyg 14:586-591

873 Yogore MG, Noble GA, Cabrera BD (1957) Studies on paragonimiasis, II. The morphology of some of the larval stages of Paragonimus in the Philippines. PJS 86:47-69
874 Yogore MG, Schanfield MS (1981) Immunoglobin allotypes in Filipinos. I. Lack of association between Schistosoma japonica infection and immunoglobulin allotypes. J Immunogenetics 8:419-423
875 Yogore MG, Tangco AF (1953) Notes on sparganosis in a Filipino. AMP 9:301-303
876 Yokogawa M, Cabrera BD, Kobayashi M, Hata H, Tazaki T, Araki K (1979) Comparative studies on Philippine Paragonimus westermani and Japanese Paragonimus westermani. AMP 15:119-131
877 Yokogawa M, Kobayashi M, Niimura M, Hata H, Tazaki T, Araki K (1979) (Immuno-serological studies on paragonimiasis in the Philippines). Jpn J Parasitol 28(1,Suppl):10
878 Yokogawa M, Kojima S, Kobayashi M, Hata H, Kano A, Cabrera BD (1977) Studies on the migration route of the larvae of Paragonimus westermani (Philippine strain) in rats. Jpn J Parasitol 26:37
879 Yokogawa M, Niimura M, Kobayashi M, Hata H, Tazaki T, Kojima S, Cabrera BD (1978) Comparative studies on paragonimiasis in the Philippines and in Japan. 4th Int Congr Parasitol, Short Communications C:95
880 Yutuc LM (1951) Observations on Manson's tapeworm Diphyllobothrium erinacei Rudolphi, 1819, in the Philippines. PJS 80:33-51
881 Yutuc LM (1951) The occurrence of Gnathostoma spinigerum (Nematoda) in a 33-day old pup. J Parasitol 37:329
882 Yutuc LM (1957) Arthropod-borne parasites of the domestic animals in the Philippines. Proc 8th Pacific Sci Congr, Quezon City, 3A:1511-1538
883 Yutuc LM (1968) Four helminths of the dog observed at breakthrough but failed to gain a foothold in greater Manila area with a consideration of the deterrent factors. Philipp J Vet Med 7:84-91
884 Yutuc LM, Cosio HF (1954) The incidence and frequency distribution of parasitic worms of naturally infected cats. Thapar Commemoration Volume, pp 305-308
885 Zhong Huilan, Cabrera BD, He Lianyin, Xu Zhibiao, Lu Baolian, Cao Weiji, Gao Peizhi (1982) Study of lung flukes from Philippines. A preliminary report. Scientia Sinica 25:521-526

LUZON
San Jose
Baler
Palayan
Cabanatuan
San Fernando
Malolos
Quezon City
MANILA
Pasig
Southern
Sta. Cruz
Tagaytay
San Pablo
Lipa
Lucena
Batangas City
POLILLO
Daet
CATANDUANES
Naga
Pili
Virac
Boac
Callapan
Tagalog
MARINDUQUE
Legazpi City
Bicol
Sorsogon
Romblon
Catarman
Masbate
Calbayog
Eastern
Catbalogan
MASBATE
BILIRAN
Naval
Visayas
Tacloban
Ormoc
LEYTE
Kalibo
Roxas
PANAY
Western
Cadiz
Silay
CEBU
San José de Buenavista
Iloilo City
Bacolod
San Carlos
Toledo
Danao
Jordan
Bago
Mandaue
La Carlota
Canlaon
Cebu City
Lapu-Lapu
Visayas
Central
Maasin
NEGROS
Bais
Visayas
Tagbilaran
BOHOL
Dumaguete
Mambajao
Siquijor
Northern
Gingoog
Dapitan
Cagayan de Oro
Dipolog
Oroquieta
Mindanao
Ozamis
Iligan
17
16
30
24
29
22
33
35
34
25
32
26
27
37
31
36
52
53
51
48
38
40
39
49
42
45
41
50
43
44
46
47
62
59
64
63

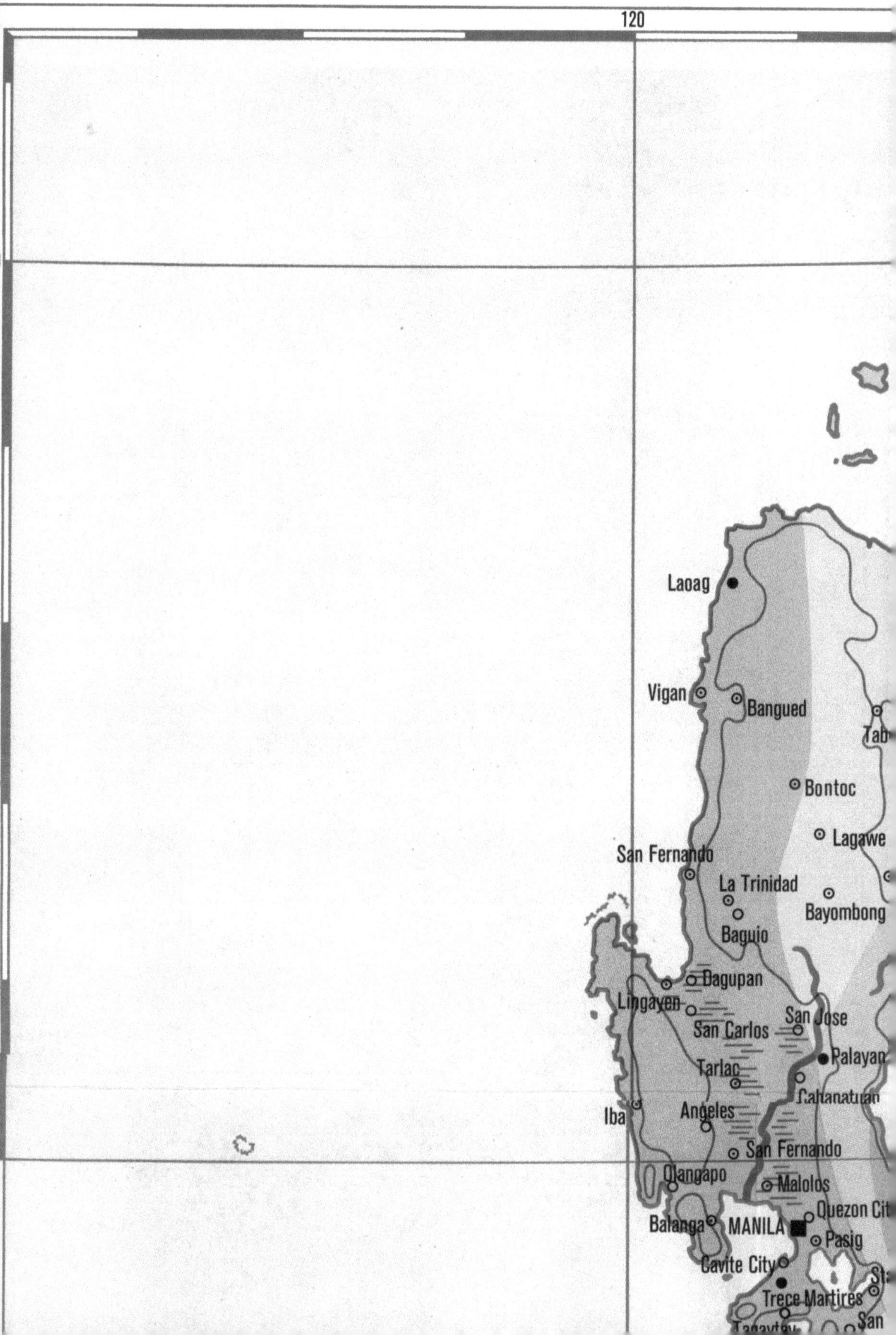
120
Laoag
Vigan
Bangued
Bontoc
Lagawe
San Fernando
La Trinidad
Bayombong
Baguio
Dagupan
Lingayen
San Carlos
San Jose
Palayan
Tarlac
Cabanatuan
Iba
Angeles
San Fernando
Olongapo
Malolos
Balanga
MANILA
Pasig
Cavite City
Trece Martires

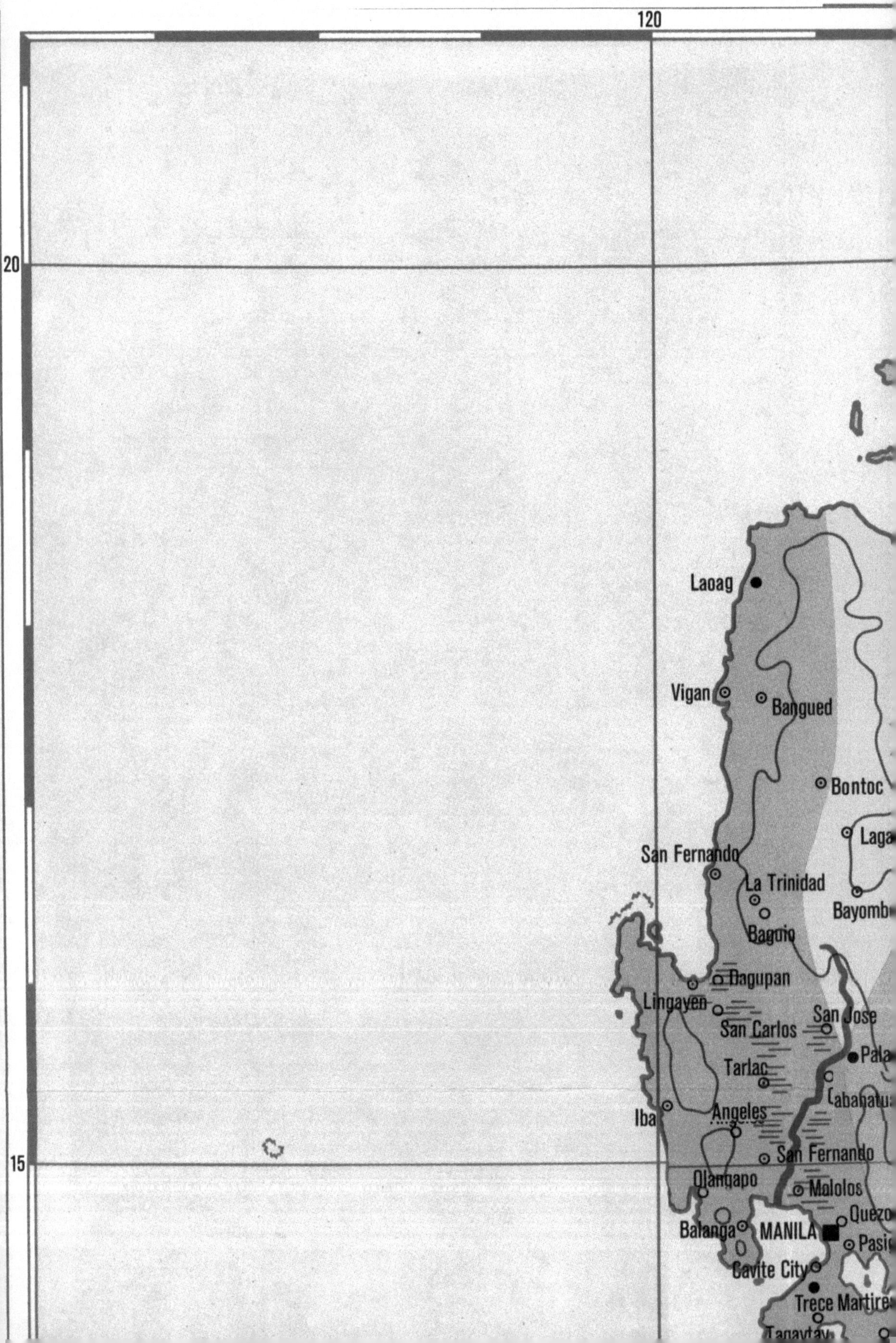
120
20
15
Laoag
Vigan
Bangued
Bontoc
San Fernando
La Trinidad
Baguio
Bayomb
Dagupan
Lingayen
San Carlos
San Jose
Tarlac
Iba
Angeles
San Fernando
Olangapo
Malolos
Balanga
MANILA
Cavite City
Trece Martires
Tagaytay

MIX
Papier aus verantwortungsvollen Quellen
Paper from responsible sources
FSC® C105338

If you have any concerns about our products,
you can contact us on
ProductSafety@springernature.com

In case Publisher is established outside the EU,
the EU authorized representative is:
Springer Nature Customer Service Center GmbH
Europaplatz 3, 69115 Heidelberg, Germany

Printed by Libri Plureos GmbH
in Hamburg, Germany